Biotechnology for Silkworm Crop Enhancement

Raviraj V Suresh • Soumen Saha •
Khasru Alam
Editors

Biotechnology for Silkworm Crop Enhancement

Tools and Applications

Springer

Editors
Raviraj V Suresh
Central Sericultural Research and Training Institute
Berhampore, West Bengal, India

Khasru Alam
Central Sericultural Research and Training Institute
Berhampore, West Bengal, India

Soumen Saha
Raiganj University
Raiganj, West Bengal, India

ISBN 978-981-97-5060-3 ISBN 978-981-97-5061-0 (eBook)
https://doi.org/10.1007/978-981-97-5061-0

This Springer imprint is published by the registered company Springer Nature Singapore Pte Ltd.
The registered company address is: 152 Beach Road, #21-01/04 Gateway East, Singapore 189721, Singapore

Foreword

I am pleased to observe that sericulture is gaining ground in Indian universities and that the sericulturist is ever so eager to adapt newer approaches and findings to comprehend the discipline. The advances in this discipline indeed are so rapid that by the time a book comes out of the press the need for another book on the subject with a fresh orientation becomes apparent. This work, therefore, is most welcome to students of sericulture as it embodies the latest approach toward the functioning of the sericulture cultivation without discarding much of the useful knowledge in sericulture gained during the last 60 years or so.

I congratulate the author Dr. Soumen Saha who is my student and an accomplished biotechnologist for producing a wonderful work which evidently supersedes all other Indian books in the field. The book will be found suitable for college students both at undergraduate and postgraduate levels.

Parthadeb Ghosh

Cytogenetics and Plant Biotechnology
Research Unit, Department of Botany
University of Kalyani,
Kalyani, India

Preface

This scientific publication, *Biotechnology for Silkworm Crop Enhancement: Tools and Applications*, aims to introduce readers to the latest developments in biotechnological methods to improve the cultivation of silkworms. The text focuses on implementing biotechnological approaches, ranging from basic to advanced techniques, enhancing silk production, increasing abiotic stress resilience, improving disease tolerance, and developing resistance in silkworm crops.

With the current climate crisis and growing demand for high-quality silk, biotechnological applications have become essential in addressing the urgent need to increase silk production. The field of sericulture has made significant breakthroughs, such as identifying DNA markers, linkage association, genome-wide association studies, generating mutants, and introducing transgenic silkworms. Developing silkworms with improved cocoon yield and disease resistance is crucial to world's efforts to boost silk production.

This comprehensive text covers 19 chapters written by pioneers and academic experts. The book covers fundamental principles and recent advancements and delves into various methodologies employed in seri-biotechnology for improving silkworm crops. These methodologies include Next-generation sequencing for DNA sequencing, using molecular markers to enhance abiotic stress tolerance (specifically high temperature and humidity conditions), expression analysis, RNA interference, gene knockout approaches to bolster disease resistance, and transcriptomics to enhance economically significant parameters such as silk content.

The book explains these topics in detail and includes contemporary research appraisals, extensive discussions, and an evaluation of the benefits and risks associated with the use of biotechnological tools. This publication serves as an invaluable reference for researchers and academics in biotechnology, molecular biology, and sericulture while also serving as an informative starting point for budding researchers. Furthermore, as mentioned earlier, the book is intended for educators, students, and professionals engaged in the disciplines.

This publication serves as an invaluable reference for researchers and academics in the fields of biotechnology, molecular biology, and sericulture while also serving as an informative starting point for budding researchers. Furthermore, the book is intended for educators, students, and professionals engaged in the aforementioned disciplines.

Berhampore, West Bengal, India — Raviraj V Suresh
Raiganj, West Bengal, India — Soumen Saha
Berhampore, West Bengal, India — Khasru Alam

Acknowledgments

I am particularly indebted to Dr. Raviraj V Suresh and Dr. Khasru Alam who edited and contributed greatly through reviewing text of each of the chapters and offering many constructive comments and suggestions; without their help this book would not have been completed. I am grateful and thankful to both of them from the bottom of my heart. I wish to express my sincere appreciation to the former and current faculty members of my laboratory.

I am thankful to Prof. Parthadeb Ghosh, Cytogenetics and Plant Biotechnology Research Unit, Department of Botany, University of Kalyani, for writing the Foreword.

I also gratefully acknowledge the constant support and encouragement given by Dr. Jula S. Nair, Director, CSR&TI, Berhampore.

I am very grateful to Prof. Dipak Kumar Roy, Vice Chancellor, Raiganj University, for their constant support.

I am thankful to all the contributors and subject experts including Professors N.B. Ramachandra, G. Subramanya, U.C. Lavania, and Debasis Chakrabarty, among others, from the inner core of my heart for instantly agreeing to my proposal and sparing time from their routine and hectic schedule to finalize and submit the chapters assigned to them within the time frame.

Thanks are also due to my wife Mrs. Shreyasi Chaki Saha and daughter Srijanya Saha for their help in various ways.

Contents

An Introduction to Biotechnology Driven Advances for Silkworm Improvement and Sustainable Perspectives 1
V. Sivaprasad, Prashanth A. Sangannavar, and Kusuma Lingaiah

Silkworm Genomics: A Novel Tool in Silkworm Crop Improvement 21
Raviraj V Suresh, Soumen Saha, Nalavadi Chandrakanth, Khasru Alam, Anil Pappachan, and Shunmugam Manthira Moorthy

An Insight into Transcriptomics of the Mulberry Silkworm, *Bombyx mori*: A Review 33
K. Lingaiah, L. Satish, V. S. Raviraj, S. M. Moorthy, and V. Sivaprasad

Proteomics of Silkworm, *Bombyx mori* L.: Recent Progress and Future Prospectus 45
L. Kusuma, L. Satish, V. S. Raviraj, S. M. Moorthy, and V. Sivaprasad

Epigenomics: A Way Forward from Classical Approach 55
Ranjini, Raviraj, S. Manthira Moorthy, and S. Gandhi Doss

Applications of Marker Assisted Selection in Silkworm Breeding for Abiotic Stress Tolerance 69
Nalavadi Chandrakanth, Raviraj V Suresh, Mallikarjuna Gadwala, and Shunmugam Manthira Moorthy

Mutation Breeding in *Bombyx mori*: Current Trends and Future Avenues 97
Nalavadi Chandrakanth, Khasru Alam, M. S. Ranjini, and Shunmugam Manthira Moorthy

Application of Marker-Assisted Selection in Silkworm Breeding for Disease Resistance 109
L. Satish, L. Kusuma, Raviraj V Suresh, S. M. Moorthy, and V. Sivaprasad

Engineered Disease Resistance Silkworm Using Genome Editing 119
Katsuhiko Ito, Pooja Makwana, and Kamidi Rahul

Biotechnological Approaches in Wild Silk Culture 133
Kaiho Kaisa, Jigyasha Tiwari, D. S. Mahesh, Suraj Shah, and Kallare P. Arunkumar

An Overview of Climatic and Genetic Influences on the Emergence of *Antheraea* spp. 147
Shuddhasattwa Maitra Mazumdar, Nabanita Banerjee, Khasru Alam, Rupa Harsha, and T. Selvakumar

The Journey of Biotechnology in Tasar Sericulture: Past Experiences, Current Strategies, and Future Horizons 167
Immanual Gilwax Prabhu, Vikas Kumar, and Narisetty Balaji Chowdary

Silkworm Databases and Research Tools: A Comprehensive Guide for Advancing Sericulture Research 219
Megha Murthy, V. S. Raviraj, Anu Sonowal, and Jula S. Nair

The Application of Biostatistical Techniques in Silkworm Breeding and Improvement 239
Rahul Banerjee, Manjunatha Gyarehalli Rangappa, Ritwika Das, Tauqueer Ahmad, Pradip Kumar Sahu, P. A. Sangannavar, S. Manthira Moorthy, and V. Sivaprasad

Dipteran Parasitoid-Silkworm Interaction: Application of Genomic and Proteomic Tools in Host-Parasitoid Communication 257
Pooja Makwana, Jula S. Nair, and Appukuttan Nair R. Pradeep

Biotechnological Approaches for the Diagnosis of Silkworm Diseases 267
Mihir Rabha, Khasru Alam, K. Rahul, and A. R. Pradeep

Implications of Bioassay in Biotechnology with Relevance to Silkworm Breeding 281
M. S. Ranjini, N. Chandrakanth, G. R. Manjunath, K. Suresh, K. B. Chandrashekar, and S. Gandhi Doss

Application of Sericin in Food Industries and Coating of Fruits and Vegetables 289
M. A. Ravindra, Azad Gull, Dhaneshwar Padhan, N. Chandrakanth, V. Sobhana, Amit Kumar, Y. Thirupathaiah, and S. Gandhi Doss

Biomedical Applications of Silkworm Sericin 303
Sayannita Das and Amitava Mandal

Editors and Contributors

About the Editors

Raviraj V Suresh a distinguished scientist in Central Silk Board, has a decade of experience in genomics and biotechnology; Dr. Raviraj V S is an expert in silkworm biotechnology, having spent 5 years specializing in this field. He has been a faculty member at the prestigious Department of Genetics and Genomics at the University of Mysore, India. Dr. Raviraj's expertise includes a range of cutting-edge techniques, such as next-generation sequencing, marker-assisted selection, micromanipulation in mice and human embryos, and gene expression studies in various model organisms. Currently, as a scientist at the Central Sericultural Research Training Institute, he focuses on developing silkworm breeds that are tolerant to abiotic stress through genomics approaches. Dr. Raviraj has successfully identified genetic markers in human and insect model systems, which have helped in the development of diseases and abiotic stress-tolerant silkworm breeds and hybrids. With numerous research articles published in international journals with high impact factors, he is a life member of the Indian Science Congress and the National Academy of Sericultural Sciences, India. Dr. Raviraj is also serving as Associate Editor in the *Journal of Sericulture Science*, India, and has acted as a referee in many international journals. His primary interest lies in identifying candidate genes and their molecular mechanism at the transcript level, which is crucial for developing new silkworm breeds and hybrids.

Soumen Saha is currently Assistant Professor of Sericulture at Raiganj University, Raiganj, India. He obtained his M.Sc. from the University of Mysore, India, and his Ph.D. from the University of Kalyani, India. He has received a prestigious UGC National Fellowship from Govt. of India during his Ph.D. and post-doc program. Dr. Saha has successfully guided 03 research scholar to a Ph.D. degree, four students continuing their Ph.D. degree, and has published more than 50 articles in international and national journals. He has also successfully completed one minor research project. His specializations include molecular biology, silkworm breeding, genome sequencing, plant tissue culture, bioinformatics, secondary metabolites, and marker-assisted breeding research. He is also a reviewer of several national and international journals.

Khasru Alam is a highly esteemed Scientist-C at the Central Silk Board, a prominent branch of the Ministry of Textiles in the Government of India. Dr. Alam holds a prestigious postgraduate degree in sericulture from the University of Mysore and completed his Doctor of Philosophy (Ph.D.) thesis on Tasar silkworm (*Antheraea mylitta D*) molecular diversity from the University of Raiganj, West Bengal, India. Currently, he is a vital scientist member of the crop protection section and has expertise in the field at the Central Sericultural Research and Training Institute, a top-tier R&D organization of the Central Silk Board, Government of India.

Dr. Khasru Alam has made significant contributions through his research papers and book chapters published in both national and international books and journals. Additionally, he has been an active participant in various national and international conferences and symposiums. Furthermore, his exceptional contributions to the field of sericulture have earned him a coveted membership in the esteemed National Academy of Sericulture Sciences, India.

In recognition of his outstanding achievements in the field of sericulture, Dr. Alam was sponsored by the International Sericultural Commission to present his groundbreaking paper on the application of deep learning in sericulture at the University of Agriculture and Veterinary Medicine, Cluj, Napoca, Romania, in 2022. Aside from his research and development work, Mr. Alam is also an esteemed scientist who provides guidance to postgraduate diploma students in sericultural sciences. His extensive knowledge and expertise in the field of sericulture make him a valuable asset to the Central Silk Board and the Ministry of Textiles.

Contributors

Tauqueer Ahmad ICAR-Indian Agricultural Statistics Research Institute, New Delhi, India

Khasru Alam Bivoltine Breeding Laboratory, Central Sericultural Research & Training Institute, Central Silk Board, Ministry of Textiles, Government of India, Berhampore, West Bengal, India

Kallare P. Arunkumar Central Muga Eri Research and Training Institute (CMER&TI), Central Silk Board, Ministry of Textiles, Government of India, Jorhat, Assam, India

Nabanita Banerjee Department of Zoology, The University of Burdwan, Burdwan, West Bengal, India

Rahul Banerjee ICAR-Indian Agricultural Statistics Research Institute, New Delhi, India

K. B. Chandra Shekar Central Sericultural Research & Training Institute, Central Silk Board, Ministry of Textile, Government of India, Mysuru, Karnataka, India

Nalavadi Chandrakanth Central Sericultural Research & Training Institute, Central Silk Board, Ministry of Textiles, Government of India, Mysuru, Karnataka, India

Narisetty Balaji Chowdary Central Tasar Research and Training Institute, Central Silk Board, Ministry of Textiles, Government of India, Ranchi, Jharkhand, India

Ritwika Das ICAR-Indian Agricultural Statistics Research Institute, New Delhi, India

Sayannita Das Molecular Complexity Laboratory, Department of Chemistry, Raiganj University, Raiganj, West Bengal, India

Surjalata Devi Silkworm Breeding and Genetics Lab, Central Sericultural Research and Training Institute, Central Silk Board, Ministry of Textiles, Government of India, Berhampore, West Bengal, India

Mallikarjuna Gadwala Central Sericultural Research & Training Institute, Central Silk Board, Ministry of Textiles, Government of India, Mysuru, Karnataka, India

S. Gandhi Doss Central Sericultural Research & Training Institute, Central Silk Board, Ministry of Textile, Government of India, Mysuru, Karnataka, India

Azad Gull Central Sericultural Research & Training Institute, Central Silk Board, Ministry of Textile, Government of India, Mysuru, Karnataka, India

Rupa Harsha Department of Zoology, Balurghat College, Balurghat, West Bengal, India

Katsuhiko Ito Department of Science of Biological Production, Graduate School of Agriculture, Tokyo University of Agriculture and Technology, Tokyo, Japan

Oshin Joshi Silkworm Breeding and Genetics Lab, Central Sericulture Research & Training Institute, Central Silk Board, Ministry of Textiles, Government of India, Berhampore, West Bengal, India

Kaiho Kaisa Central Muga Eri Research and Training Institute (CMER&TI), Central Silk Board, Ministry of Textiles, Government of India, Jorhat, Assam, India

Amit Kumar Central Sericultural Research & Training Institute, Central Silk Board, Ministry of Textile, Government of India, Mysuru, Karnataka, India

Vikas Kumar Central Tasar Research and Training Institute, Central Silk Board, Ministry of Textiles, Government of India, Ranchi, Jharkhand, India

L. Kusuma Central Sericulture Research & Training Institute, Central Silk Board, Ministry of Textiles, Government of India, Mysuru, Karnataka, India

D. S. Mahesh Central Muga Eri Research and Training Institute (CMER&TI), Central Silk Board, Ministry of Textiles, Government of India, Jorhat, Assam, India

Pooja Makwana Central Sericultural Research & Training Institute, Central Silk Board, Ministry of Textiles, Government of India, Berhampore, West Bengal, India

Amitava Mandal Molecular Complexity Laboratory, Department of Chemistry, Raiganj University, Raiganj, West Bengal, India

G. R. Manjunath Central Silk Board, Research Coordination Section, Central Office, Ministry of Textile, Government of India, Bangaluru, Karnataka, India

Shuddhasattwa Maitra Mazumdar Central Silk Board, Ministry of Textiles, Government of India, Kathikund, Jharkhand, India

S. M. Moorthy Central Silk Board, Ministry of Textiles, Government of India, Bangalore, Karnataka, India

Megha Murthy UCL Queen Square Institute of Neurology, London, UK

Jula S. Nair Silkworm Breeding and Genetics Lab, Central Sericultural Research and Training Institute, Central Silk Board, Ministry of Textiles, Government of India, Berhampore, West Bengal, India

Dhaneshwar Padhan Central Sericultural Research & Training Institute, Central Silk Board, Ministry of Textile, Government of India, Mysuru, Karnataka, India

Anil Pappachan National Silkworm Seed Organization, Central Silk Board, Ministry of Textiles, Government of India, Bangalore, Karnataka, India

Immanual Gilwax Prabhu Central Tasar Research and Training Institute, Central Silk Board, Ministry of Textiles, Government of India, Ranchi, Jharkhand, India

Appukuttan Nair R. Pradeep Central Sericultural Research & Training Institute, Central Silk Board, Ministry of Textile, Government of India, Berhampore, West Bengal, India

Mihir Rabha Central Sericultural Research and Training institute, Central Silk Board, Ministry of Textile, Government of India, Berhampore, West Bengal, India

Kamidi Rahul Central Sericultural Research & Training Institute, Central Silk Board, Ministry of Textiles, Government of India, Berhampore, West Bengal, India

Manjunatha Gyarehalli Rangappa Central Silk Board, Ministry of Textiles, Government of India, Bengaluru, Karnataka, India

M. S. Ranjini Central Sericultural Research & Training Institute, Central Silk Board, Ministry of Textiles, Government of India, Mysuru, Karnataka, India

M. A. Ravindra Central Sericultural Research & Training Institute, Central Silk Board, Ministry of Textile Government of India, Mysuru, Karnataka, India

Soumen Saha Cytogenetics & Plant Biotechnology Unit, Department of Sericulture, Raiganj University, Raiganj, West Bengal, India

Pradip Kumar Sahu Bidhan Chandra Krishi Viswavidyalaya, Mohanpur, West Bengal, India

Prashanth A. Sangannavar Central Silk Board, Ministry of Textiles, Government of India, Bengaluru, Karnataka, India

L. Satish Silkworm Breeding and Molecular Biology Laboratory II, Central Sericultural Research and Training Institute, Central Silk Board, Ministry of Textiles, Government of India, Mysuru, Karnataka, India

T. Selvakumar Basic Tasar Silkworm Seed Organization, Central Silk Board, Ministry of Textiles, Government of India, Bilaspur, Chhattisgarh, India

Suraj Shah Central Muga Eri Research and Training Institute (CMER&TI), Central Silk Board, Ministry of Textiles, Government of India, Jorhat, Assam, India

V. Sivaprasad Central Silk Board, Ministry of Textiles, Government of India, Bengaluru, Karnataka, India

V. Sobhana Central Sericultural Research & Training Institute, Central Silk Board, Ministry of Textile, Government of India, Mysuru, Karnataka, India

Anu Sonowal Silkworm Breeding and Genetics Lab, Central Sericultural Research and Training Institute, Central Silk Board, Ministry of Textiles, Government of India, Berhampore, West Bengal, India

K. Suresh Central Sericultural Research & Training Institute, Central Silk Board, Ministry of Textile, Government of India, Berhampore, West Bengal, India

Raviraj V Suresh Silkworm Breeding and Genetics Lab, Central Sericulture Research & Training Institute, Central Silk Board, Ministry of Textiles, Government of India, Berhampore, West Bengal, India

Y. Thirupathaiah Central Sericultural Research & Training Institute, Central Silk Board, Ministry of Textile Government of India, Mysuru, Karnataka, India

Jigyasha Tiwari Central Muga Eri Research and Training Institute (CMER&TI), Central Silk Board, Ministry of Textiles, Government of India, Jorhat, Assam, India

An Introduction to Biotechnology Driven Advances for Silkworm Improvement and Sustainable Perspectives

V. Sivaprasad, Prashanth A. Sangannavar, and Kusuma Lingaiah

Abstract

Silkworms have been cultivated for centuries across the world, primarily for silk production. The relentless pursuit of enhanced silk yield, quality, and survival was augmented by the application of biotechnological tools and resources to the otherwise traditional and conventional industry. Genome sequencing of *Bombyx mori* and the databases available on the public domain have been quite useful in integrating molecular approaches with conventional breeding methods for evolving silkworm hybrids with higher productivity and quality and sustainable yields for adverse climatic conditions. Over the years, various research groups have identified a large number of genes/proteins differentially expressed or associated with disease resistance/tolerance and also quite a few quantitative trait loci (QTLs), which especially leads to the development of *Bombyx mori* breeds/hybrids by gene expression/manipulation. The development of molecular diagnostic methods for various silkworm diseases has also impacted the sustainability in the field. These efforts have ushered a new era of precision and innovation in silkworm improvement for specific or desired trait(s) such as silk productivity and quality; tolerance to diseases, high temperature, and relative humidity; and general survival and adaptability. The research collaborations are proving to be vital in translating the outputs from basic science in mulberry and non-mulberry silkworms to counter biotic and abiotic stress into the commercially beneficial applications. Functional genomics hold vast potential to improve the quality of produce by offering directions into gene mining, gene drive, genetic manipulation, and molecular breeding. Currently, the impact of biotechnological advances

V. Sivaprasad (✉) · P. A. Sangannavar
Central Silk Board, Bengaluru, India

K. Lingaiah
Central Sericulture Research & Training Institute, Central Silk Board, Mysuru, India

R. V Suresh et al. (eds.), *Biotechnology for Silkworm Crop Enhancement*,
https://doi.org/10.1007/978-981-97-5061-0_1

in enhancing silk production is in transformative phase and needs continuous efforts as the potential benefits have far-reaching implications in improving silk productivity/quality and generation of novel biomaterials. The most plausible approaches for the development of silkworm breeds with improved silk quality, disease resistance/tolerance, and sustainable productivity include marker-assisted selection (MAS) and transgenic and genome editing tools. This chapter provides an overview of significant biotechnological advances made/undertaken to develop novel and pathbreaking silkworm genetic resources and also perspectives for greater interventions utilizing functional genomics, molecular-assisted breeding, genome editing tools, etc. paving the way for sustainable sericulture.

Keywords

Biotechnology tools · Genome sequencing · Marker-assisted selection · Functional genomics · Integrated breeding approaches · *Bombyx mori* · Non-mulberry silkworms

Introduction

Various genera of silkworms are exploited commercially for the production of natural and versatile silk fibers across the world. Among them, the most prominent one is highly domesticated mulberry silkworm (*Bombyx mori*), and others include wild (forest-based) saturniid moths especially muga (*Antheraea assamensis*), tropical tasar (*Antheraea mylitta*), Chinese oak tasar (*Antheraea pernyi*), Japanese wild (*Antheraea yamamai*), and semidomesticated eri (*Samia ricini*). Asian countries such as China, Japan, India, Thailand, Korea, Vietnam, etc. are well known for cultivation of sericigenous insects and the respective host plants since time immemorial. Silk produced by *B. mori* is characterized for its durability and strength, while the one produced by wild silk moths is distinct with characteristic luster, color, and textures. Besides the huge economic value of silk filaments and silk fabric, these silkworms are playing an important role as model insect systems for genetical, biological, molecular, and medical studies as they have richest mutation repertoires in almost every aspect of morphology, physiology, and development (Goldsmith et al. 2005; Sivaprasad et al. 2022). The silk industry boasts rich heritage interwoven with the traditional practitioners in rural communities and plays a pivotal role in generating gainful employment opportunities for socioeconomic empowerment and quality livelihoods. The global demand for silk is ever increasing for textile purposes, while the recent applications of silk biomaterials for biomedical, cosmetic, and industrial applications provide ample scope for sustaining the silk production.

Biotechnology Advances and Silkworm Improvement

Biotechnology tools such as transcriptomics, functional genomics, marker-assisted breeding, quantitative trait loci mapping, genomic selection, and genetic engineering (transgenesis and gene editing) have enabled targeted modification of silkworm genomes to instill desired quantitative and qualitative traits of importance, beyond the conventional characteristics. Researchers have successfully introduced genes to enhance silk fiber strength, color, and resistance to pathogens to develop superior silkworm strains. Molecular biology approaches in tandem with traditional genetical methods are accelerating the process of cocoon crop improvement as breeders could make informed decisions. Genetic and genomic approaches have wider applicability in sericulture, and the advent of genome sequencing technology has resulted in rapid advances. Genetic differences between the individuals are manifested in differences in DNA sequences, biochemical characteristics (e.g., protein structure or isoenzyme properties), physiological properties (e.g., abiotic stress resistance or growth rate), and morphological characters (Rao and Hodgkin 2002). The distribution and variability of microsatellite sequences in a species genome can elucidate its genetic history from the standpoint of evolution and artificial selection (Prasad et al. 2005). Microsatellites form a suitable genetic marker for analyzing pedigree, population structure, genome variation, evolutionary process, and fingerprinting purposes (Abdul-Muneer 2014). Interspecific comparisons of microsatellite loci have repeatedly shown that the loci are longer and more variable in the species from which they are derived (the focal species) than homologous loci in other (non-focal) species (Hutter et al. 1998).

The rapid advancement of information technology in life sciences (artificial intelligence, image analysis, machine and deep learning, etc.) along with genomic technologies and bioinformatics (NGS, ChIP-seq, RNA-seq, RAD-seq, QTL mapping, GBS, etc.) has resulted in interdisciplinary research to incite massive data generation and analysis. Genome editing tools such as zinc finger nucleases (ZFNs), transcription activator-like effector nucleases (TALENs), and clustered regularly interspaced palindromic repeats (CRISPR)/CRISPR-associated (Cas) system through targeted deletions, insertions, and substitutions of genetic sequences have revolutionized genetic research with broad applications in areas such as medicine, agriculture, and biotechnology (Moon et al. 2019). Emerging biotechnological tools such as genome editing tools hold immense promise for precise and rapid trait manipulation facilitating the creation of transgenic lines. The ability to modify specific genes with unprecedented accuracy offers a game-changing approach for tailoring the silkworm characteristics.

Mulberry Silkworm, *Bombyx mori* L.

Mulberry silkworm, *B. mori*, is the most commercially exploited sericigenous lepidopteran insect feeding on intensely cultivated mulberry leaves across the sericultural countries. *B. mori* is highly domesticated and monophagous. Their genetic resources are quite large, and several countries conserve around 3000 including land races, elite breeds, and improved breeding lines exhibiting distinct economic traits and are of diverse origin in tropical and temperate countries. Around 400 described mutants have been mapped to >200 loci comprising 28 linkage groups with a ~ 432 Mb genome size. Elaborate and systematic breeding programs are conducted to develop productive hybrids continuously using conventional, molecular, and integrated methods (Sivaprasad et al. 2022). All sericultural countries are endowed with highly popular silkworm hybrids for commercial raw silk production. Trait syngeny and genetic manipulations are quite effective and are widely used by farmers for rearing silkworms by the farmers. Genome investigations on *B. mori* are quite important for breeding and genetic studies, as well as for isolating valuable genes and promoters for comparative genomics (Mita et al. 2004; Yamamoto et al. 2006). Mulberry silkworms are completely dependent on humans for their survival and propagation.

The genome resequencing of 40 silkworms (29 phenotypically and geographically diverse domesticated silkworm lines and 11 wild silkworms) employing Illumina-Solexa sequencing strategy revealed the presence of ~14 million and 13 million SNPs in the domesticated and wild varieties, respectively, and also clearly differentiated the SNPs, indels, and structural variations between wild and domesticated silkworms. Several highly polymorphic SNPs (117) were identified and validated from 121 SNP positions genotyped for 4840 sites across the 40 silkworm genomes (Xia et al. 2009).

The high-quality genome assembly of *B. mori* performed employing gene prediction based on the new genome assembly utilizing available mRNA and protein sequence data revealed 16,880 gene models with an N50 of 2154 bp representing more accurate coding sequences and gene sets than the old ones (Kawamoto et al. 2019). The recently published high-resolution *B. mori* pan genome indicated high density of genomic variants representing 7308 new and 4260 (22%) core genes including 3,432,266 nonredundant structure variations (SVs). The study also identified hundreds of genes and SVs involved in artificial selection, particularly in silkworm domestication and breeding. Further, two genes responsible for economic (silk yield and silk fineness) and two ecologically adaptive traits (egg diapause and aposematic coloration) were also identified (Tong et al. 2022).

Multiple transcription factors are involved in transcriptional regulation of silk protein genes during *B. mori* development. A clear genomic understanding of the silkworm biology has been revealed by transcriptome analysis of silkworm gene expression patterns in multiple tissues (Gan et al. 2011; Xia et al. 2007), different developmental stages (Xu et al. 2012), host responses to infection with pathogenic microorganisms (Zhou et al. 2013), 20-hydroxyecdysone (20E) regulation of innate immunity and glycolysis (Tian et al. 2015), and dosage compensation in the sex chromosome (Zha et al. 2009). Further, the RNA-seq studies have identified silkworm-specific miRNAs, differential expression of miRNA-star and their corresponding miRNAs, TE associated miRNAs, and also candidate targets of miRNAs (Liu et al. 2010; Xia et al. 2014).

Microarrays and RNA-seq methods have been extensively applied to study the transcriptome analysis owing to the availability of the silkworm genome sequence (Xia and Xiang 2013). Silk glands are specialized in the synthesis of several secretory proteins. Expression of genes encoding the *B. mori* silk proteins in silk glands with strict territorial and developmental specificities is regulated by several transcription factors. Quantitative PCR analysis suggested that the expression level of *Bmsage* in a high silk strain was higher as compared to a lower silk strain on day 3 of fifth instar larvae (Zhao et al. 2014). De novo transcriptome of wild mulberry silkworm, *B. mandarina* (Cheng et al. 2015), resulted in the identification of 883 and 425 unigenes that were upregulated in the MSG and 425 in PSG, respectively. The transcriptomic analysis of transgenic silkworm strain revealed overexpression of Ras1CA in posterior silk gland, which improves fibroin production and thereby silk production (Ma et al. 2014). Comparative transcriptome analyses on silk glands of six silk moths represented the genetic basis of silk structure and coloration (Dong et al. 2015). Lower quantities of silk production were correlated with 779 upregulated and 9 downregulated transcripts (Wang et al. 2016). Zhou et al. (2020) in a comparison of co-expression networks in silk gland emphasized the causes of silk yield increases during domestication of silkworm evident with BGIBMGA001462 gene involvement in the transport pathway and BGIBMGA000074 in the ribosome pathway. Both genes co-expressed with Fib-H in the domestic silkworm and relatively high gene expression levels were detected in the silk gland indicating that silk-encoding genes are involved in the enhanced silk yields following silkworm domestication. Eleven QTLs related to cocoon yield traits mapped on to seven chromosomes were identified, wherein the expression of KWMTBOMO04917 was higher in the wild silkworm than in the domestic silkworms. KWMTBOMO12906, encoding DNA polymerase epsilon, was highly expressed in the domestic silkworms, and its expression in the posterior silk gland positively correlated with cocoon yields (Fang et al. 2020). A single-cell transcriptomic atlas developed by Ma et al. (2022) unraveled the developmental trajectory and gene expression status of silk gland cells. Emphasis was laid on the marker genes in the regulation of silk gland development and silk protein synthesis including the heterogeneity of silkworm silk gland cells and their gene expression dynamics (Tang et al. 2007).

Transcriptomic analysis of *B. mori* developmental features was studied, and a total of 12,254 assembled transcripts from the wing disc at L5D6, PP, and P0 stages were identified. High levels of expression of cuticle proteins and chitin-related genes reflect active chitin metabolism, which is related to morphological and structural changes during the metamorphosis from wing disc to pupal wing (Ou et al. 2014). Integument transcriptome analysis of wild-type strain Dazao and the mutant strains +/bd and bd/bd showed a total of 254 novel transcripts revealing 28 DEGs contributing to bd larval melanism and remarkably higher expression of 15 cuticular protein genes in the bd/bd mutant (Wu et al. 2016). Comparative integument transcriptome analysis identified genes associated in the pattern formation in three allelic mutants of *B. mori* and eight common genes involved in the construction of multiple twin-spot marking patterns in the three mutants (Ding et al. 2020). Transcriptome analysis of silkworm reproduction under dimethoate stress indicated that pesticide exposure significantly slowed down the development of silkworm eggs (Zheng et al. 2022).

By combining transcriptomic analysis and population genomics with QTLs, Fang et al. (2020) identified 14 domestication-related genes in the transcriptome of the fifth instar larval silk glands of *B. mori* and *B. mandarina*. The domestication processes were reconstructed, and selective key events were identified in 137 representative silkworm strains by sequencing. The results represented an evolutionary scenario in which silkworms may have been initially domesticated in China followed by independent spreads along the Silk Road. The development of most local strains from diverse ancestral origin further improved for modern silk production in Japan and China. The gene indicators predicted key roles for nitrogen and amino acid metabolism contributing to the promotion of silk production and circadian-related genes for adaptation (Xiang et al. 2018).

The detection and exploitation of naturally occurring DNA sequence polymorphisms have wide potential applications in crop improvement programs for varietal and parentage identification. Several molecular markers such as random amplified polymorphic DNA (RAPD), simple sequence repeats (SSRs), and inter simple sequence repeats (ISSRs) were used for the identification of the *B. mori* silkworm breeds (Awasthi et al. 2008; Reddy et al. 1999). The research on markers utilized for silkworm improvement programs is elucidated in detail by Sivaprasad et al. (2022).

The creation of a transgenic silkworm, *B. mori*, through gene transfer using a vector derived from the piggyBac transposon by Tamura et al. (2000) has resulted in rapid advancement of transgenic insects for various desired traits. Currently, transgenic silkworms have been engineered for the purpose of synthesizing recombinant human proteins such as procollagen, cytokines, and monoclonal antibodies within their larvae, pupae, cocoons, and silk glands. Further, transgenic silkworm research has been utilized successfully incorporating favorable traits into silk and its related products. *B. mori*-spider silk serves as a notable example significantly transforming the silk industries. The transgenesis process is standardized in the cultivated silkworm, and over a thousand transgenic silkworm lines have been created so far including the most important ones exhibiting tolerance to BmNPV (Subbaiah et al. 2013) and fluorescent silk (Suzuki 2019; Tomita 2019). But the commercial exploitation of these transgenic lines is entangled with regulatory issues and environmental complications.

Rapid advances in genome editing tools have enabled efficient gene knockout experiments in a wide variety of organisms including silkworms. Positional cloning of silkworm mutant, multi lunar (L) with twin-spot markings on sequential segments, revealed that cis-regulatory change in Wnt1 is responsible for the spot patterning (Yamaguchi et al. 2013). Heritable genome editing with CRISPR/Cas9 was carried out for four loci (Bm-ok, BmKMO, BmTH, and Bmtan) which showed 16.7–35.0% of mutation frequencies at specific sites could be induced by direct microinjection of specific guide RNA and Cas9-mRNA into silkworm embryos. Design of sgRNA, microinjection of a customized sgRNA/Cas9-mRNA mixture, and an effective crossing strategy representing the three key components essential for successful acquisition of a homozygous mutant through genome editing (Wei et al. 2014). *B. mori* CRISPR system was employing commonly used SpCas9, which recognizes only the target sites with NGG PAM sequence (Ma et al. 2017). Genome editing of bmGSTu2 (gene disruption) revealed a decrease in median lethal dose values to an organophosphate insecticide and a decrease in acetylcholine levels in silkworms (Yamamoto et al. 2018). The targeted gene disruption techniques in *B. mori* have been extensively studied by Baci et al. (2021). Multiple gene editing methodologies are available for the silkworm including the binary transgene-based techniques to generate mutants and delivery of CRISPR/Cas9 system via DNA-free ribonucleoproteins. T7 promoter, widely used in the ribonucleoprotein-based method for production of sgRNAs in vitro, needs a 5′ GG motif for efficient initiation (Zou et al. 2021).

BmBLOS2 marker gene controls the development of uric acid granules in the larval epidermis through ZFNs effectively inducing somatic and germline mutations (Takasu et al. 2013). This was further confirmed by Zhu et al. (2017) that this marker can be employed to confirm the efficacy of various genome editing tools in *B. mori* and other Lepidoptera insects. Application of TALENs in *B. mori* by directly micro-injecting mRNA from two synthetic TALEN pairs into embryos was reported by Ma et al. (2012) with significantly higher and diverse efficiency inducing mosaic and heritable mutations. *CRISPR/Cas9* system has been achieved successfully in *B. mori* through various delivery strategies. Ma et al. (2014) knocked out Bmfib-H of and reported a smaller and empty silk gland, abnormally developed posterior silk gland cells, an extremely thin cocoon that contains only sericin proteins, and a slightly heavier pupae in the genome-edited silkworms. Further, these results also implicated that removal of endogenous Bmfib-H protein could significantly increase the expression level of exogenous protein. The genome editing of BmFibH gene in order to unravel its role in the development of silk gland was carried out by Cui et al. (2018) with an efficiency of only 12.5% hatchings; however, the unhatched eggs also indicated that all embryos were genetically edited. Knockout of BmFibH gene resulted in the formation of naked pupa or thin cocoons, and silkworms exhibiting naked pupae did not survive. Inactivation of BmFibH indicated that several other genes are involved in the processes of degradation such as the autophagy represented by upregulation.

W chromosome-based genetic-sexing system combining TALENs and CRISPR/Cas9 technologies in *B. mori* has also been successful. One strain with TALEN-mediated, W-specific Cas9 expression driven by silkworm germ cell-specific nanos (nos) promoter and another strain with U6-derived single-guide RNA (sgRNA) expression targeting transformer 2 (tra2), an essential gene for silkworm embryonic development, were developed. The F1 hybrids exhibited complete female-specific lethality during embryonic stages (Zhang et al. 2018). Inactivation of exogenous and endogenous genes through base-editing-induced nonsense mutations was made possible with an efficiency of up to 66.2% by employing BE3 as an important knockout tool (Li et al. 2018).

The identification of a null mutation in BmorCPH24 (gene encoding a cuticular protein) in mutant larvae through spatiotemporal analyses indicated that BmorCPH24 is expressed in the larval epidermis postecdysis, whereas RNAi-mediated knockdown and CRISPR/Cas9-mediated knockout of BmorCPH24 produced the abnormal body shape and inhibited pigment typical of the mutant phenotype (Xiong et al. 2017). Analysis of silkworm molecular breeding potential of CRISPR/Cas9 systems for white egg 2 genes suggested that silkworm molecular breeding is possible and would be a very effective way to shorten the breeding period than the traditional breeding process (Park et al. 2019). CRISPR/Cas9-mediated knockout of *Bombyx* methoprene tolerant 1 (Met1), JH receptor, and a putative domestication gene caused developmental retardation in the brain, unlike precocious pupation of the cuticle (Cui et al. 2021). Phenomic characterization of mutant lines identified a large set of genes responsible for visual phenotypic or economically valuable trait changes and identified KWMTBOMO12902 as a strong candidate gene for breeding applications in sericulture industry (Ma et al. 2024).

Genome databases are valuable resources and essential for better understanding of functional genomics, comparative genomics, and molecular evolution as well as gene discovery, transposable elements, gene organization, and genome variability among the insect species and sericigenous insects in particular. Over 180,000 silkworm ESTs from independent projects are available in public database, and the two largest database projects were constructed by Mita et al. (2004) and Cheng et al. (2004). Subsequently, Japanese and Chinese groups independently achieved *B. mori* WGS at 3X and 5.9X coverage level (The International Silkworm Genome Consortium 2008; Yamamoto et al. 2008; Mita et al. 2004; Xia et al. 2004), which formed a strong foundation for the genome-wide functional analysis in silkworms. Silk database (SilkDB) is an integrated representation of six draft genome sequences, cDNAs, EST clusters, transposable elements (TEs), and functional annotations of genes with assignments to InterPro domains and Gene Ontology (GO) terms for *B. mori* (Wang et al. 2005). *KAIKObase* (Ver 4) was built using the chromosomal-level genome assembly of *B. mori* including a genetic map viewer and genome viewer (Kawamoto et al. 2019).

Non-mulberry Silkworms

The cultivation of non-mulberry silkworms especially tasar, oak tasar, and muga are predominantly reliant on forest trees involving outdoor rearing practices, whereas the eri silk production involves indoor rearing processes through the utilization of castor/cassava leaves. Non-mulberry silkworms exhibit polyphagous feeding behavior unlike the mulberry silkworm and primarily belong to the Saturniidae family with the largest group of sericigenous insects including notable genera such as *Antheraea*, *Philosamia*, *Samia*, and *Attacus*.

Muga Silkworm, *Antheraea assamensis* Helfer

The muga silkworm, *A. assamensis* (n = 15), is polyphagous endemic to India (Assam and other northeastern states) producing lustrous golden-yellow silk and is priced five times over the mulberry silk. Larval color polymorphism (green, blue, almond, yellow) is a distinctive trait, and the known eco-strains are Harujukia (Harubhangia), Barjukia (Barbhangia), and Lebang (Lebang). The green-colored muga silkworms are predominant and cultivated across the year. Few wild muga accessions, viz., RMRS Aa00–1, RMRS-Aa00–2, RMRS-Aa00–3, and RMRS-Aa00–4, have been collected from natural forests and characterized. Only two muga silkworm breeds (CMR-1 and CMR-2) were developed utilizing wild germplasm resources by Central Muga Eri Research & Training Institute, Lahdoigarh, Assam (India). The integrated approach involving sericulture and forest department along with local communities aims to safeguard the rich diversity of muga silkworms in their natural habitats.

A. assamensis has a narrow genetic base and its mono-racial status with little variability in the entire population is a constraint for the development of new breeds/hybrids with desired traits such as diapausing strains and stress or disease tolerant lines. The only choice is to evolve transgenic lines through the application of transgenic or gene editing techniques. Arunkumar et al. (2009) developed several useful microsatellite markers for analyzing the genetic structure and phylogenetic status of *A. assamensis* through ESTs and genomic libraries from ten polymorphic loci. Further, 6 populations were genotyped using 13 informative markers to examine the population structure and genetic variation, and relatively high level of genetic diversity was observed in WWS-1 population that originated from the West Garo Hills, Meghalaya (Arunkumar et al. 2012).

The research on wild silkworm silk proteins is scant, and only low-coverage EST data are available for *A. assama*, *S. cynthia*, and *A. mylitta*. A comprehensive transcriptomics of the silk glands of Saturniidae silk moths (*A. yamamai*, *A. pernyi*, *A. assama*, *A. selene*, *S. cynthia*, and *R. newara* along with *B. mori*) was made by Dong et al. (2015) to identify the genetic basis for the variances in silk quality. The fifth instar larval transcriptome of *A. assamensis* could provide critical information on candidate genes and signaling pathways (Hasnahana et al. 2017).

The immune-related genes (AMPs) from the fat body mRNA of *S. c. ricini* that are activated/repressed by microbial injection with *E. coli* were demonstrated by the EST frequencies of two important hemolymph storage protein genes, *ScSP1* and *ScSP2* (Meng et al. 2008). The conserved iron-binding proteins, ferritins, mainly involved in the immune system, detoxification, and iron storage were studied through transcriptome, and ScFerHCH is probably crucial for *S. c. ricini*'s iron storage, antioxidation, and innate immunological protection against infections (Yu et al. 2017).

Muga silk moth silk proteins and genome are the least understood among saturniid moths presenting a unique challenge in the field of genomics. The exploration of muga silk moth transcriptome is very significant as it could shed more light on the genetic makeup of this unique species and pave way for deeper understanding of its biology and genetics. First ever complete *A. assamensis* mitogenome was made through NGS and contrasted with the other known lepidopterans for improved comprehension of the evolutionary and comparative biology (Singh et al. 2017). The genome sequencing and assembly of Indian golden silk moth, *A. assamensis*, was reported recently providing further insights into the genetic architecture, insights on evolutionary history, and genetic perspectives revealing unique characteristics as compared to the other silk moths (Dubey et al. 2023). The study utilized two sequencing platforms and three library strategies revealing a 501.8 Mb genome with notable features. Further, the genome analysis identified genes related to silk fiber production, feeding behavior, and duplicated gene families emphasizing the role of negative selection and highlighted the shared and unique genes in *A. assamensis*. The study also explored the evolution of key insect gene families such as GST, ABC transporter, and CYP450. The WGS of muga silkworm is a significant genomic resource for future molecular interventions in improving *A. assamensis*.

Even though a number of databases have been created for *B. mori*, lack of genetic resources in wild silk moths is the root cause for scant data in the public domain. WildSilkbase is an EST database of three saturniid silk moths (*Antheraea assama*, *A. mylitta*, and *Samia cynthia ricini*) with 57,113 ESTs grouped and assembled into 10,019 singletons and 4019 contigs (Arunkumar et al. 2008). *SilkBase* is an integrated database containing transcriptome and genomic resources of *B. mori*, *B. mandarina*, *Trilocha varians*, *Ernolatia moorei*, and *S. ricini* with an easy-to-use integrated genome browser (Kawamoto et al. 2022). *SilkwormBase* is a comprehensive genetic resource database with information about 608 genes and 481 wild silkworm strains, transcriptomes, and genes (Priyadharshini and Maria 2019). *VanyaSilkbase* is a web resource with transcriptome and genomic information (DEGs, SNPs) of wild silk moths (*S. ricini*, *A. yamamai*, *A. assamensis*, *A. pernyi*, and *A. mylitta*) with a genome browser (https://vanyasilkbase.cmerti.res.in/). Very few attempts have been made for the establishment of transgenic wild silkworms. Incorporation of diapause trait into the muga silkworm, using transgenic methods, was suggested to mitigate seed crop losses during unfavorable seasons and enhance the muga silk production (Takeda et al. 2014).

Eri Silkworm, *Samia ricini* Donovan

S. ricini (n = 13) is a polyphagous multivoltine silkworm and is semidomesticated for the production of eri silk (white or brick-red) known for its hardiness, disease tolerance, and thermal properties better than wool and highly nutritious pupae as staple food in India's northeastern region, Thailand, and other countries. Eri silkworm is considered as a useful pharmacological model. Among the eight known ecoraces of eri silkworm are Borduar, Titabar, Dhanubhanga, Khanapara, Sile, Nongpoh, Mendipathar, and Kokrajhar based on larval markings, color, and place of collection; but only few of them are commercially exploited due to superior economic traits. Twelve strains of eri silkworm have been identified based on larval markings, viz., White Plain (WP), Green Plain (GP), Yellow Plain (YP), White

Dotted (WD), Green Dotted (GD), Yellow Dotted (YD), White Semi Zebra (WSZ), Green Semi Zebra (GSZ), Yellow Semi Zebra (YSZ), White Zebra (WZ), Green Zebra (GZ), and Yellow Zebra (YZ). These strains are maintained through recurrent selection over the generations under standard laboratory conditions.

The genetic resources of *S. ricini* are better than *A. assamensis*, and 26 ecoraces have been identified (Lee et al. 2020). They are capable of hybridizing with wild *Samia* species found throughout Asia to produce viable offsprings (Lee and Shimada 2024). Hybridization between the wild (*S. canningi*) and cultivated eri (*S. ricini*) silkworms and their utility have gained immense importance for a deeper understanding of the genetic diversity, characteristics, and potential applications. Unlike *B. mori*, wild *Samia* species such as *S. canningi*, *S. pryeri*, *S. walkeri*, etc. could be exploited to generate eri silkworm hybrids. Eri ecoraces commonly used by the farmers for cocoon production are low-yielding (5–8 kg/100 dfls). A high-yielding eri silkworm race, C2, was developed for commercial exploitation with a yield potential of 13.5 kg/100 dfls, higher fecundity, and shell weight.

The whole genome sequence of *S. ricini* (scaffolds: 155; size: 458 Mb; N50 length: 21 Mb) was determined to predict 16,702 protein-coding genes to enhance genetic research in eri silkworms (Lee et al. 2021). The genomic information was updated, and all the contigs were assigned to respective linkages leading to the construction of a chromosome-scale genome assembly (NCBI Acc. No. GCA_014132275.2), which also facilitated forward genetic analysis. Positional cloning of blue larval pigmentation trait (1Mbp) was located on chromosome 14. Later CRISPR/Cas9-mediated knockout of candidate genes was undertaken to identify the responsible gene. This genetic information would be quite useful to accelerate the genomic selection and breeding of *S. ricini* (Lee and Shimada 2024). Gene controlling pupal dormancy is known to be linked to Z chromosome of *S. cynthia*, which could be fixed in the populations in a cross between the two species. Successful genome editing system utilizing transcription activator-like effector nucleases (TALENs) was developed in *S. ricini* (Lee et al. 2018; Lee and Shimada 2024) and created several gene knockout lines including mutants with long diapause character.

Tropical Tasar Silkworm, *Antheraea mylitta* Drury

The tropical tasar silkworm, *A. mylitta*, has wider distribution and is abundantly available in tropical regions, while its temperate counterpart, oak tasar silkworm (*A. proylei*), is cultivated only in limited areas. Improvements in tasar silk production and its quality necessitate development of improved races/breeds. Consequently, exploration, cataloguing, conservation, and characterization of genetic resources or ecoraces are crucial for commercial utilization by breeders and geneticists. Currently, 44 ecoraces of tasar silkworm have been reported with the majority in Jharkhand, Chhattisgarh, and Odisha. Significant phenotypic variation is observed in naturally grown cocoons among and within the tasar ecoraces. Specific breeding lines such as Double Laying Silk Daba (DLSD), Plus Laying Silk Daba (PLSD), and Average Laying Silk Daba (ALSD) have been developed for the genetic improvement of widely cultivated Daba ecorace (bivoltine and trivoltine). Further, two tropical tasar silkworm breeds, viz., BDR-10 and CTR-14, exhibiting higher fecundity have been developed and are currently under evaluation trials.

The tasar ecoraces thrive in their natural habitat by feeding on Sal tree leaves, which the tribals collect cocoons from the forest for their livelihood. These wild ecoraces contribute approximately 15–20% of the total silk production showcasing excellent behavioral and quantitative traits under in situ conditions as compared to captivity. However, they are not easily manageable, and their characteristics tend to degrade with human interference resulting in low survival. Proper survey and collection efforts have not been conducted for an extended period raising concerns about the potential extinction of ecoraces due to deforestation, habitat loss, and climate change effects as natural populations serve as a repository of genes and alleles developed over centuries through natural evolution.

The Central Tasar Research & Training Institute (CTRTI), Ranchi, Jharkhand, developed conservation models for Modal (Odisha), Raily (Chhattisgarh), and Laria (Jharkhand) ecoraces. These models are implemented by state governments to multiply in the fringe forest areas providing livelihoods to tribal communities as non-timber forest produce. In the fast depletion of natural populations due to persistent ecological disturbance, the most crucial element for sustainability is the conservation of wild tasar silk moth diversity. A systematic multiplication program of wild ecoraces (in situ conservation) is supported by the Central Silk Board. An integrated strategy incorporating simultaneous in situ and ex situ conservation of tasar ecoraces, involvement of forest departments and tribal communities, and exploration of semen cryopreservation and artificial insemination techniques is the need of the hour for addressing the decline of biological diversity. Lack of breeds for hotter zones remains a significant constraint in enhancing the tasar silk production, which makes imperative for genetic and molecular characterization of ecoraces and initiatives to promote their conservation and development of productive and high-yielding thermotolerant tasar breeds.

Shallow sequencing of *A. mylitta* genome was undertaken and yielded 65 GB sequence data with 35.9% GC content. Comparative genomic analysis indicated substantial similarities with 77.4% *A. mylitta* genome aligning to *A. yamamai* and only 11.3% with *B. mori* suggesting a closer genetic relationship with *A. yamamai*. Following the above genetic information, de novo whole genome sequencing was undertaken employing PacBio and Illumina technologies, and the sequence data was submitted to NCBI for providing accessibility to the researchers (NCBI Acc. No. JACEII000000000). The PacBio platform contributed 10.22 GB data. Concurrently, sequencing of 10X chromium libraries via the Illumina HiSeq X 10 system utilizing 2x150 pair-end chemistry generated 106,443.28 MB with a GC content of 37.68% and a high-quality score (Q30) of 93.305%. The genomic data assembly resulted in an assembled genome size of approximately 707.76 MB with a remarkable genome coverage of 184.6x incorporating 20,891 scaffolds with an N50 length of 5,182,261 base pairs and 33,698 contigs with an N50 contig length of 136,984 base pairs. The finished genome completeness was estimated at 97.50%.

Gene annotation was conducted utilizing a suite of bioinformatics tools leading to the annotation of 25,544 genes. A detailed BLAST search comparison between the proteins of *A. mylitta* and *A. yamamai* revealed significant similarity with 86.27% of *A. mylitta* proteins showing hits with *A. yamamai* proteins. Orthology analysis conducted identified 14,317 ortholog groups among *B. mori*, *A. yamamai*, and *A. mylitta* further elucidating the evolutionary relationships and functional genomics across these species. This comprehensive genomic characterization of *A. mylitta* not only enhances the understanding of its genetic architecture but also contributes valuable insights into the evolutionary biology of sericigenous insects, functional genomics, and gene discovery for productive traits.

Bioinformatic analysis revealed significant differences in gene expression between diapause and non-diapause *A. mylitta* pupae implicating key genes involved in metabolic suppression, hormonal pathways, stress response, and environmental adaptation. An intricate interplay between environmental cues and hormonal regulation dictates the diapause status (Prabhu et al. 2024).

Chinese Oak Tasar Silkworm, *Antheraea Pernyi*

Southern China is the origin of Chinese oak tasar silkworm, *Antheraea pernyi* (n = 49), which is cultivated widely for its silk. The temperate tasar or oak tasar silk that originated from China was introduced in India in the 1960s. On the other hand, the temperate tasar moth, *Antheraea roylei* (n = 30, 31, 32), is found in sub-Himalayan region of India. A fertile interspecific hybrid, *Antheraea proylei*, was evolved through a cross between the Chinese *A. pernyi* and its Indian counterpart. Very meager quantities of oak tasar silk is produced in the sub-Himalayan belt of India, covering states such as Jammu and Kashmir, Manipur, Himachal Pradesh, Uttarakhand, Meghalaya, Mizoram, and Nagaland utilizing *A. proylei*, *A. roylei*, *A. frithi*, and *A. pernyi*. Currently, oak tasar industry in India is under the revival through strategic measures and interventions for its resurgence and sustainability. Oak tasar ecoraces being genetically distinct necessitate in situ conservation to facilitate their multiplication, which is crucial for future breeding programs.

Molecular phylogeny of silk moths reveals the origin of domesticated silk moth, *B. mori*, from Chinese *B. mandarina* and paternal inheritance of *A. proylei* mitochondrial DNA (Arunkumar et al. 2006). The prothoracic gland transcriptomes of *B. mori* and *A. pernyi* share an expressed gene profile demonstrating that orthologs in non-model organisms may be identified using RNA-seq technology (Bian et al. 2019).

Japanese Wild Tasar Silkworm, *Antheraea yamamai*

A. yamamai is a native of Japan and is a 31-chromosome Japanese oak tasar silkworm. The silk produced from these insects has unique characteristics such as thickness, compressive elasticity, and chemical resistance and is used to make highly valued silk fabrics.

Illumina and PacBio sequencing platforms were used to generate 147G base pairs for the construction of the *A. yamamai* genome with 210-fold coverage. The assembled genome of *A. yamamai* was 656 Mb (>2 kb) with 3675 scaffolds, and the N50 length is 739 Kb with a 34.07% GC ratio. Identified repeat elements covered 37.33% of the total genome, and the completeness of the constructed genome assembly was estimated to be 96.7%. A total of 15,481 genes were identified based on the gene prediction results obtained from 3 different methods (ab initio, RNA-seq-based, known gene-based) and manual curation. These results provide valuable genomic information, which will help enrich the understanding of the molecular mechanisms relating to not only specific phenotypes such as wild silk itself but also the genomic evolution of Saturniidae (Kim et al. 2017a).

The mitochondrial genome of *A. yamamai* is 15,341 bp and encodes 13 protein-coding genes, 2 ribosomal RNA genes, and 22 transfer RNA genes (Sun et al. 2021). Sequence comparison identified 22 SNVs in the *A. yamamai* mitochondrial genomes between the Chinese and Korean populations, indicating a low intraspecific variation. Phylogenetic analyses with maximum-likelihood and Bayesian inference methods revealed a close relationship between *A. yamamai* and *Antheraea frithi* and supported the relationship among *Antheraea* species (((*A. yamamai* + *A. frithi*) + *A. pernyi*) + *A. assamensis*).

Population genetic characterization of *A. yamamai* (Lepidoptera: Saturniidae) in South Korea was determined using ten novel microsatellite markers and two mitochondrial DNA gene sequences (COI and ND4) by Kim et al. (2017b). The two mtDNA gene sequences revealed very low total genetic variation and, consequently, low geographic variation validating the use of more variable molecular markers. Genotyping and population-based FIS, FST, and RST and global Mantel tests suggested that the *A. yamamai* populations were overall well-interconnected. Nevertheless, STRUCTURE analyses using microsatellite data and mtDNA sequences indicated the presence of two genetic pools.

The gene structure and mRNA expression of *A. yamamai* cuticle protein gene AyCP12 (1107 bp) were determined by Kim et al. (2005), consisting of an intron and 2 exons coding for a 112 amino acid polypeptide (12.163 KDa and a pI of 4.4). The AyCP12 protein contained a type-specific consensus sequence identifiable in other insect cuticle proteins, and the deduced amino acid sequence of the AyCP12 cDNA is most homologous to *A. pernyi* cuticle protein ApCP13 (82% protein sequence identity). Northern blot analysis revealed that AyCP12 showed the epidermis-specific expression.

Sustainable Perspectives

Enhancement of productivity improvement in case of mulberry silkworm, *B. mori*, has been fully realized almost, while challenges such as crop loss due to infectious diseases, mortality due to abiotic stresses (high temperature, high humidity, radiation), adaptation to climate change, and novel biomaterial resources are yet to be realized. Hence, continuous research is expected to provide better adaptive breeds/ hybrids, mitigative strategies, and provide much sought after biomaterials with novel properties. On the other hand, much needed research has to be initiated on all fronts in case of wild silkworms, and there is tremendous scope for improvement of wild silk production. The current status is very appalling, and immediate efforts are essential to catch up *B. mori* research. The wild silkworm conservation efforts aiming to ensure sustainable use in breeding programs are essential for fostering the development of high-productivity, disease-resistant, and climate-resilient breeds/ varieties and unravelling the genomic information.

The major fields of biotechnology such as genome editing, transgenesis, and omics technologies have seen extraordinary advancements in the last 20 years. The much farmed *B. mori* has seen all and is being researched further, while very few attempts have been made with wild silkworms. The comprehensive understanding of wild silkworm genomes and continued efforts on their functional genomics, marker-assisted breeding, genome editing, and epigenetics would provide valuable insights into the genetic basis of key traits akin to *B. mori*. Detailed genomic information is instrumental in implementing targeted breeding strategies and genetic improvements, ultimately leading to the sustainable development of wild silk moths. The utilization of novel biotechnologies to tackle enduring problems in silk farming is crucial as it is expected to give valuable contributions for general improvement of silk industry through enhanced survival, productivity, and quality. The integration of genomic, transcriptomic, and biotechnological approaches should result in profound understanding of the molecular mechanisms underlying silk production, insect-host plant interactions, and the overall adaptability of silkworms to environmental changes. The following perspectives are quite essential for the sustainable development of silk industry:

- Genomic studies on wild silk silkworms
- Functional genomics on genes rendering silk productivity/quality characteristics
- Identification of genes for resistance/tolerance to infectious pathogens and abiotic stress
- Meaningful conservation of ecoraces and wild silk species
- Genome editing approaches for developing silkworm breeds with desired traits
- Adoption of silkworm improvement programs utilizing AI techniques
- Development of field-oriented disease diagnosis kits
- Development of diapausing eri and muga silkworms
- Development of disease resistant/tolerant seri-genetic resources
- Pyramiding of tolerance to biotic and abiotic factors in silkworms
- New silk biomaterials with novel properties

References

Abdul-Muneer PM (2014) Application of microsatellite markers in conservation genetics and fisheries management: recent advances in population structure analysis and conservation strategies. Genet Res Int 691759:1–11

Arunkumar KP, Metta M, Nagaraju J (2006) Molecular phylogeny of silkmoths reveals the origin of domesticated silkmoth, *Bombyx mori* from Chinese *Bombyx mandarina* and paternal inheritance of *Antheraea proylei* mitochondrial DNA. Mol Phylogenet Evol 40(2):419–427

Arunkumar KP, Tomar A, Daimon T, Shimada T, Nagaraju J (2008) WildSilkbase: an EST database of wild silkmoths. BMC Genomics 9:338

Arunkumar KP, Kifayathullah L, Nagaraju J (2009) Microsatellite markers for the Indian golden silkmoth, *Antheraea assama* (Saturniidae: Lepidoptera). Mol Ecol Resour 9:268–270

Arunkumar KP, Sahu AK, Mohanty AR, Awasthi AK, Pradeep AR, Urs SR, Nagaraju J (2012) Genetic diversity and population structure of Indian Golden Silkmoth (*Antheraea assama*). PLoS One 7(8):e43716. https://doi.org/10.1371/journal.pone.0043716

Awasthi AK, Kar PK, Srivastava PP, Rawat N, Vijayan K, Pradeep AR, Urs SR (2008) Molecular evaluation of bivoltine and mutant silkworm (*Bombyx mori* L.) with RAPD, ISSR and RFLP STS markers. Indian J Biotechnol 7:188–194

Baci GM, Cucu AA, Giurgiu AI, Muscă AS, Bagameri L, Moise AR (2021) Advances in editing silkworms (*Bombyx mori*) genome by using the CRISPR-cas system. Insects 13(1):28

Bian HX, Chen DB, Zheng XX, Ma HF, Li YP, Li Q (2019) Transcriptomic analysis of the prothoracic gland from two lepidopteran insects, domesticated silkmoth *Bombyx mori* and wild silkmoth *Antheraea pernyi*. Sci Rep 9(1):5313. https://doi.org/10.1038/s41598-019-41864-0

Cheng TC, Xia QY, Qian JF, Liu C, Lin Y, Zha XF, Xiang ZH (2004) Mining single nucleotide polymorphisms from EST data of silkworm, *Bombyx mori*, inbred strain Dazao. Insect Biochem Mol Biol 34(6):523–530

Cheng T, Fu B, Wu Y, Long R, Liu C, Xia Q (2015) Transcriptome sequencing and positive selected genes analysis of *Bombyx mandarina*. PLoS One 10:e0122837. https://doi.org/10.1371/journal.pone.0122837

Cui Y, Zhu Y, Lin Y, Chen L, Feng Q, Wang W, Xiang H (2018) New insight into the mechanism underlying the silk gland biological process by knocking out fibroin heavy chain in the silkworm. BMC Genom 19:215

Cui Y, Liu ZL, Li CC, Wei XM, Lin YJ, You L (2021) Role of juvenile hormone receptor Methoprene-tolerant 1 in silkworm larval brain development and domestication. Zool Res 42(5):637

Ding X, Liu J, Tong X, Wu S, Li C, Song J, Hu H, Tan D, Dai F (2020) Comparative analysis of integument transcriptomes identifies genes that participate in marking pattern formation in three allelic mutants of silkworm, *Bombyx mori*. Functional & Integrative Genomics 20(3):223–235

Dong Y, Dai F, Ren Y, Liu H, Chen L, Yang P (2015) Comparative transcriptome analyses on silk glands of six silkmoths imply the genetic basis of silk structure and coloration. BMC Genomics 16(1):203. https://doi.org/10.1186/s12864-015-1420-9

Dubey H, Pradeep AR, Neog K, Debnath R, Aneesha PJ, Shah SK, Sivaprasad V, Arunkumar KP (2023) Genome sequencing and assembly of Indian golden silkmoth, *Antheraea Assamensis* Helfer (Saturniidae, Lepidoptera). Genomics 116(3):110841

Fang SM, Zhou QZ, Yu QY, Zhang Z (2020) Genetic and genomic analysis for cocoon yield traits in silkworm. Sci Rep 10:5682. https://doi.org/10.1038/s41598-020-62507-9

Gan L, Liu X, Xiang Z, He N (2011) Microarray-based gene expression profiles of silkworm brains. BMC Neurosci 12:1–14

Goldsmith MR, Shimada T, Abe H (2005) The genetics and genomics of the silkworm, *Bombyx mori*. Annu Rev Entomol 50:71–100

Hasnahana Chetia HC, Debajyoti Kabiraj DK, Deepika Singh DS, Mosahari PV, Suradip Das SD, Pragya Sharma PS, Kartik Neog KN, Swagata Sharma SS, Jayaprakash P, Utpal Bora UB

(2017) *De novo* transcriptome of the muga silkworm, *Antheraea assamensis* (Helfer). Gene 611:54–65. https://doi.org/10.1016/j.gene.2017.02.021

Hutter CM, Schug MD, Aquadro CF (1998) Microsatellite variation in *Drosophila melanogaster* and *Drosophila simulans*: a reciprocal test of the ascertainment bias hypothesis. Mol Biol Evol 15(12):1620–1636

International Silkworm Genome Consortium (2008) The genome of a lepidopteran model insect, the silkworm *Bombyx mori*. Insect Biochem Mol Biol 38(12):1036–1045

Kawamoto M, Jouraku A, Toyoda A, Yokoi K, Minakuchi Y, Katsuma S, Fujiyama A, Kiuchi T, Yamamoto K, Shimada T (2019) High-quality genome assembly of the silkworm, *Bombyx mori*. Insect Biochem Mol Biol 107:53–62. https://doi.org/10.1016/j.ibmb.2019.02.002

Kawamoto M, Kiuchi T, Katsuma S (2022) SilkBase: an integrated transcriptomic and genomic database for *Bombyx mori* and related species. Database 2022:baac040. https://doi.org/10.1093/database/baac040

Kim BY, Park NS, Jin BR, Lee BH, Seong SI, Hwang JS (2005) A cuticle protein gene from the Japanese oak silkmoth, *Antheraea yamamai*: gene structure and mRNA expression. Biotechnol Lett 27:1499–1504

Kim SR, Kim KY, Jeong JS, Kim MJ, Kim KH, Choi KH, Kim I (2017a) Population genetic characterization of the Japanese oak silkmoth, *Antheraea yamamai* (Lepidoptera: Saturniidae), using novel microsatellite markers and mitochondrial DNA gene sequences. Gen Mol Res 16(2):gmr16029608

Kim SR, Kwak W, Kim H, Caetano-Anolles K, Kim KY, Kim SB (2017b) Genome sequence of the Japanese oak silk moth, *Antheraea yamamai*: the first draft genome in the family Saturniidae. *Giga*. Science 7:GIX113

Lee J, Shimada T (2024) The rise and future of genetics in *Samia ricini*. Key note address: 08. In: International conference on silkworm seed industry: opportunities and future prospects, Bangalore, India

Lee J, Kiuchi T, Kawamoto M, Shimada T, Katsuma S (2018) Accumulation of uric acid in the epidermis forms the white integument of *Samia ricini* larvae. PLoS One 13(10):e0205758

Lee J, Nishiyama T, Shigenobu S, Yamaguchi K, Suzuki Y, Shimada T (2020) The genome sequence of *Samia ricini*, a new model species of lepidopteran insect. Mol Ecol Resour 21:327–339

Lee J, Nishiyama T, Shigenobu S, Yamaguchi K, Suzuki Y, Shimada T, Kiuchi T (2021) The genome sequence of *Samia ricini*, a new model species of lepidopteran insect. Mol Ecol Resour 21(1):327–339

Li Y, Ma S, Sun L, Zhang T, Chang J, Lu W, Chen X, Liu Y, Wang X, Shi R, Zhao P (2018) Programmable single and multiplex base-editing in *Bombyx mori* using RNA-guided cytidine deaminases. G3: Genes, Genomes. Genetics 8(5):1701–1709

Liu S, Li D, Li Q, Zhao P, Xiang Z, Xia Q (2010) MicroRNAs of *Bombyx mori* identified by Solexa sequencing. BMC Genomics 11:148

Ma S, Zhang S, Wang F, Liu Y, Liu Y, Xu H, Xia Q (2012) Highly efficient and specific genome editing in silkworm using custom TALENs. PLoS One 7(9):e45035. https://doi.org/10.1371/journal.pone.0045035

Ma L, Ma Q, Li X, Cheng L, Li K, Li S (2014) Transcriptomic analysis of differentially expressed genes in the Ras1 CA-overexpressed and wild type posterior silk glands. BMC Genomics 15:1–13

Ma S, Liu Y, Liu Y, Chang J, Zhang T, Wang X, Shi R, Lu W, Xia X, Zhao P, Xia Q (2017) An integrated CRISPR *Bombyx mori* genome editing system with improved efficiency and expanded target sites. Insect Biochem Mol Biol 83:13–20

Ma Y, Zeng W, Ba Y, Luo Q, Ou Y, Liu R (2022) A single-cell transcriptomic atlas characterizes the silk-producing organ in the silkworm. Nat Commun 13:3316. https://doi.org/10.1038/s41467-022-31003-1

Ma S, Zhang T, Wang R, Wang P, Liu Y, Chang J, Xia Q (2024) High-throughput and genome-scale targeted mutagenesis using CRISPR in a nonmodel multicellular organism, *Bombyx mori*. Genome Res 34(1):134–144. https://doi.org/10.1101/.08.02.551621

Meng Y, Omuro N, Funaguma S, Daimon T, Kawaoka S, Katsuma S (2008) Prominent down-regulation of storage protein genes after bacterial challenge in eri-silkworm, *Samia cynthia ricini*. Arch Insect Biochem Physiol 67(1):9–19. https://doi.org/10.1002/arch.20214

Mita K, Kasahara M, Sasaki S, Nagayasu Y, Yamada T, Kanamori H, Namiki N, Kitagawa M, Yamashita H, Yasukochi Y et al (2004) The genome sequence of silkworm, *Bombyx mori*. DNA Res 11:27Ð35

Moon SB, Kim DY, Ko JH, Kim YS (2019) Recent advances in the CRISPR genome editing tool set. Exp Mol Med 51(11):1–11

Ou J, Deng HM, Zheng SC, Huang LH, Feng QL, Liu L (2014) Transcriptomic analysis of developmental features of *Bombyx mori* wing disc during metamorphosis. BMC Genomics 15:1–15

Park JW, Yu JH, Kim SB, Kim SW, Kim SR, Choi KH (2019) Analysis of silkworm molecular breeding potential using CRISPR/Cas9 systems for white egg 2 gene. Int J Ind Entomol 39(1):14–21

Prabhu GI, Nidhi S, Vikas S, Gadad H, Pandey JP, Mittal V, Chowdary NB (2024) Expression and functional analysis of genes related to diapause in *Antheraea mylitta* D. In: Intlernational conference on silkworm seed industry: opportunities and future prospects, Bangalore, India (T3S1:OP-06)

Prasad MD, Muthulakshmi M, Madhu M, Archak S, Mita K, Nagaraju J (2005) Survey and analysis of microsatellites in the silkworm, *Bombyx mori*: frequency, distribution, mutations, marker potential and their conservation in heterologous species. Genetic 169:197–214

Priyadharshini P, Maria JA (2019) Silkworm databases and its applications. J Int Acad Res Multidiscip 7:11

Rao RV, Hodgkin T (2002) Genetic diversity and conservation and utilization of plant genetic resources. Plant Cell Tissue Organ Cult 68:1–19

Reddy KD, Abraham EG, Nagaraju J (1999) Microsatellites in the silkworm, *Bombyx mori*: abundance, polymorphism, and strain characterization. Genome 42:1057–1065

Singh D, Kabiraj D, Sharma P, Chetia H, Mosahari PV, Neog K, Bora U (2017) The mitochondrial genome of Muga silkworm (*Antheraea assamensis*) and its comparative analysis with other lepidopteran insects. PLoS One 12(11):e0188077. https://doi.org/10.1371/journal.pone.0188077

Sivaprasad V, Chandrakanth N, Moorthy SM (2022) Genetics and genomics of *Bombyx mori* L. In: Genetic methods and tools for managing crop pests. Springer, pp 127–209

Subbaiah EV, Royer C, Kanginakudru S, Satyavathi VV, Babu AS, Sivaprasad V (2013) Engineering silkworms for resistance to baculovirus through multigene RNA interference. Genetics 193:63–75

Sun SW, Huang JC, Liu YQ (2021) The complete mitochondrial genome of the wild silkmoth *Antheraea yamamai* from Heilongjiang, China (Lepidoptera: Saturniidae). Mitochondrial B Resour 6:2209–2211

Suzuki MG (2019) Increased applicability of genetically modified silkworms leads to innovation in sericulture and sericology. In: 6th APSERI, Mysuru, India, pp 12–16

Takasu Y, Sajwan S, Daimon T, Osanai-Futahashi M, Uchino K, Sezutsu H, Zurovec M (2013) Efficient TALEN construction for *Bombyx mori* gene targeting. PLoS One 8(9):e73458

Takeda M, Kobayashi J, Kamimura M, Prakash PJ (2014) Introduction of diapause trait into the muga silkmoth, *Antheraea assamensis*: a scientific collaboration over the Himalayas. Int J Wild Silkmoth Silk 18:39–54

Tamura T, Thibert C, Royer C, Kanda T, Eappen A, Kamba M (2000) Germline transformation of the silkworm *Bombyx mori* L. using a piggyBac transposon-derived vector. Nat Biotechnol 18(1):81–84

Tang J, Li W, Zhang X, Zhou C (2007) The gene expression profile of *Bombyx mori* silkgland. Gene 396:369–372

Tian HL, Wang FG, Zhao JR, Yi HM, Wang L, Wang R (2015) Development of maizeSNP3072, a high-throughput compatible SNP array, for DNA fingerprinting identification of Chinese maize varieties. Mol Breed 35:1–11. https://doi.org/10.1007/s11032-015-0335-0

Tomita S (2019) Development of new silk materials with GM silkworms. In: 6th APSERI, Mysuru, India, p 158

Tong X, Han MJ, Lu K, Tai S, Liang S, Liu Y, Hu H, Shen J, Long A, Zhan C, Ding X (2022) High resolution silk worm pan-genome provides genetic insights into artificial selection and ecological adaptation. Nat Commun 13(1):5619

Wang J, Xia Q, He X (2005) SilkDB: a knowledgebase for silkworm biology and genomics. Nucleic Acids Res 33:399–402

Wang S, You Z, Feng M, Che J, Zhang Y, Qian Q, Komatsu S, Zhong B (2016) Analyses of the molecular mechanisms associated with silk production in silkworm by iTRAQ-based proteomics and RNA sequencing-based transcriptomics. J Proteome Res 15(1):15–28

Wei W, Xin H, Roy B, Dai J, Miao Y, Gao G (2014) Heritable genome editing with CRISPR/Cas9 in the silkworm. Bombyx mori PLoS One 9(7):e101210

Wu S, Tong X, Peng C, Xiong G, Lu K, Hu H (2016) Comparative analysis of the integument transcriptomes of the black dilute mutant and the wild-type silkworm *Bombyx mori*. Sci Rep 6(1):26114

Xia Q, Cheng D, Duan J, Wang G, Cheng T, Zha X, Liu C, Zhao P, Dai F, Zhang Z (2007) Microarray-based gene expression profiles in multiple tissues of the domesticated silkworm, *Bombyx mori*. Genome Biol 8(8):R162-10.1186/gb-2007-8-8-r162

Xia Q, Zhou Z, Lu C, Cheng D, Dai F, Li B, Zhao P, Zha X, Cheng T, Chai C (2004) A draft sequence for the genome of the domesticated silkworm (*Bombyx mori*). Science 306:1937–1940

Xia Q, Guo Y, Zhang Z, Li D, Xuan Z, Li Z (2009) Complete resequencing of 40 genomes reveals domestication events and genes in silkworm (*Bombyx)*. Science 326(5951):433–436

Xia Q, Xiang Z (eds) (2013) The Genome of Silkworm. Science Press, Beijing

Xia Q, Li S, Feng Q (2014) Advances in silkworm studies accelerated by the genome sequencing of *Bombyx mori*. Annu Rev Entomol 59(1):513–536

Xiang H, Liu X, Li M, Zhu YN, Wang L, Cui Y (2018) The evolutionary road from wild moth to domestic silkworm. Nat Ecol Evol 2(8):1268–1279. https://doi.org/10.1038/s41559-018-0593-4

Xiong G, Tong X, Gai T, Li C, Qiao L, Monteiro A (2017) Body shape and coloration of silkworm larvae are influenced by a novel cuticular protein. Genetics 207(3):1053–1066

Xu Q, Lu A, Xiao G, Yang B, Zhang J, Li X, Guan J, Shao Q, Beerntsen BT, Zhang P, Wang C (2012) Transcriptional profiling of midgut immunity response and degeneration in the wandering silkworm. Bombyx mori PLoS One 7(8):e43769. https://doi.org/10.1371/journal.pone.0043769

Yamaguchi J, Banno Y, Mita K, Yamamoto K, Ando T, Fujiwara H (2013) Periodic Wnt1 expression in response to ecdysteroid generates twin-spot markings on caterpillars. Nat Commun 4(1):1857

Yamamoto K, Narukawa J, Kadono-Okuda K, Nohata J, Sasanuma M, Suetsugu Y (2006) Construction of a single nucleotide polymorphism linkage map for the silkworm, *Bombyx mori*, based on bacterial artificial chromosome end sequences. Genetics 173(1):151–161

Yamamoto K, Nohata J, Kadono-Okuda K, Narukawa J, Sasanuma M, Sasanuma S, Minami H, Shimomura M, Suetsugu Y, Banno Y, Osoegawa K, J. de Jong P, Goldsmith MR, Mita K (2008) A BAC-based integrated linkage map of the silkworm. Bombyx mori Genome Biol, 9: R21.

Yamamoto K, Higashiura A, Hirowatari A, Yamada N, Tsubota T, Sezutsu H, Nakagawa A (2018) Characterisation of a diazinon-metabolising glutathione S-transferase in the silkworm *Bombyx mori* by X-ray crystallography and genome editing analysis. Sci Rep 8(1):16835

Yu HZ, Zhang SZ, Ma Y, Fei DQ, Li B, Yang LA (2017) Molecular characterization and functional analysis of a ferritin heavy chain subunit from the Eri-silkworm, *Samia cynthia ricini*. Int J Mol Sci 18(10):2126

Zha X, Xia Q, Duan J, Wang C, He N, Xiang Z (2009) Dosage analysis of Z chromosome genes using microarray in silkworm, *Bombyx mori*. Insect Biochem Mol Biol 39:315–321

Zhang Z, Niuc B, Jic D (2018) Silkworm genetic sexing through W chromosomelinked, targeted gene integration. PNAS 115(35):8752–8756

Zhao XM, Liu C, Li QY, Hu WB, Zhou MT, Nie HY, Zhang YX, Peng ZC, Zhao P, Xia QY (2014) Basic helix-loop-helix transcription factor Bmsage is involved in regulation of fibroin H-chain gene via interaction with SGF1 in *Bombyx mori*. PLoS One 9(4):e94091

Zheng X, Liu F, Shi M (2022) Transcriptome analysis of the reproduction of silkworm (*Bombyx mori*) under dimethoate stress. Pestic Biochem Physiol 183:105081

Zhou Y, Gao L, Shi H, Xia H, Gao L et al (2013) Microarray analysis of gene expression profile in resistant and susceptible *Bombyx mori* strains reveals resistance-related genes to nucleopolyhedrovirus. Genomics 101:256–262

Zhou R, Li S, Liu J, Wu H, Yao G, Sun Y, Chen Z, Li W, Du Y (2020) Up-regulated FHL2 inhibits ovulation through interacting with androgen receptor and ERK1/2 in polycystic ovary syndrome. EBioMedicine 52:102635

Zhu GH, Peng YC, Zheng MY, Zhang XQ, Sun JB, Huang Y (2017) CRISPR/Cas9 mediated BLOS2 knockout resulting in disappearance of yellow strips and white spots on the larval integument in *Spodoptera litura*. J Insect Physiol 103:29–35

Zou YL, Ye AJ, Liu S, Wu WT, Xu LF, Dai FY, Tong XL (2021) Expansion of targetable sites for the ribonucleoprotein-based CRISPR/Cas9 system in the silkworm *Bombyx mori*. BMC Biotechnol 2:1–9

Silkworm Genomics: A Novel Tool in Silkworm Crop Improvement

Raviraj V Suresh, Soumen Saha, Nalavadi Chandrakanth, Khasru Alam, Anil Pappachan, and Shunmugam Manthira Moorthy

Abstract

For thousands of years, silkworms have been bred for their silk production, which is a crucial industry in many countries. Scientists have made significant progress in improving the production of silk through the study of silkworm genomics. This involves examining the entire genetic material of silkworms, including their DNA sequences, gene functions, and genetic variations among different strains. In 2004, the genome of the silk moth *Bombyx mori* was sequenced, a groundbreaking achievement in the field of insect genomics. This milestone has allowed researchers to identify key genes involved in silk production, development, immunity, and other essential biological processes. Silkworm genomics has revolutionized the field of sericulture by providing a deeper understanding of the genetic basis of silk production and other essential biological processes in silkworms. Genomics has also facilitated the development of molecular breeding strategies to improve desired traits in silkworms more efficiently. Silkworms are susceptible to various diseases that can significantly impact silk production. By identifying genes that confer resistance, scientists can develop silkworm strains with improved disease resistance, thereby reducing economic

R. V Suresh (✉) · K. Alam
Bivoltine Breeding Laboratory, Central Sericultural Research & Training Institute, Central Silk Board, Berhampore, India

S. Saha
Department of Sericulture, Raiganj University, Raiganj, West Bengal, India

N. Chandrakanth
Bivoltine Breeding Laboratory, Central Sericultural Research & Training Institute, Central Silk Board, Mysuru, India

A. Pappachan · S. M. Moorthy
National Silkworm Seed Organization, Central Silk Board, Bangalore, India

R. V Suresh et al. (eds.), *Biotechnology for Silkworm Crop Enhancement*, https://doi.org/10.1007/978-981-97-5061-0_2

losses. The knowledge gained from studying the silkworm genome has significant implications for crop improvement, allowing researchers to develop silk with enhanced quality and quantity, breed silkworms with improved disease resistance, and explore new avenues for molecular breeding strategies. The continuous advancements in genomic technology and analytical techniques promise even more exciting discoveries and innovations in the future of silkworm research and sericulture industry. This chapter provides insights into the silkworm genomics and candidate genes identified using genomics for applications in silkworm breeding and inducing disease resistance, as well as its role in improving silk fiber.

Keywords

Genomics · Molecular breeding · Disease resistance · Crop improvement

Introduction

Silkworm *Bombyx mori*, a small insect, with vital economic and agricultural significance, was domesticated approximately 5000 years ago from its ancestors, *Bombyx mandarina* (Tong et al. 2022). Presently, more than 1000 inbred lines of the domesticated silkworm are maintained and propagated worldwide. The domesticated silkworm has coexisted with humans for thousands of years, helping to develop the sericulture business and giving us materials for clothing and art (Yang et al. 2021). Currently, 20 million Chinese farmers grow mulberries and raise silkworms; the country's yearly output of silkworm cocoons makes up around 80% of the global output and has been the world leader for more than 30 years (Tong et al. 2022). In, India Sericulture provides income for approximately 8 million farmers which is the second leading silk producer in the world (https://inserco.org/en/statistics).

Silkworms have been studied extensively as model organisms to uncover the genetic principles of insect physiology. In the early 1900s, genetic and physiological research on silkworms led to several significant basic discoveries in hormones, pheromones, brain anatomy, physiology, and insect genetics (Goldsmith et al. 2005). In the 2000s, there was a notable acceleration in biotechnological research such as comprehensive genome level investigations which led to the creation of the silkworm genome's draft map, fine map, epigenetic map, and variation atlas (Xia et al. 2014). To keep pace with the rapidly changing productive forces and methods, new breeding methods and a fresh way of thinking are desperately needed to speed up the transformation and advancement of sericulture. The development of silkworm genetic breeding has been crucial in the advancement of the sericulture-based silk industry (Ma and Xia 2017).

Silkworm breeding has advanced quickly as a result of the consecutive discoveries of a number of fundamental principles and theories, including "The Origin of Species," Mendelian inheritance, population inheritance, and quantitative inheritance. The potential for raising production through conventional breeding techniques has run out after a protracted period of investigation. The discipline of silkworm breeding is continuously utilizing new technologies and innovative approaches due to the rapid advancement of technology (Xia et al. 2014). In 1991, Goldsmith of Rhode Island University in the United States suggested the international silkworm molecular breeding plan. Molecular marker maps of silkworms, including AFLP, RFLP, RAPD, SSR, and SNP, were created as part of this endeavor (Goldsmith et al. 2005). Apart from notable advancements in BmNPV (*Bombyx mori* nucleopolyhedrovirus) resistance, fluorescence identification of silkworm cocoons, and breeding of specialty varieties, overall progress in genetic breeding of silkworms has been relatively slow in the over 50 years since hybrids were made popular. Proteomes, metabolomes, genetic variation maps, gene expression chips, and genome maps have all been examined one after the other. Certain aims have been effectively converted into silkworm genetic breeding work through the use of silkworm transgenic technology. Many uses for transgenic silkworms have been reported, including the synthesis of medications, collagen, and antibodies (Tada et al. 2015; Tomita 2011). The silk material itself has garnered significant interest for a variety of biological applications (Mori and Tsukada 2000; Aigner et al. 2018; Farokhi et al. 2018 & Holland et al. 2019). To fully utilize silkworm with genetic engineering technology, precise genomic and genetic information will therefore be required (Yang et al. 2021). Thus, a silk genome resource database can serve as the foundation for studies on silkworms and also make study in other domains easier. The advent and rapid advancement of genome editing technologies in recent times have invigorated the field of insect genetic breeding. One of the first insects to accomplish genome editing is *Bombyx mori*. This chapter focuses on the primary and recent accomplishments in genetics and genomic technologies related to silkworm crop improvement.

Silkworm Genetics

Genetics plays a major role in the development of an organism from embryo to adult. Genetics is a vast field of study involving many components (Fig. 1). Genetics plays a major role in evolution of any organism from larger mammals to tiny moths. This section highlights the genetic content of the silk moth from its journey from wild moth to domestication. The domestication of silkworms is a topic of debate and has garnered significant attention. Researchers are interested in whether *Bombyx mori* originated from a single group of *Bombyx mandarina* or multiple groups with differing voltinism (Arunkumar et al. 2006; Li et al. 2015). Japanese researchers initiated the first studies on the silkworm genetics, which are available in their native language and are not easily understood due to language barrier. The first compilation of monograph on silkworm genetics which excluded molecular studies was investigated by Tazima and Ohuma (1995). *Bombyx mori* follows the WZ/ZZ chromosome system of sex determination, which is seen in lepidopteran insects, where males are homogametic (ZZ) and the females are heterogametic (ZW) (Xu et al. 2017). Genome-wide SNP-based neighbor-joining phylogenetic tree and principal component analysis indicate a clear genetic separation between domesticated and

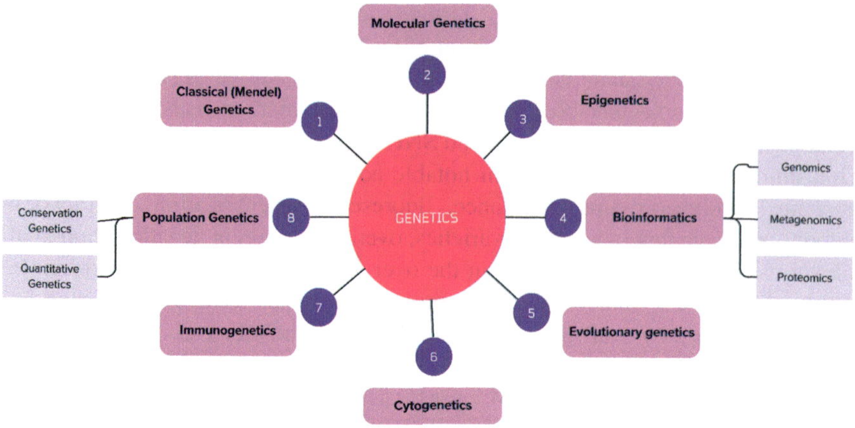

Fig. 1 Representation of the components of basic genetics

wild silkworm populations. *Bombyx mandarina*, the wild silkworm, offers ample opportunities for genetic manipulation and can help us understand many other bombycids, covering many aspects such as domestication, feeding patterns, nutrition uptake, and evolution. Additionally, *B. mandarina* has shed light on the underlying genetic mechanisms behind several cocoon yield traits in *B. mori*, as evidenced by studies conducted by Goncu and Parlak (2008), Gu et al. (2009, 2010, 2012), and Fang et al. (2020).

Silkworm Genome

The Bombycidae family under the Lepidoptera order is where most species of silk moths are found. This family has about 160,000 species, and a majority of them are agriculture pests. According to d'Alencon et al. (2010), *Bombyx mori* is the primary model species, and it was the first Lepidoptera to be fully sequenced.

Genomics can be divided into many components (Fig. 2) from comparative genomics to functional genomics. These genomic components have varied utilities in silkworm development. The first silkworm (*Bombyx mori*) genome was sequenced through a collaborative effort between Chinese researchers from Southwest University in Chongqing, China, and Japanese researchers from the University of Tokyo. The genome sequencing was completed, and the results were published in 2004. The silkworm's haploid genome (~450 Mb) is one-seventh the size of the human genome and 3.4 times larger than *Drosophila melanogaster*'s. The primary data needed for its genetic research was made available in 2004 with the original genome assembly of the silkworm carried out in the strain p50T that was disclosed by Mita et al. (2004) and Xia et al. (2004). The silkworm genome is estimated to contain around 14,623 protein-coding genes. These genes are responsible for producing proteins essential for various biological processes. The genes in the

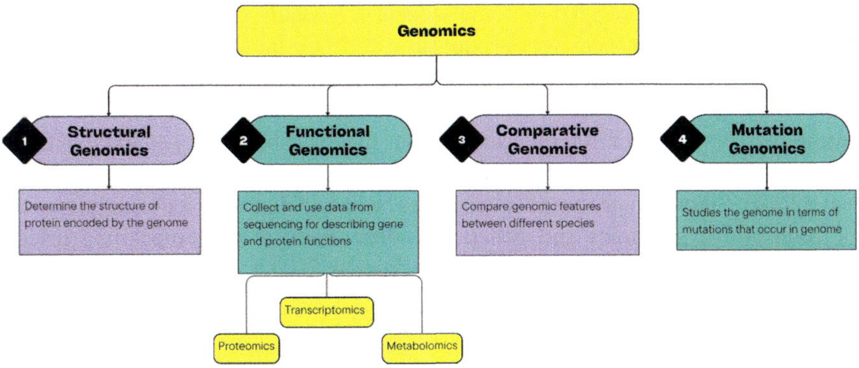

Fig. 2 Representation of components of genomics and its functions

Table 1 Overview of Single gene functions

Gene	Function	Authors
BmE2F1 transcription factor	Reduction in number of silk gland cells by 7.68% and silk yield by 22%	Tong et al. (2022)
BmE2F1 CRISPR-cas9 knockout	Reduction in number of silk gland cells by 23% and silk yield by 16%	Tong et al. (2022)
BmE2F1 transgenic overexpression	High yield fine silk strain	Tong et al. (2022)
BmChit β-GlcNAcase gene	L phenotype	Tong et al. (2022)
BmFibH	Mechanical properties of silk fiber, heritable dominance, cocoon-pupae balance, and secretion mechanism	Wang and Nakagaki (2014)
BmPTTH	Severe developmental delay at both the larval and pupal stages	Uchibori-Asano et al. (2019)
BmOrco	Larva feeding and adult mating behaviors impaired in the Orco mutant	Liu et al. (2017)
BmSage,	Embryonic development of silk gland	Xin et al. (2015)
Awd (abnormal wing disc) Fng (fringe)	Deformed wings	Ling et al. (2015)
BmSxl, Bmtra2, BmImp, BmImpM, BmPSI, and BmMasc	Sex determination	Xu et al. (2017)
Cas9 mRNA& BmWnt1-specific sgRNA	Wnt-deficient phenotype-embryos could not hatch and showed severe defects in body segmentation and pigmentation	Zhang et al. (2015)
BmTCTP	Controls larval growth and development	Liu et al. (2018)

silkworm genome are involved in functions such as silk production, development, immunity, and other physiological processes. Few of the important genes having significant impacts in the sericulture field are represented in Table 1. In the year 2008, the International Silkworm Genome Consortium announced that there are two distinct strain assemblies from Chinese and Japanese research groups that were integrated to improve the p50T genome sequence. A highly contiguous assembly of the p50T genome sequences was produced in 2019 by the use of BAC sequences and PacBio long reads (Kawamoto et al. 2019). This species boasts a comprehensive array of genetic resources, including molecular linkage maps generated from various markers and over 400 reported mutants that have been mapped to more than 200 loci. These loci are distributed across 28 linkage groups as reported by Mita et al. (2004). Silkworm genome sequence has been established as a useful reference for comparative genomics and genetics studies aimed at comprehending insect diversity and other related aspects (Boregowda et al. 2021).

Silkworm Pan-Genome

The pan-genome is the complete set of genes found in a species, which includes both the genes that are commonly shared among individuals and the genes that vary between the species. Pan-genome approach takes us beyond single reference genome information. This idea has a spectrum of applications in genomics, evolutionary biology, and the study of species-specific adaptations.

Recently, a research group headed by Tong employed the pan-genome approach using the third-generation sequencing, where they re-sequenced 1078 silkworm strains and assembled long-read genomes for 545 representatives of silkworms. Their analysis revealed that the silkworm genome harbors a high density of genomic variants, which are responsible for the diversity among the silkworm population. Furthermore, the researchers identified 7308 new genes, which were previously unknown, and 4260 (22%) core genes that are conserved, which function in gene regulation. In addition, they uncovered 3,432,266 nonredundant structure variations (SVs) in the genome that contribute to the genetic diversity of the silkworm population. Moreover, the researchers revealed hundreds of genes and SVs that may have played a role in the artificial selection (domestication and breeding) of silkworms. This information is of great significance for the silkworm breeding industry, as it provides insight into the genetic basis of the desirable traits that have been selected for over the years. Four genes were highlighted, of which two of them (*BmE2F1* and *BmChit β-GlcNAcase*) were found responsible for economic traits silk yield and silk fineness and two ecologically adaptive traits (Table 1), gene *BmTret1-like for* egg diapause and gene *Wnt-1* for aposematic coloration. These genomic resources will aid functional genomic studies and breeding improvements for silkworms at a population scale.

Intervention of Genomics in Silkworm Breeding Program

A large number of breeds/hybrids are developed in India, China, Japan, and some temperate countries like Bulgaria. The productivity of the breeds/hybrids highly depends on the climatic conditions for commercial exploitation throughout the year. As we have approached the era of climate change, silkworm breeds resurgent to the climate change or elevated temperature and humidity are the need of the hour for utilization of silkworm breeds throughout the year. Genomic applications have been a great utility in development of silkworm breeds/hybrids by identifying specific genetic markers in the genome of the silkworm. The identification of genetic markers (SSRs, SNPs, and QTLs) and advanced technologies such as TALEN and CRISPR/Cas9, GWAS, transcriptome, proteome, microbiome, and metabolome associated with many adaptive traits such as tolerance to heat, temperature, humidity, and other stress conditions have yielded in developing many breeds/hybrids which are tolerant to temperature and humidity (Chandrakanth et al. 2015; Moorthy et al. 2016). These genetic markers were utilized in developing new bivoltine breeds/hybrids which not only have high productivity but possess survival potential at elevated stress conditions. Recently under trial, bivoltine breed such as TT21 X TT56 double hybrid has the best survival fitness at elevated temperature. These breeds utilized genetic markers such S0806 and S0816 as a major component for developing adaptiveness to the elevated stressed conditions (Table 2).

Congenic Breeds

The tropical silkworm gene pool is characterized by high heterogeneity and an increased genetic load, which is further compounded by the accumulation of spontaneously mutated genes. To enable consistent expression of heterosis in a congenial environment, pure/syngeneic lines that are more homozygous at the molecular

Table 2 Representation of silkworm breeds/hybrids developed by utilizing gene markers and other advanced technologies

Sl. No.	Breed name	Bivoltine/ multivoltine	Markers used	Application
1	TT21 X TT56	Bivoltine	S0803 and S0816—SSR marker	Tolerant to high temperature
2	SK6R	Bivoltine	Nsd-2 resistant allele marker	BmBDV resistance
3	SK7R	Bivoltine	Nsd-2 resistant allele marker	BmBDV resistance
4	WB1HH	Bivoltine	S0803 and S0816—SSR marker Pyx-3 and Pyx-4—SNP marker	Tolerant to high temperature and humidity
5	N5HH	Bivoltine	S0803 and S0816—SSR marker	Tolerant to high temperature and humidity
6	B.Con4HH	Bivoltine	S0803 and S0816—SSR marker	Tolerant to high temperature and humidity
7	Jingsong	Bivoltine	Antivirus transgenic-CRISPR/Cas9	Tolerant to high temperature
8	Nistari	Multivoltine	Genome editing of nsd-2 gene	BmBDV resistant

level are recommended. Congenic breeds have been developed through a new breeding approach that involves the introgression of traits controlled by multiple genes from donor syngeneic lines to recipient syngeneic lines. This approach has been employed to develop improved silkworm breeds such as MCon1, MCon4, BCon1, and BCon4.

MCon1 and MCon4 are polyvoltines with high cocoon shell that were developed by introgressing high shell from a syngeneic bivoltine line, JPN (donor), to a multivoltine line, CB5 (recipient) and D6P being the donor and M6DPC-M being the recipient. Similarly, BCon1 and BCon4 are bivoltines with high survival that were developed by introgressing multigenic traits (high survival) from a syngeneic multivoltine line, CB5 (donor), to a bivoltine syngeneic line, JPN (recipient), with D6P being the recipient and M6DPC-LM being the donor. The breeding approach employed in developing these congenic breeds has shown great promise in improving silkworm breeds, resulting in breeds with desirable traits and characteristics.

Genome Application for Disease Resistance

Bivoltine raw silk production is not yet to its potential in tropical climates due to various factors that impact the host plant, mulberry varieties, and silkworm hybrids. These factors include rearing management, disinfection, and fluctuating climate. Although researchers have identified and developed bivoltine silkworm breeds over several years, these breeds are not sturdy enough to withstand weather changes or prevalent diseases. Moreover, the newly productive silkworm breeds also face the challenge of poor-quality mulberry leaves throughout the year due to adverse environmental conditions. To overcome these challenges, there is a need to shift breeding strategies and approaches, along with improved silkworm-rearing practices. Climate-related factors such as temperature, humidity, and nutrient content in mulberry leaves may cause diseases or crop failures, particularly during adverse seasons. Hence, it is necessary to develop robust silkworm breeds using molecular tools such as marker-assisted selection (MAS) and transgenic and genome editing to address these problems defying traditional approaches (Sivaprasad and Chandrasekharaiah 2008). The diseases which rampantly harm the silkworm crops are BmNPV (nuclear polyhedrosis), BmDNV (*Bombyx mori densovirus*), and BmIFV (infectious flacherie). These diseases possess several symptoms and spread rapidly once infected in tropical conditions.

Scientists have identified breeds naturally immune to diseases such as Nistari and Pure Mysore and attempted to develop resistance by backcrossing strategies. However, these breeds did not achieve the full potential of resistance. Recently, Japanese and Indian biotechnologists have made a significant discovery about silkworms. They have identified five major genes that determine the resistance of silkworms to certain viruses, particularly the BmNPV virus. These genes are responsible for metabolism, including the *Bombyx mori putative deoxynucleoside kinase mRNA*, *Acyl-CoA desaturase (desat3) mRNA*, *Bombyx mori phenol UDP-glucosyltransferase mRNA*, and *Bombyx mori PKG-Ib mRNA*. After several attempts and utilizing a large number of genetic markers, these genes were introduced to a productive silkworm breed CSR2 by backcross. As a result, researchers developed resistant strains of silkworms, namely, MASN4, MASN6, and MASN7. These resistant strains have exhibited enhanced tolerance for BmNPV, up to 67%.

BmIFV infection is usually associated with BmDNV, and its tolerance is controlled by polygenes. This disease is less effective or has no effect at high temperature (37 °C). FV1 X FV12 and FV12 X FV13 are the two hybrids identified in India for its productivity and survival.

The resistance to BmDNV1 is dependent on both dominant and recessive genes. Two resistance genes have been identified—*Nid-1* and *nsd-1* genes—which provide protection against BmDNV1, and *nsd-2* gene, which provides protection against BmDNV2 (Kadono-Okuda et al. 2014). The mutated *nsd-2* gene confers resistance against BmDNV2 (Table 2). Homozygous recessive nsd-1 provides resistance against BmDNV1. The non-susceptibility to BmDNV1 is determined by two genes: *nsd1*, located on chromosome 21, and *Nid-1*, located on chromosome 17. *Nid-1* is epistatic to +*nsd-1*, which is an allelic and dominant gene to *nsd-1* (Eguchi et al. 1986, 1991, 1998, 2007). Silkworms achieve absolute non-susceptibility to BmDNV1 through two independent genes: *nsd-1* and *Nid-1*.

Development of multiviral disease-resistant breeds: Studies reported by Satish et al. 2023 at CSRTI, Mysore a pioneering Institute in India, have led in the development of silkworm hybrids viz. FV23 x FV225 and FV225 x FV212 by combining conventional breeding methods with virus-challenge studies. These breeds were developed by utilizing marker-assisted selection utilizing the available genetic resources and markers already identified for BmDNV1, BmNPV, and BmIFV.

Researchers have not only limited the genomic studies to silkworm but they have created transgenic silkworm models to understand the problems related to human diseases and developmental studies (Table 3). Hence, silkworm genome models can be considered as one of the best models available to understand basic genetics and answer some human health complications.

Table 3 Silk fiber's functionalization and diversification were made possible by genetically engineered technology

Sl. no.	Functionalization and diversification	Authors
1	Silk containing human collagen	Adachi et al.
2	Silk containing acidic and alkaline fibroblast growth factors	Wang Fengand Kambe
3	Fluorescent silk with antibacterial function	Li Zhen et al.
4	Improvement in calcium-binding capacity of silk fiber by expressing calcium-binding protein in transgenic silkworms	Wang Shaohua
5	Silkworms resistant to BmNPV	Isobe
6	Human type III collagen in the silk glands	Tomita

Conclusion

Sericulture is the primary economy of many small and large farmers across the world. The unraveling of the genome of silkworms paved the way for the identification of several genetic markers that have been utilized for developing silkworm breeds/hybrids that are highly productive and disease-resistant. These techniques should be further integrated with conventional breeding techniques to increase the immune response of the silkworm and enhance productivity from silkworm cocoon weight, shell, shape, silk fiber quality, tolerance to abiotic stress, and disease resistance. Biotechnological tools such as gene markers, allele markers, gene expression studies, gene editing, and transgenic model development, have become a boon in the silkworm industry. These innovations have significantly improved the protection of the silkworm and therby increasing the profit of the sericulture industry.

References

Aigner TB, DeSimone E, Scheibel T (2018) Biomedical applications of recombinant silk-based materials. Adv Mater 30:1704636

Arunkumar KP, Metta M, Nagaraju J (2006) Molecular phylogeny of silkmoths reveals the origin of domesticated silkmoth, Bombyx mori from Chinese Bombyx mandarina and paternal inheritance of Antheraea proylei mitochondrial DNA. Molecular Phylogenetics and Evolution 40(2):419–427

Boregowda MH (2021) Silkworm genomics: current status and limitations. Advances in Animal Genomics. Academic Press, pp 259–280

Chandrakanth N, Moorthy SM, Ponnuvel KM, Sivaprasad V (2015) Identification of microsatellite markers linked to thermotolerance in silkworm by bulk segregant analysis and in silico mapping. Genetika 47(3):1063–1078

d'Alençon E, Sezutsu H, Legeai F, Permal E, Bernard-Samain S, Gimenez S et al (2010) Extensive synteny conservation of holocentric chromosomes in Lepidoptera despite high rates of local genome rearrangements. Proceedings of the National Academy of Sciences 107(17):7680–7685

Eguchi R, Shimazaki A, Shibukawa A (1986) Genetic analysis of neatness defects by diallel cross in the silkworm, Bombyx mori L. Acta Sericologia 137:29–36

Eguchi R, Ninaki O, Hara W (1991) Genetical analysis on non-susceptibility to densonucleosis virus in silkworm, Bombyx mori. J Seric Sci Jpn 60:384–389

Eguchi R, Hara W, Simazaki A, Hirota K, Ichiba M, Ninagi O, Nagayasu K (1998) Breeding of the silkworm race Taisei non susceptible to a densonucleosis virus type 1. J Seric Sci Jpn 67:361–366

Eguchi R, Nagayasu K, Ninagi O, Hara W (2007) Genetic analysis on the dominant non-susceptibility to densonucleosis virus type 1 in the silkworm, Bombyx mori. Sanshi-Kontyu Biotec 2007:159–163

Fang S-M, Zhou Q-Z, Yu Q-Y, Zhang Z (2020) Genetic and genomic analysis for cocoon yield traits in silkworm. Nat Sci Reports 10:5682

Farokhi M, Mottaghitalab F, Fatahi Y (2018) Overview of silk fibroin use in wound dressings. Trends Biotechnol 36:907–922

Goldsmith MR, Shimada T, Abe H (2005) The genetics and genomics of the silkworm, *Bombyx mori*. Annu Rev Entomol 50:71–100

Goncu E, Parlak O (2008) Someautophagic and apoptotic features of programmed cell death in the anterior silk glands of the silkworm, Bombyx mori. Autophagy 4:1069–1072

Gu SH, Lin JL, Lin PL, Chen CH (2009) Insulin stimulates ecdysteroidogenesis by prothoracic glands in the silkworm, Bombyx mori. Insect Biochem Mol Biol 39:171–179

Gu SH, Lin JL, Lin PL (2010) PTTH-stimulated ERK phosphorylation in prothoracic glands of the silkworm, Bombyx mori: role of Ca2+/calmodulin and receptor tyrosine kinase. J Insect Physiol 56:93–101

Gu SH, Yeh WL, Young SC, Lin PL, Li S (2012) TOR signalling is involved in PTTH-stimulated ecdysteroidogenesis by prothoracic glands in the silkworm, Bombyx mori. Insect Biochem Mol Biol 42:296–303

Holland C, Numata K, Rnjak-Kovacina J (2019) The biomedical use of silk: past, present, future. Adv Healthc Mater 8:1800465

Kadono-Okuda K, Ito K, Murthy GN, Sivaprasad V, Ponnuvel KM (2014) Molecular mechanism of densovirus resistance in silkworm. Bombyx mori. Sericologia 54(1):1–10

Kawamoto M, Jouraku A, Toyoda A, Yokoi K, Minakuchi Y, Katsuma S, Shimada T (2019) High-quality genome assembly of the silkworm, Bombyx mori. Insect Biochem Mol Biol 107:53–62. https://doi.org/10.1016/j.ibmb.2019.02.002

Li Z, Jiang Y, Cao G, Li J, Xue R, Gong C (2015) Construction of transgenic silkworm spinning antibacterial silk with fluorescence. Mol Biol Rep 42:19–25

Ling L, Ge X, Li Z, Zeng B, Xu J, Chen X, Shang P, James AA, Huang Y, Tan A (2015) MiR-2 family targets awd and fng to regulate wing morphogenesis in Bombyx mori. RNA Biol 12:742–748. https://doi.org/10.1080/15476286.2015.1048957

Liu Y, Wan S, Liu JUN, Zou Y, Liao S (2017) Antioxidant activity and stability study of peptides from enzymatically hydrolyzed male silkmoth. J Food Process Preserv 41(1):e13081

Liu Y, Zhang H, He F et al (2018) Combined toxicity of chlorantraniliprole, lambda-cyhalothrin, and imidacloprid to the silkworm Bombyx mori (Lepidoptera: Bombycidae). Environ Sci Pollut Res 25:22598–22605. https://doi.org/10.1007/s11356-018-2374-7

Ma SY, Xia QY (2017) Genetic breeding of silkworms: from traditional hybridization to molecular design. Yi Chuan. 39(11):1025–1032. https://doi.org/10.16288/j.yczz.17-103

Mita K, Kasahara M, Sasaki S, Nagayasu Y, Yamada T, Kanamori H (2004) The genome sequence of silkworm, Bombyx mori. DNA Res 11(1):27–35. https://doi.org/10.1093/dnares/11.1.27

Mori H, Tsukada M (2000) New silk protein: modification of silk protein by gene engineering for production of biomaterials. Rev Mol Biotechnol 74:95–103

Moorthy SM, Chandrakanth N, Krishnan N (2016) Inheritance of heat stable esterase in near isogenic lines and functional classification of esterase in silkworm Bombyx mori. Invertebrate Survival Journal 13(1):1–10

Satish L, Kusuma L, Shery AMJ, Moorthy SM, Manjunatha GR, Sivaprasad V (2023) Development of productive multi-viral disease-tolerant bivoltine silkworm breeds of Bombyx mori (Lepidoptera: Bombycidae). Applied Entomology and Zoology 58(1):61–71

Sivaprasad V, Chandrasekharaiah, (2008) Biotechnological approaches for the breeding of silk-worms for disease resistance. 6th Mulberry Silkworm Breeders Meet, Mysore, India, pp 41–43

Tada M, Tatematsu KI, Ishii-Watabe A et al (2015) Characterization of anti-CD20 monoclonal antibody produced by transgenic silkworms (Bombyx mori). mAbs 7:1138–1150

Tazima Y, Ohuma A (1995) Preliminary experiments on the breeding procedure for synthesizing a high temperature resistant commercial strain of the silkworm, Bombyx mori L. Japan Silk Sci Res Inst 43:1–16

Tomita M (2011) Transgenic silkworms that weave recombinant proteins into silk cocoons. Biotechnol Lett 33:645–654

Tong X, Han MJ, Lu K et al (2022) High-resolution silkworm pan-genome provides genetic insights into artificial selection and ecological adaptation. Nat Commun 13:5619. https://doi.org/10.1038/s41467-022-33366-x

Uchibori-Asano M, Jouraku A, Uchiyama T, Yokoi K, Akiduki G, Suetsugu Y, Kobayashi T, Ozawa A, Minami S, Ishizuka C et al (2019) Genome-wide identification of tebufenozide resistant genes in the smaller tea tortrix, Adoxophyes honmai (Lepidoptera:Tortricidae). Sci Rep 9(1):4203. https://doi.org/10.1038/s41598-019-40863-5

Wang Y, Nakagaki M (2014) Editing of the heavy chain gene of Bombyx mori using transcription activator like effector nucleases. Biochem Biophys Res Commun 450:184–188

Xia Q, Zhou Z, Lu C, Cheng D, Dai F, Li B, Biology Analysis Group et al (2004) A draft sequence for the genome of the domesticated silkworm (Bombyx mori). Science 306(5703):1937–1940. https://doi.org/10.1126/science.1102210

Xia Q, Li S, feng, Q. (2014) Advances in silkworm studies accelerated by the genome sequencing of Bombyx Mori. Annu Rev Entomol 59:513–536

Xin H-H, Zhang D-P, Chen R-T, Cai Z-Z, Lu Y, Liang S, Miao Y-G (2015) Transcription factor bmsage plays a crucial role in silk gland generation in silkworm, Bombyx mori. Arch Insect Biochem Mol Biol 90:59–69. https://doi.org/10.1002/arch.21244

Xu J, Zhan S, Chen S, Zeng B, Li Z, James AA, Tan A, Huang Y (2017) Sexually dimorphic traits in the silkworm, Bombyx mori, are regulated by doublesex. Insect Biochem Mol Biol 80:42–51. https://doi.org/10.1016/j.ibmb.2016.11.005

Yang C-c, Yokoi K, Yamamoto K, Jouraku A (2021) An update of KAIKObase, the silkworm genome database. Database 2021:baaa099. https://doi.org/10.1093/database/baaa099

Zhang Q, Hua X, Yang Y et al (2015) Stereoselective degradation of flutriafol and tebuconazole in grape. Environ Sci Pollut Res 22:4350–4358. https://doi.org/10.1007/s11356-014-3673-2

An Insight into Transcriptomics of the Mulberry Silkworm, *Bombyx mori*: A Review

K. Lingaiah, L. Satish, V. S. Raviraj, S. M. Moorthy, and V. Sivaprasad

Abstract

Rapid advancements have been made in the field of structural and functional genomics using the available silkworm genome sequence data. Annotated genome sequences, silkworm genomic maps, and transcriptome data are potentially helpful resources for the improvement of silkworm breeding programs. Of the various genomic methodologies, transcriptome analysis is of increasing significance in comprehending the altered manifestation of genetic variations toward a particular characteristic or disease. Understanding the transcriptome is crucial for deciphering the functional components of the genome since it represents the entire collection of RNA transcripts that a genome produces at any specific given time. Numerous transcriptomic research on the silkworm *Bombyx mori* have identified the transcription factors related to disease resistance and susceptibility, phoxim-induced resistance, alternative splicing events, silk structure and coloration, and other aspects which could be used, via high-throughput genotyping techniques, to create better variety or trait-specific genetic resources. Therefore, functional genomics thus offers guidance for future gene mining, genetic modification, or breeding efforts to meet the different challenges for boosting productivity and quality of silk generated and further leading to significantly increase the quality of raw silk production.

K. Lingaiah (✉) · L. Satish
Silkworm Breeding and Molecular Biology Laboratory II, Central Sericultural Research and Training Institute, Central Silk Board, Ministry of Textiles, Government of India, Mysuru, Karnataka, India

V. S. Raviraj
Central Sericultural Research and Training Institute, Central Silk Board, Ministry of Textiles, Government of India, Berhampore, West Bengal, India

S. M. Moorthy · V. Sivaprasad
Central Silk Board, Bengaluru, India

R. V Suresh et al. (eds.), *Biotechnology for Silkworm Crop Enhancement*,
https://doi.org/10.1007/978-981-97-5061-0_3

Keywords

Silkworm · Transcriptome · Functional genomics · Trait · DEGs

Introduction

Silkworm, or *Bombyx mori*, is a valuable model insect for biological, molecular, and genetic research, because of its potential applications in the silk industry and medical technology (Goldsmith et al. 2005). It also has the richest mutation repertoire in nearly every area of morphology, physiology, and development (Sivaprasad et al. 2022). The silkworm genome data is important for breeding and genetic research, for identifying important genes and promoters, and for comparative genomics, which is why research into it started a few years ago. Due to human selection, silkworms have developed an absolute reliance on humans for existence; over 1000 domesticated variants of the species are available globally (Yamamoto et al. 2006). With a genome size of approximately 432 million base pairs and over 400 identified mutations that have been mapped to over 200 loci, this species possesses highly developed genetic resources. In tropical and temperate regions, an estimated 3000 silkworm genotypes with a variety of origins are preserved as ecotypes and distinct inbred lines (Nagaraja and Nagaraju 1995). Because of its abundant genetic resources, *B. mori* is a perfect reference species for Lepidoptera, which makes it easier to conduct basic and comparative genomic research and develop innovative genome-based methodologies for sericulture (Mita et al. 2003). The public database contains more than 180,000 silkworm ESTs from independent projects. Separately, the Chinese and Japanese groups reached the WGS at a coverage level of 5.9X and 3X, respectively (The International Silkworm Genome Consortium 2008; Mita et al. 2003; Xia et al. 2004). Xia et al. (2009) conducted a genome resequencing of 40 silkworms, comprising 11 wild silkworms and 29 phenotypically and geographically diverse domesticated silkworm lines. The results showed that the domesticated and wild varieties had approximately 14 million and 13 million SNPs, respectively, and that the genome complexity of the two types of silkworms differed significantly in terms of SNPs, indels, and structural variations (Xia et al. 2009).

The high-quality genome assembly of silkworm *Bombyx mori* is used in gene prediction based on the newly assembled genome utilizing existing mRNA and protein sequence data, which showed 16,880 gene models with a N50 of 2154 bp, which indicates more precise coding sequences and gene sets (Kawamoto et al. 2019). A significant density of genomic variants representing 7308 novel genes, 4260 (22%) core genes, and 3,432,266 nonredundant structural variations (SVs) was revealed by the silkworm's most recent high resolution pan genome study (Tong et al. 2022). A number of genes and SVs implicated in artificial selection, mainly in domestication and silkworm breeding, were also found in the study. Additionally, two genes linked to two environmentally adapted traits, viz., egg diapause and aposematic coloration, as well as two commercially important traits like silk production and silk fineness were found.

Transcriptome Analysis of Silkworm

Throughout the silkworm development, a variety of transcription factors are involved in the transcriptional control of the genes that code for silk proteins. Because transcriptome analysis can decipher biological pathways, molecular mechanisms, and functional aspects of the genome, it can shed light on functional genomics (Li et al. 2012; Wang et al. 2009). Comprehension on the functional components of the genome requires a complete set of the transcriptome, which is the total collection of RNA transcripts that the genome produces at any given time. Thus, RNA-sequencing, or RNA-seq, is a crucial technique for transcriptome analysis based on deep sequencing technology (Li et al. 2016; Wang et al. 2009). Several transcriptome analyses pertaining to silkworm gene expression patterns in multiple tissues (Gan et al. 2011; Xia et al. 2007) and at different developmental stages (Xu et al. 2012), responses of host to infectious pathogens (Zhou et al. 2013), 20-hydroxyecdysone regulation of innate immunity and glycolysis (Tian et al. 2010), and dosage compensation (Zha et al. 2009) and reproduction (Zheng et al. 2022) have yielded a detailed understanding of the silkworm genome. Additionally, the ensuing RNA-seq research has discovered miRNAs unique to silkworms, miRNAs related with TE, and miRNAs that exhibit differential expression of miRNA-star and their corresponding miRNAs (Liu et al. 2010; Xia et al. 2014).

Silk Gland Transcriptomics

Based on the availability of the silkworm genome sequence data, microarrays and RNA-seq techniques have been widely used to study the transcriptome analysis (Xia et al. 2014). The silk glands of silkworm are specialized in the synthesis of several secretory proteins. Numerous transcription factors control the expression of the genes that code for the silk proteins in *B. mori* silk glands, which have tight developmental and territorial restrictions. According to quantitative PCR study, on day 3 of the larval fifth instar, the expression level of Bmsage was higher in a high silk strain than in a lower silk strain (Zhoa et al. 2014). The genetic basis of silk structure and pigmentation was shown by comparative transcriptome analysis on the silk glands of six silk moth species (Dong et al. 2015). A comparative transcriptome analysis between domestic and wild silkworms identified 32 differentially expressed genes (DEGs), of which 16 were upregulated in domestic as opposed to wild silkworms. While the two domestic silkworms' upregulated DEGs were linked to the effective biosynthesis and secretion of silk proteins, the two wild silkworms' upregulated DEGs were assumed to play significant roles in inducing pathogen tolerance and environmental adaptation (Fang et al. 2015). The causes of the rise in silk yield throughout the domestication of silkworms were highlighted by Zhou et al. (2020) in a study comparing the co-expression networks in the silk gland. This was demonstrated by the participation of the BGIBMGA001462 gene in the transport route and BGIBMGA000074 in the ribosome pathway. Both genes were shown to have relatively high levels of gene expression in the silk gland and to co-express with Fib-H in the domestic silkworm networks of genes that code for silk. This further suggested that the increase in silk yield following silkworm domestication is due to genes associated to silk manufacturing. According to Cheng et al. (2015), de novo transcriptome sequencing for *B. mandarina* revealed a total of 883 and 425 unigenes that were upregulated in the middle and posterior silk glands, respectively. Wang et al. (2016) found 788 transcripts, of which 779 were upregulated and 9 were downregulated; the nine downregulated transcripts were associated with reduced silk production. Ionotropic kainite 5, LIM-binding protein, and likely cytochrome P450 were annotated on five of these nine transcripts (BGIBMGA008713, BGIBMGA004648, BGIBMGA008727, BGIBMGA012206, BGIBMGA012704, BGIBMGA000056, BGIBMGA012202, BGIBMGA007487, and BGIBMGA009567); the remaining four transcripts were left unannotated. Additionally, two strains of L10 and JS were examined using RNA-seq and quantitative polymerase chain reaction to ascertain the divergence of the silk gland, which regulates silk biosynthesis in silkworms. The results of the study revealed that the differentially expressed genes (DEGs) were enriched in three pathways, which were primarily connected to the processing and biosynthesis of proteins. Hu et al. (2016) used RNA-sequencing and comparative transcriptome analysis to identify 2178 DEGs in Nd-PSG and WT-PSG. The majority of these DEGs were linked to cellular stress responses, such as autophagy, the ubiquitin-proteasome system, and the heat shock response, and they were significantly upregulated in Nd-PSG, indicating that mutant FibH was changing cellular homeostasis and activating adaptive responses in the posterior silk gland cells.

Zhu et al. (2022) investigated the comparative silk gland transcriptomes of wild, early domestic, and enhanced silkworms. The study also focused on the changes in gene expression compared to early domestication as a result of current breeding techniques and the evolution of silkworm cocoon silk. Further, differences of 2711 DEGs between early domesticated and improved silkworms and 2264 DEGs between improved and wild silkworms were documented. In contrast to the 1671 DEGs unique to the improved silkworm, only 158 DEGs were found in wild silkworms. GDAP2, a significant putative improvement gene from upregulated DEGs, was also found to be involved in silk behavior and the general robustness of the enhanced silkworm.

Developmental Stages and Transcriptomes

Li et al. (2012) analyzed the whole transcriptome profile of *B. mori* at various developmental stages, leading to the discovery of 5428 new exons. The deterioration of the silkworm midgut and the synthesis of innate immunity-related proteins during the wandering stage were discovered by the transcriptional profiling of the midgut immune response. Using a transgenic silkworm strain, Ma et al. (2014) performed transcriptome study and discovered that overexpressing $Ras1^{CA}$ in the posterior silk gland enhances the production of fibroin, which in turn increases the production of silk. Transcriptome analysis of *B. mori* developmental characteristics found 12,254 assembled transcripts from the wing disc during the L5D6, PP, and P0 stages. Elevated expression of cuticle proteins and genes associated with chitin signifies the active metabolism of chitin in connection with the morphological and structural alterations that occur during the transformation from the wing disc to the pupal wing (Ou et al. 2014). Wang et al. (2015) conducted transcriptome profiling of the silkworm spinneret and Filippi's gland and found that >11,000 genes were enriched in cuticle proteins and ion transport-related genes in the spinneret, with a putative transporter identified as BGIBMGA013432.

An analysis of the expression of genes in the midgut of *B. mori* larvae during the molting stage revealed a significant difference in the expression of the gene encoding ecdysoneoxidase between the feeding and molting stages (Yang et al. 2016). Additionally, the expression of genes such as trypsin-like serine protease, BGIBMGA001320-TA, BGIBMGA003605-TA, and BGIBMGA010023-TA was either downregulated entirely or partially upon entering the molting stage. Consequently, this served as the catalyst for alterations that most likely reduced the physiological functioning of the midgut by downregulating genes related to metabolism and transport (Yang et al. 2016).

An analysis of the lipid transcriptome of the wild-type strain Dazao, as well as the mutant strains +/bd and bd/bd, revealed 254 novel transcripts in total. Additionally, it was discovered that 28 genes were differentially expressed and contributed to bd larval melanism, while 15 genes related to cuticular proteins were notably more expressed in the bd/bd mutant (Wu et al. 2016). With expression of 1509 DEGs between H9 and P50 silkworm strains, Yang et al. (2018) analyzed the transcriptome analysis of the colleterial gland suggesting its functional role in mucous secretion. Of these, 1001 genes were upregulated and 508 genes were downregulated in P50 compared with H9.

Ding et al. (2020) conducted a comparative investigation of integument transcriptomes and found eight common genes involved in the building of numerous twin-spot marking patterns in the three allelic mutants of the silkworm, *B. mori*, as well as genes involved in marking pattern development in the three mutants. Sixty-eight DEGs between LC/+ and + LC/+LC, 188 DEGs between LCa/+ and + LCa/+Lca, and 336 differentially expressed genes (DEGs) between L and Dazao (wild type which exhibits normal markings) were identified, along with 8 common DEGs among 3 pairwise comparisons. These included Kruppel-like factor, TATA-binding protein, protein patched, UDP glycosyltransferase, an unknown secreted protein, and three cuticular proteins (Ding et al. 2020).

Shen et al. (2022) conducted a thorough transcriptome sequencing investigation of the anterior-posterior axis gene expression atlas, revealing that the posterior midgut differed transcriptionally from the anterior and middle midgut. The results also demonstrated the regional specialization of hormone regulation, chitin metabolism, transmembrane transport, and digestive enzyme synthesis in various midgut regions, and they implied that homeobox transcription factors are essential for transcriptional differences throughout the midgut. The study examined transcriptome variations in the head and midgut tissues of male and female silkworms that were given either artificial diets or fresh mulberry leaves. The results showed that the male and female silkworms' head tissues had 2969 and 3427 DEGs, and the midgut tissues contained 923 and 619 DEGs. The roles in these DEGs were represented by digestion and absorption of nutrients and silkworm innate immunity (Li et al. 2022).

Yokoi et al. (2021) conducted an RNA-seq analysis of ten major silkworm larvae tissues, establishing a detailed reference transcriptome data in silkworm tissues. The study identified the almost complete sequence of the sericin-I gene and reported several transcripts with distinct expression profiles. Furthermore, tight genomic clusters were represented by the highly expressed transcripts in the midgut, which may have originated from tandem gene duplication. A thorough transcriptome analysis of 12 *B. mori* tissues, including testis tissue, showed that 1705 testis-specific genes and 20,962 genes were widely expressed (Kakino et al. 2023).

Diapause and Transcriptomics in Silkworm

Studying the transcriptome of silkworm diapause eggs at 20 and 48 h after oviposition, researchers found 6402 DEGs overall at 48 h compared to 20 h after oviposition. This finding adds to our understanding of the role of the oxidative phosphorylation pathway in diapause as well as its involvement in ribosome-related metabolism and hydrogen transport (Gong et al. 2020). Based on Liu et al. (2021), transcriptome analysis of bivoltine silkworm embryos cultured at diapause-inducing (25 °C) and non-diapause-inducing (15 °C) temperatures during the blastokinesis (BK) and head pigmentation (HP) phases revealed changes in miRNA abundance, target genes of differentially expressed miRNAs under temperature-treated embryos, and identification of 411 known miRNAs along with 71 novel miRNAs during the 2 phases with 108 and 74 distinct miRNA expression patterns for each of the 2 periods.

Transcriptomic Studies Under Different Conditions

In silkworm-based research, transcriptome technology is widely used to evaluate changes in physiological and biochemical processes at the transcriptional level under various circumstances. After treating silkworms with 7.5 μM H_2S, a transcriptome analysis showed 1200 DEGs, of which 977 represented upregulation and 223 represented downregulation. Hydrogen sulfide represents a novel signaling molecule required essentially for regulatory functions involving growth, the cardiovascular system, oxidative stress, and inflammation in several organisms. According to Zhang et al. (2021), these DEGs were functionally involved in the transport pathway, cellular community, metabolism of carbohydrates, and immune-associated signal transduction.

According to Zheng et al. (2022) transcriptome investigation of silkworm reproduction under dimethoate stress, exposure to dimethoate considerably slowed down the development of silkworm eggs. Eleven QTLs linked to cocoon yield features were found by Fang et al. (2020) and mapped to seven chromosomes in silkworms.

The expression of KWMTBOMO04917 was found to be higher in wild silkworms than in domestic ones. KWMTBOMO12906 is the DNA polymerase gene. In domestic silkworms, epsilon was significantly expressed, and there was a positive correlation between its expression in the posterior silk gland and cocoon yield.

Transcriptome study of *B. mori* selenium-treated fat body tissue was carried out by Jiang et al. (2020), which identified changes in gene expression in response to selenium therapy in silkworm fat bodies treated with 50 μM and 200 μM of selenium and found 912 DEGs (371 upregulated and 541 downregulated) and 1420 DEGs (1078 upregulated and 342 downregulated). According to the study, reduced selenium concentrations control nutrition and encourage high selenium levels, and antioxidant pathways demonstrated the detoxifying function in silkworms.

A transcriptome investigation of the silkworm brain under stress from fasting revealed 330 DEGs, emphasizing the complex interactions between the availability of nutrients and other biological systems related to immunity, metabolism, hormones, and illnesses (Li et al. 2023).

Transcriptomic Studies and Silkworm Diseases

Silkworm is susceptible to infectious diseases caused by various microbial pathogens. Viral diseases are prevalent in almost all the sericultural areas in India and account for 15–20% silkworm crop losses. Resistance of silkworm to nuclear polyhedrosis and infectious flacherie viruses is controlled by polygenes (Watanabe 1986), while resistance of silkworm to densonucleosis is controlled by single recessive gene (Kadono-Okuda et al. 2014). Several transcriptome based studies in *B. mori* have helped to understand the complexity of the silkworm diseases, thereby throwing light on several genes and pathways contributing toward the elucidation of disease resistance or susceptibility. Transcriptome analysis of *B. mori* bidensovirus (BmBDV) infected midgut of silkworm revealed 334 upregulated and 272 downregulated DEGs with upregulation of DEGs involved in structural constituents of cuticle, antioxidant, and immune system processes. The study also emphasized on the antioxidant genes governing host cells from virus induced oxidative stress, being significantly upregulated after BmBDV infection (Sun et al. 2020).

Variations in expression of the genes of *B. mori* upon infection to a densonucleosis virus BmDNV-Z were studied utilizing 2 *B. mori* near-isogenic lines, Jingsong and Jingsong.nsd-Z NIL, highly susceptible and completely resistant to BmDNV-Z, and identified 151 DEGs among which 11 genes were significantly upregulated in the midgut of the Jingsong.nsd-Z.NIL strain following BmDNV-Z infection (Bao et al. 2008).

In order to understand the molecular mechanism of silkworm immune response against the fungal infection, high-throughput transcriptome sequencing analysis of the silkworm larvae during early response against *B. bassiana* infection was carried out and identified 1430 DEGs including 960 upregulated and 470 downregulated which were implicated in biological processes, such as defense and response, signal transduction, phagocytosis, regulation of gene expression, RNA splicing, biosynthesis and metabolism, and protein transport (Hou et al. 2014). DEGs generated from transcriptomes of differentially resistant silkworm strains following AcMNPV infection in *B. mori* results revealed that DEGs in p50 were concentrated in the metabolic pathway and apoptosis pathway and identified BmTex261 involved in the apoptosis pathway (Ding et al. 2021). Transcriptome analysis of strain-specific and gender-specific response of silkworm to BmNPV infection identified 18 genes that showed differential expression in the NB after BmNPV inoculation among which 14 were upregulated and 4 down-regulated, suggesting their possible role in eliciting BmNPV resistance. Further, the transcript abundance of eight genes was higher in males in comparison with females except for one gene (He et al. 2021). In conclusion, Although high-quality genome sequence of the silkworm genome is available, the utilization of this information, in combination with extensive transcriptomic analyses, will exemplify our understandings of the role of crucial genes invovled in specific pathways and mechanism in understading the traits. The transcriptome data will significantly contribute towards accelerating the understandings of the future functional genomics studies on the silkworm as well as other related species.

Conclusion

Although much improved high-quality genome sequence of the silkworm genome is available, the utilization of this information, in combination with transcriptome analyses data, will exemplify our understandings of the role of crucial genes involved in specific pathways and mechanism in understanding the traits. The transcriptome data will significantly contribute toward accelerating the understandings of the future functional genomics studies on the silkworm as well as other related organisms.

References

Bao Y, Li M, Zhao Y, Ge J, Wang C, Huang Y, Zhang C (2008) Differentially expressed genes in resistant and susceptible *Bombyx mori* strains infected with a densonucleosis virus. Insect Biochem Mol Biol 38(9):853–861. https://doi.org/10.1016/j.ibmb.2008.06.004

Cheng T, Fu B, Wu Y, Long R, Liu C, Xia Q (2015) Transcriptome sequencing and positive selected genes analysis of Bombyx mandarina. PLoS One 10(3):e0122837

Ding X, Liu J, Tong X, Wu S, Li C, Song J, Hu H, Tan D, Dai F (2020) Comparative analysis of integument transcriptomes identifies genes that participate in marking pattern formation in three allelic mutants of silkworm, *Bombyx mori*. Funct Integr Genomics 20(2):223–235. https://doi.org/10.1007/s10142-019-00708-w

Ding X, Wang X, Kong Y, Zhao C, Qin S, Sun X, Li M (2021) Comparative transcriptome analysis of *Bombyx mori* (Lepidoptera) larval Hemolymph in response to *Autographa californica* nucleopolyhedrovirus in differentially resistant strains. Processes 9:1401. https://doi.org/10.3390/pr9081401

Dong Y, Dai F, Ren Y, Liu H, Chen L, Yang P, Liu Y et al (2015) Comparative transcriptome analyses on silk glands of six silkmoths imply the genetic basis of silk structure and coloration. BMC Genomics 16(1):203. https://doi.org/10.1186/s12864-015-1420-9

Fang S, Hu B, Zhou Q, Yu Q et al (2015) Comparative analysis of the silk gland transcriptomes between the domestic and wild silkworms. BMC Genomics 16:60

Fang SM, Zhou QZ, Yu QY, Zhang Z (2020) Genetic and genomic analysis for cocoon yield traits in silkworm. Sci Rep 10(1):5682. https://doi.org/10.1038/s41598-020-62507-9

Gan L, Liu X, Xiang Z, He N (2011) Microarray-based gene expression profiles of silkworm brains. BMC Neurosci 12:8

Goldsmith MR, Shimada T, Abe H (2005) The genetics and genomics of the silkworm *Bombyx mori*. Annu Rev Entomol 50:71–100

Gong J, Zheng X, Zhao S, Yang L, Xue Z, Fan Z, Tang M (2020) Early molecular events during onset of diapause in silkworm eggs revealed by transcriptome analysis. Int J Mol Sci 21(17):6180. https://doi.org/10.3390/ijms21176180

He S, Xu J, Fan Y, Zhu F, Chen K (2021) Transcriptomic analysis of strain-specific and gender-specific response of silkworm to BmNPV infection Inv. Sur J 18:98–107

Hou C, Qin G, Liu T, Geng T, Gao K, Pan Z et al (2014) Transcriptome analysis of silkworm, *Bombyx mori*, during early response to *Beauveria bassiana* challenges. PLoS One 9(3):e91189. https://doi.org/10.1371/journal.pone.0091189

Hu W, Chen Y, Lin Y, Xia Q (2016) Developmental and transcriptomic features characterize defects of silk gland growth and silk production in silkworm naked pupa mutant. Insect Biochem Mol Biol 111:103175. https://doi.org/10.1016/j.ibmb.2019.05.010

International Silkworm Genome Consortium (2008) The genome of a lepidopteran model insect, the silkworm Bombyx mori. Insect Biochem Mol Biol 38(12):1036–1045. https://doi.org/10.1016/j.ibmb.2008.11.004

Jiang L, Peng LL, Cao YY, Thakur K, Hu F, Tang SM, Wei ZJ (2020) Transcriptome analysis reveals gene expression changes of the fat body of silkworm (*Bombyx mori* L.) in response to selenium treatment. Chemosphere 245:125660. https://doi.org/10.1016/j.chemosphere.2019.125660

Kakino K, Mon H, Ebihara T, Hino M, Masuda A, Lee JM, Kusakabe T (2023) Comprehensive transcriptome analysis in the testis of the silkworm, *Bombyx mori*. Insects 14:684. https://doi.org/10.3390/insects14080684

Kawamoto M, Jouraku A, Toyoda A, Yokoi K, Minakuchi Y, Katsuma S et al (2019) High-quality genome assembly of the silkworm, *Bombyx mori*. Insect Biochem Mol Biol 107:53–62. https://doi.org/10.1016/j.ibmb.2019.02.002

Li Y, Wang G, Tian J, Liu H, Yang H, Yi Y, Wang J, Shi X et al (2012) Transcriptome analysis of the silkworm (*Bombyx mori*) by high-throughput RNA sequencing. PLoS One 7(8):e43713. https://doi.org/10.1371/journal.pone.0043713

Li G, Qian H, Luo X, Xu P, Yang J, Liu M, Xu A (2016) Transcriptomic analysis of resistant and susceptible *Bombyx mori* strains following BmNPV infection provides insights into the antiviral mechanisms. Int J Genom 2:1–10. https://doi.org/10.1155/2016/2086346

Li J, Chen C, Zha X (2022) Midgut and head transcriptomic analysis of silkworms reveals the physiological effects of artificial diets. Insects 13(3):291. https://doi.org/10.3390/insects13030291

Li Y, Wang X, Dong H, Xia Q, Zhao P (2023) Transcriptomic analysis of starvation on the silkworm brain. Insects 14:658. https://doi.org/10.3390/insects14070658

Liu S, Li D, Li Q, Zhao P, Xiang Z, Xia Q (2010) MicroRNAs of *Bombyx mori* identified by Solexa sequencing. BMC Genomics 11:148

Liu L, Zhang P, Gao Q, Feng X, Han L, Zhang F, Bai Y, Han M, Hu H, Dai F et al (2021) Comparative transcriptome analysis reveals bmo-miR-6497-3p regulate circadian clock genes during the embryonic diapause induction process in bivoltine silkworm. Insects 12(8):739. https://doi.org/10.3390/insects12080739

Ma L, Ma Q, Li X, Cheng L, Li K, Li S (2014) Transcriptomic analysis of differentially expressed genes in the Ras1(CA)-overexpressed and wildtype posterior silk glands. BMC Genomics 15:182

Mita K, Morimyo M, Okano K, Koike Y, Nohata J, Kawasaki H, Kadono-Okuda K et al (2003) The construction of an EST database for *Bombyx mori* and its application. Proc Natl Acad Sci 100(24):14121–14125

Nagaraja GM, Nagaraju J (1995) Genome fingerprinting of the silkworm, *Bombyx mori* using random arbitrary primers. Electrophoresis 16:1633–1638

Okuda KK, Ito K, Murthy GN, Sivaprasad V, Ponnuvel KM (2014) Molecular mechanism of densovirus resistance in silkworm, *Bombyx mori*. Sericologia 54(1):1–10

Ou J, Deng H, Zheng S, Huang L, Feng Q, Liu L (2014) Transcriptomic analysis of developmental features of *Bombyx mori* wing disc during metamorphosis. BMC Genomics 15:820

Shen Y, Zeng X, Chen G, Wu X (2022) Comparative transcriptome analysis reveals regional specialization of gene expression in larval silkworm (*Bombyx mori)* midgut. Insect Sci 29(5):1329–1345. https://doi.org/10.1111/1744-7917.13001

Sivaprasad V, Chandrakanth N, Moorthy SM (2022) Genetics and genomics of *Bombyx mori* L. In: Genetic methods and tools for managing, Crop pests. Springer, pp 127–209

Sun Q, Guo H, Xia Q, Jiang L, Zhao P (2020) Transcriptome analysis of the immune response of silkworm at the early stage of *Bombyx mori* bidensovirus infection. Dev Comp Immunol 106:103601. https://doi.org/10.1016/j.dci.2019.103601

Tian L, Guo E, Wang S, Liu S, Jiang R, Cao Y et al (2010) Developmental Regulation of Glycolysis by 20-hydroxyecdysone and Juvenile Hormone in Fat Body Tissues of the Silkworm, Bombyx mori. J Mol Cell Biol 2(5):255–263

Tong X, Han MJ, Lu K et al (2022) High-resolution silkworm pan-genome provides genetic insights into artificial selection and ecological adaptation. Nat Commun 13:5619. https://doi.org/10.1038/s41467-022-33366-x

Wang Z, Gerstein M, Snyder M (2009) RNA-Seq a revolutionary tool for transcriptomics. Nat Rev Genet 10:57–63

Wang H, Wang L, Wang Y et al (2015) High yield exogenous protein HPL production in the *Bombyx mori* silk gland provides novel insight into recombinant expression systems. Sci Rep 5:13839. https://doi.org/10.1038/srep13839

Wang S, You Z, Feng M, Che J et al (2016) Analyses of the molecular mechanisms associated with silk production in silkworm by iTRAQ-based proteomics and RNA sequencing-based Transcriptomics. J Proteome Res 15:15–28

Watanabe H (1986) Resistance to the silkworm, Bombyx mori to viral infection. Agric Ecosyst Environ 15:131–139

Wu S, Tong X, Peng C, Xiong G, Lu K, Hu H, Tan D et al (2016) Comparative analysis of the integument transcriptomes of the black dilute mutant and the wild-type silkworm Bombyx mori. Sci Rep 6:26114

Xia Q, Zhou Z, Lu C, Cheng D, Dai F, Li B, Zhao P, Zha X et al (2004) A draft sequence for the genome of the domesticated silkworm (*Bombyx mori*). Science 306(5703):1937–1940. https://doi.org/10.1126/science.1102210

Xia Q, Cheng D, Duan J, Wang G, Cheng T et al (2007) Microarray-based gene expression profiles in multiple tissues of the domesticated silkworm, *Bombyx mori*. Genome Biol 8:R162

Xia Q, Guo Y, Zhang Z, Li D, Xuan Z, Li Z, Dai F et al (2009) Complete resequencing of 40 genomes reveals domestication events and genes in silkworm (*Bombyx*). Science 326(5951):433–436. https://doi.org/10.1126/science.1176620

Xia Q, Li S, Feng Q (2014) Advances in silkworm studies accelerated by the genome sequencing of *Bombyx mori*. Annu Rev Entomol 2014(59):513–536

Xu Q, Lu A, Xiao G, Yang B, Zhang J, Li X et al (2012) Transcriptional profiling of midgut immunity response and degeneration in the wandering silkworm, *Bombyx mori*. PLoS One 7(8):e43769. https://doi.org/10.1371/journal.pone.0043769

Yamamoto K, Narukawa J, Kadono-Okuda K, Nohata J, Sasanuma M, Suetsugu Y, Banno Y, Fujii H, Goldsmith MR, Mita K (2006) Construction of a single nucleotide polymorphism linkage map for the silkworm, *Bombyx mori*, based on bacterial artificial chromosome end sequences. Genetics 173(1):151–161

Yang B, Huang W, Zhang J, Xu Q, Zhu S, Zhang Q, Beerntsen BT et al (2016) Analysis of gene expression in the midgut of *Bombyx mori* during the larval molting stage. BMC Genomics 17(1):866. https://doi.org/10.1186/s12864-016-3162-8

Yang L, Gao Q, Dai J, Yuan G, Wang L, Qian C et al (2018) Comparative transcriptome analysis of silkworm, *Bombyx mori* colleterial gland suggests their functional role in mucous secretion. PLoS One 13(5):e0198077. https://doi.org/10.1371/journal.pone.0198077

Yokoi K, Tsubota T, Jouraku A, Sezutsu H, Bono H (2021) Reference transcriptome data in silkworm *Bombyx mori*. Insects 12:519. https://doi.org/10.3390/insects12060519

Zha X, Xia Q, Duan J, Wang C, He N, Xiang Z (2009) Dosage analysis of Z chromosome genes using microarray in silkworm, *Bombyx mori*. Insect Biochem Mol Biol 39:315–321

Zhang R, Cao YY, Du J, Thakur K, Tang SM, Hu F, Wei ZJ (2021) Transcriptome analysis reveals the gene expression changes in the silkworm (*Bombyx mori*) in response to hydrogen sulfide exposure. Insects 12(12):1110. https://doi.org/10.3390/insects12121110

Zheng X, Liu F, Shi M et al (2022) Transcriptome analysis of the reproduction of silkworm (*Bombyx mori*) under dimethoate stress. Pesticide Biochem Physiol 183:105081

Zhoa M, Liu C, Li Q, Hu W, Zhou M et al (2014) Basic helix loop helix transcription factor *Bmsage* is involved in regulation of fibroin H chain gene via interaction with SGF1 in *Bombyx mori*. PLoS One 9(4):e94091

Zhou Y, Gao L, Shi H, Xia H, Gao L et al (2013) Microarray analysis of gene expression profile in resistant and susceptible *Bombyx mori* strains reveals resistance-related genes to nucleopolyhedrovirus. Genomics 101:256–262

Zhou QZ, Fu P, Li SS, Zhang CJ, Yu QY, Qiu CZ et al (2020) A comparison of co-expression networks in silk gland reveals the causes of silk yield increase during silkworm domestication. Front Genet 27(11):225. https://doi.org/10.3389/fgene.2020.00225

Zhu K, Chen Y, Chen L, Xiang H (2022) Comparative silk transcriptomics illuminates distinctive impact of artificial selection in silkworm modern breeding. Insects 13:1163. https://doi.org/10.3390/insects13121163

Proteomics of Silkworm, *Bombyx mori* L.: Recent Progress and Future Prospectus

L. Kusuma, L. Satish, V. S. Raviraj, S. M. Moorthy, and V. Sivaprasad

Abstract

Recent advances in the proteomic analysis of silkworm, *Bombyx mori*, in different tissues and organs have paved way for better understanding of the regulation of various biological activities. Comparative proteomic analyses have led to the identification of crucial proteins and varied expression during metamorphosis like storage proteins and 30 K proteins from the hemolymph. The identification of tissue specific proteins of silkworm larvae, pupae, and moths has unraveled various metabolic pathways related to metabolism-related disease resistance and susceptibility, molting, spatial and temporal expression of enzymes and activities, mucins, chitin-binding proteins, chitin deacetylases, serpins, serine proteases, etc. Adopting technologies like 2D gel electrophoresis and MALDI-TOF-MS and expression of silk gland specific proteins, developmental stage specific proteins, antimicrobial proteins, and immune related proteins have been identified, which have revolutionized the understanding of complex protein profiles in silkworm. Further, silk fibroin has a number of outstanding properties, not only for textile applications but also for biomaterials. The high tensile strength and toughness is suitable for by-products/biomaterials. *B. mori* silkworm silk fibroin is the versatile fiber that could be regenerated for utilization in industrial and biomedical applications. Structure-property relationships are among the most important goals in silk industry to understand the basic mechanism of fibroin structure and

L. Kusuma · L. Satish (✉)
Silkworm Breeding and Molecular Biology Laboratory II, Central Sericultural Research and Training Institute, Central Silk Board, Mysuru, Karnataka, India

V. S. Raviraj
Central Sericultural Research and Training Institute, Central Silk Board, Berhampore, West Bengal, India

S. M. Moorthy · V. Sivaprasad
Central Silk Board, Bengaluru, India

R. V Suresh et al. (eds.), *Biotechnology for Silkworm Crop Enhancement*,
https://doi.org/10.1007/978-981-97-5061-0_4

its interactions as elementary unit influences mechanical properties. These information provide way for exploring genetic markers responsible for silk mechanical properties and thereby develop silkworm breeds with improved production of silk fiber that is equivalent to that of spider silk.

Keywords

Bombyx mori · Proteomics · Silk fibroin · Protein profiling

Introduction

The total protein content of an organism constitutes its proteome complex. The proteome possesses a wide range of functions and characteristics in all organisms (Zubair et al. 2022). Among several model organisms studied, silkworm holds a unique place in biology because of the exclusive protein produced in the silk glands, the silk. Silkworm represents an ideal lepidopteran model system for biological, molecular, and genetic studies with a catalog of information available on its morphology, genetics, and genetic diversity (Nagaraju and Goldsmith 2002) and is one of the most economically important insects because of the worldwide importance of the protein—silk. The silk gland of *Bombyx mori* is an organ that is exclusively specialized for the synthesis and secretion of silk proteins (Zhang et al. 2006). The silk produced from different silkworms varies in its constituent but is basically composed of two threads (fibroins) bonded by adhesive proteins (sericins) (Gamo et al. 1977; Zhang et al. 2015). Silk fibroin is made up of major components of silk fibroin that include the heavy fibroin chain, the light fibroin chain, and P25/fibrohexamerin (Inoue et al. 2000; Tanaka et al. 1999).

Protein Structure of Fibroin

The exceptional quality bivoltine silk measures around 1500 m in length and 10–20 μm in diameter and is made up of many nanofibrils—each fiber has sericin and fibroin proteins. The silk fibroin molecule is made up of the glycoprotein P25 (30 kDa) and two chains of different lengths: the light (L) chain, which is 26 kDa, and the heavy (H) chain, which is 390 kDa. In the endoplasmic reticulum of the posterior silk gland, the H chain, L chain, and P25 combine to produce the elementary unit. This 2.3 MDa elementary unit complex consists of one fibrohexamerin (P25) molecule and six sets of H and L chains. Disulphide bonding between H and L-Chain and maintaining H_6:L_6:$P25_1$ conformation is significant for quality check of fibroin synthesis (Inoue et al. 2000, 2004).

Structure of Fibroin Heavy Chain

The amino acid sequence of *B. mori* silk fibroin heavy chain is highly repetitive. From X-ray diffraction studies, it is well-known that the hexaamino acid sequence, GAGAGS, is the main component of the typical β-sheet crystals of silk fibroin, which form the monoclinic unit cell of silk II. The crystalline segments are composed of GAGAGS sequences, and besides the GAGAGS sequences, there are several types of irregular sequences. One type of irregular sequence (GT-GT) appears 11 times in the heavy chain. Another type of irregular sequence contains GY (GY_GY). Small irregular units, GAAS, are also present. The amino acid residues at the N-terminus and C-terminus of the heavy chain are also irregular. Cys residues in each chain end are capable of covalent disulfide bonding with other Cys residues.

There is some regularity in the amino acid sequence of the fibroin heavy chain.

(1) The crystalline segments, GAGAGS, are separated by 60 segments containing GY (GY_GY) sequences. There are 13 different types of GY_GY sequences, and the 5 most abundant are as follows:

1. GY(GX)3GYGXGY(GX)3GY, 24 times.
2. GAGAGY, seven times.
3. GYGXGY (GX)3GY(GX)3GY(GX)3GY, seven times.
4. GY(GX)3GY, six times.
5. GYGXGY(GX)3GY, six times (where X is usually A and sometimes V and I residues. The positions of V and I are irregular).

GY_GY sequences are usually followed by the small GAAS segment.

GAAS segments always appear at the same positions, except for each terminus.

GAGAGS/GY_GY crystalline building blocks are usually composed of Gly, Ala, Ser, and Tyr, which are and/or may be the main components of the β-sheet crystals of silk fibroin. Val, Thr, Ile, and Phe are not major residue types but sometimes appear in GAGAGS/GY_GY blocks (Asakura 2021).

Two to eight GAGAGS/GY_GY blocks appear between 11 GT_GT irregular sequences. GT_GT irregular sequences are immediately followed by GAGAGS sequences, except for the last one. Evolutionarily conserved GT_GT irregular sequences are usually composed of 31 residues, except for the first one (32 residues) and the last one (34 residues). GT_GT irregular sequences always contain Pro residues, which can be a major factor for changing the backbone direction. The C-terminus and N-terminus are composed of completely nonrepeating amino acid residues.

Structure of L Chain and P25 Protein

Based on cDNA sequencing research, the molecular weight of the L chain is 25,800, while the molecular weight of P25 was inferred from the genomic sequence to be 25,000. There are 244 amino acids in the L chain overall and 240 in the P25 chain.

It is revealed that the L chain's secondary structure, as Chou and Fasman predicted, is abundant in β-sheets and twists. Asx, also known as aspartic acid or asparagine, accounted for 36% of the main amino acid residues in the L chain, followed by alanine (34%), serine (22%), glycine, and glutamine. In contrast, alanine makes up 33% of P25's primary amino acid makeup, followed by serine (24%), glycine (22%), and isoleucine (20%). About 47% of the amino acids in the L chain are hydrophobic side chains (Asakura 2021).

Structure of Silk I

After drying naturally without the aid of outside forces, the solid state structure of silk fibroin is known as silk I and is kept in the middle silk glands. It is a soluble form that doesn't precipitate at high concentrations and is stable and nonviscous. Silk I has both regions with a clearly defined ordered structure and regions that are random coil. These are arranged in the (AGSGAG)n sequence. According to a thorough examination of silk I's 13C solid state NMR spectra, roughly 30% of alanine and 40% of serine are Cβ. In the silk I primary structure, the computed fraction of Ala residues of AGSGA sequences is 54%, whereas in the Ser, it is 69%. Consequently, this indicated that roughly 55–58% of the AGSGAG sequence takes silk I structure (Ma et al. 2014; Wilson et al. 2000).

Structure of Silk II

The structure thought to be composed of cocoon silk is called silk II. The combination of crystalline and noncrystalline protein domains is known as silk II. Marsh et al. (1955) used X-ray crystallography to study silk II, the first structure they analyzed, and further discovered that silk II is composed of antiparallel β-sheets. Of the native silk fiber studied by NMR, 25% had β-sheet and 18% had β-turn. Marsh first offered a polar antiparallel β-sheet structure; however, Takahashi et al. (1999) proposed an anti-polar antiparallel β-sheet structure with more data available. More than 50% of the noncrystalline area silk fiber has a β-sheet structure, which is a combination of β-turn and β-sheet.

Structural Transition from Silk I to Silk II

The organized structure of silk I serves as the foundation for the production of silk II, as was previously described. The creation of β-sheet conformation in silk II is aided by the β-turn secondary structure of silk I, which already has intermolecular hydrogen bonding mediated by serine residues (Wilson et al. 2000). Additionally, a rise in salt content, a fall in pH, shear-induced alignment, and gradual dryness all contribute to the creation of silk dope (silk I) and cocoon silk (silk II). Because the

N-terminal domain contains acidic amino acid residues that trigger structural modifications, pH decrease mediates the nucleation to the silk II structure.

Proteomics of Silkworm Silk Gland

The proteome in the silk gland lumen using liquid chromatography-tandem mass spectrometry was determined in fifth day of the fifth instar to day 1 of wandering. The annotations of luminal proteins identified 868 intracellular proteins, 262 extracellular proteins, and 141 transmembrane proteins. Based on the protein function, proteins in the silk gland lumen mainly consist of fibroins, sericins, seroins, extracellular matrix proteins, protease inhibitors, enzymes, and proteins of unknown function, of which fibroins were the most abundant components. Hormone metabolism enzymes, viz., juvenile hormone esterase (JHE) and JHE-like proteins, juvenile hormone epoxide hydrolases (JHEH) and JHEH-like proteins, ecdysone oxidase (EO) and EO-like proteins, 3-dehydroecdysone 3 alpha-reductase (3DE-3α-R), and 3-dehydroecdysone 3 beta-reductase (3DE-3β-R), were identified in the silk gland lumen (Dong et al. 2016).

In another study, silk gland proteins were separated by 2D followed by in-gel digestion, and PMF identified 93 proteins tentatively as proteins including fibroin L chain and P25 were found as multiple isoforms (Zhang et al. 2006). Over 200 proteins were identified in the hemolymph of last instar silkworm larvae.

Proteomics of Silkworm Midgut

2DE and matrix-assisted laser desorption/ionization time-of-flight mass spectrometry (MALDI-TOF-MS) were used to study the midgut of fifth instar larvae. Proteins, viz., 30 K lipoprotein precursor, low-molecular-mass 30 kDa lipoprotein 21G1 precursor, low-molecular-mass 30 kDa lipoprotein 19G1, low-molecular-mass 30 kDa lipoprotein 21G1 precursor, low-molecular-mass 30 kDa lipoprotein PBMHPC-19 precursor, low-molecular-mass 30 kDa lipoprotein 19G1, mature 30 K lipoprotein, imaginal disk growth factor, Actin-depolymerizing factor1, transferrin, 60 kDa heat shock protein, mitochondrial, Protein phosphatase 1 catalytic subunit, Acireductone dioxygenase, Glyoxylate reductase/hydroxypyruvate reductase, Protein disulfide isomerise, DnaJ-like protein, Dihydrolipoamide dehydrogenase, Aspartate aminotransferase, Protein disulfide-isomerase-like protein ERp57, Ribosomal protein P0, Ubiquitin carboxyl-terminal hydrolase, Ribosomal protein P2, Heat shock protein 70, Peptidylprolyl isomerase B, 10 kDa heat shock protein, mitochondrial, Enolase, Triosephosphate isomerise, Glyceraldehyde-3-phosphate dehydrogenase, Enoyl-CoA hydratase precursor 1, Fatty acid binding protein, Apolipophorin-III precursor, Acyl-coenzyme A dehydrogenase, Acyl-coenzyme A dehydrogenase, 3-hydroxy acyl-CoA dehydrogenase, Lysophospholipase, Acetoacetyl-CoA thiolase, Phosphatidylethanolamine binding protein isoform 2, H+-transporting ATP synthase beta subunit isoform 2, H+-transporting ATP synthase beta subunit isoform 1, Abnormal wing disc-like protein, Vacuolar ATP synthase subunit F, Cytochrome c oxidase subunit Va, H+-transporting ATP synthase delta subunit, H+-transporting ATP synthase subunit d, Vacuolar ATPase B subunit, Ubiquinol-cytochrome c reductase, Vacuolar ATP synthase catalytic subunit A, ATP synthase subunit alpha, ATP synthase subunit alpha, Hydroxybutyrate CoA-transferase, putative, Arginine kinase, Ubiquinol-cytochrome c reductase core protein II, ATP synthase subunit alpha, ATP synthase subunit alpha, NADH dehydrogenase (ubiquinone) Fe-S protein 8, Immune-related Hdd13, Cyclophilin A, Cyclophilin, Glutathione S-transferase 10, Thiol peroxiredoxin, Actin A3, Actin, muscle-type A1, Actin, cytoplasmic A4, Myosin 1 light chain, Actin, cytoplasmic A4,Beta-tubulin, Myosin heavy chain, Myosin light chain 2, Muscular protein 20, Transgelin, Myosin light chain 2, Juvenile hormone diol kinase, Putative farnesoic acid O-methyl transferase, Putative farnesoic acid O-methyl transferase, Diapause hormone precursor, 27 kDa glycoprotein, p27K, Cellular retinoic acid binding protein, Stardust, NADPH oxidase, Serpin-2 were identified.

Proteomics of Silkworm Hemolymph

Several studies on the proteomics of the hemolymph of silkworm have been conducted which identified the proteins. Inter-alpha-trypsin inhibitor heavy chain H4 precursor, SP2 Storage protein Transferrin, Beta-N-acetylglucosaminidase 2, Prophenoloxidase, CI8 Chymotrypsin inhibitor, Antitrypsin isoform 1, Aminoacylase, Gelsolin, Aldose reductase, Glyoxylate reductase, Serine protease homolog 1, Ester hydrolase, Serine proteinase-like protein, 30kD Bmlp1, 30kD Bmlp3, 30kD Bmlp2, Hydroxypyruvate isomerise, 30kD Bmlp7, Bombyrin, Peptidylprolyl isomerase B, Juvenile hormone binding protein, Putative paralytic peptide-binding protein homolog, Glycerophosphoryl diester phosphodiesterase, SP1, SP2, Hemolin, Vitellogenin, 32 kDa apolipoprotein cytosolic fatty-acid binding protein, Bioclock protein Diapause time regulate protein EA4, Apoliprotein III (Hou et al. 2007; Kajiwara et al. 2009; Hou et al. 2010; Zhang et al. 2014; Xia et al. 2014)

Proteomics of Silkworm Fat Body

Silkworm fat body forms a crucial site where several intermediary metabolic processes occur and is a source of sustenance for growth throughout the life cycle. Fat body proteins are responsible for storing nutrients, providing energy, and regulating hormones and have been identified using several proteomic approaches (Wu et al. 2014). Proteome analysis of day 3 fifth instar silkworm larvae, *B. mori*, by 2D PAGE on 2D gels analyzed by capillary HPLC coupled with ion trap mass spectrometry identified five ATPase proteins (Kajiwara et al. 2009). Out of five ATPase proteins, two showed similarities to aspartate decarboxylases. Several cytoskeletal proteins, viz., actin, cofilin, villin, profiling bound to actin, tropomyosin, alpha-tubulin, and beta-tubulin, represented the galore of proteins in fat bodies of silkworm. Three calreticulin proteins such as Heat shock protein, chaperonin, and prefoldin were also characterized vastly along with lipocalin-related proteins that bind to the cytosolic fatty acids. Among the abundances of the proteins in fat body, metal binding proteins, viz., calcium-binding proteins, zinc-finger protein, and selenium binding protein, were also found to be involved in various metabolic activities (Kajiwara et al. 2009). Enzymes involved in carbohydrate metabolism such as enolase, glyceraldehyde 3-phosphate dehydrogenase, xylose isomerase, isocitrate dehydrogenase, lactate/malate dehydrogenase, and succinate dehydrogenase were also formed the rich proteome in fat bodies. Study of proteomic analysis in the fat body of silkworm by Hou et al. identified 34 proteins several of which represented involvement in metabolism and immunity among which important proteins identified were Heat shock proteins, 30 K proteins, and Actin (Hou et al. 2007). In a proteomic study based on the heat tolerant (multivoltine) and heat susceptible (bivoltine) silkworms (*Bombyx mori*) in response to heat shock through MALDI-TOF/TOF spectrometer, MS/MS and MS analysis identified eight proteins—small heat shock proteins (sHSPs) and HSP70. These were expressed similarly in both breeds, while 4 proteins were expressed exclusively in bivoltine breed and 12 protein in multivoltine breed (Moghaddam et al. 2008). Another approach involving shotgun proteomics explored 138, 217, and 86 proteins in the fat body of larvae, pupae, and moths, respectively, and comparatively increased number of metabolism-related enzymes were identified at the pupal stage than at the any other stages (Xia et al. 2014; Yang et al. 2010).

Comparative proteomic analysis of fat body on the fifth day of fifth instar was studied by Chen et al. (2015) between BmFib-H gene knockout *Bombyx mori* line (FGKO) and its wide-type Dazao, which revealed single gene knockout in silk gland triggering large-scale metabolic pathway changes in fat body. Several proteins involved in glycolysis/gluconeogenesis, pentose phosphate pathway, and glycine-serine biosynthetic pathway were found to be downregulated in the FGKO fat body, whereas those involved in ribosomal proteins, eukaryotic translation initiation factor, and elongation factor were found to be upregulated. This study was first to signify the cross talk between silk gland and fat body in silkworm, and this would lead to redistribution of nutrients in the FGKO fat body resulting in the increase of the pupal weight. The differential expression of sex-related fat body proteins during the larval-pupal developmental stages of silkworm was characterized which identified 25 sex-related proteins with 13 differential protein constituents, and the expression of fat body proteins from male and female also differed in which β-tubulin represented specific expression in female silkworms and also RACK is only expressed in the female silkworms of the fifth day of the fifth instar (Wu et al. 2014).

Crystal Structures of *B. mori* Proteins

The protein structure of *B. mori* is under studied area of research. Only few proteins have been structurally studied through crystallography. Of them are 30 kDa proteins known as *B. mori* low molecular weight lipoproteins (Bmlps)—these homologous proteins, which have a molecular weight of roughly 30 kDa, make up approximately

5% of the total plasma proteins in the hemolymph of silkworm *Bombyx mori* larvae in their fifth instar. It has been claimed that these so-called 30 K proteins are involved in the innate immune response as well as the transfer of sugar and/or fat. The crystal structure of a 30 K protein, Bmlp7, at 1.91 Å has been discovered. Two crystal structures (Bmlp3 and Bmlp7) of 30-kDa LPs from *Bombyx mori* have been reported thus far. The crystal structure of Bmlp6, a different member of the 30-kDa LP, has also been reported (Pietrzyk et al. 2013).

The initial crystal structure of a complex consisting of two storage proteins, SP2 and SP3, has been ascertained. The proteins were extracted from their native source, the hemolymph of the mulberry silkworm (*Bombyx mori* L.). Using arylphorin, a protein rich in aromatic amino acid residues from oak silkworms, as the original model, the structure was solved using molecular replacement (Hou et al. 2014).

In insects, juvenile hormones (JHs) regulate a variety of important life events. Major agricultural pests are members of the Lepidoptera order, to which JH signaling is crucially regulated by a species-specific, low molecular weight, high-affinity JH-binding protein (JHBP) in hemolymph that carries JH from the site of production to target tissues (Fujimoto et al. 2013; Suzuki et al. 2011). Therefore, it is anticipated that JHBP will provide a great target for the creation of innovative, targeted insect growth regulators (IGRs) and insecticides. A deeper comprehension of JHBP's structural biology ought to open up new avenues for the structure-based medication design of these kinds of molecules. The two molecules of 2-methyl-2,4-pentanediol (MPD), one of which is bound in the JH-binding pocket and the other in a second cavity, are complexed with the silkworm *B. mori* JHBP. The crystal structure of this complex is also unraveled.

Esterase A4 (EA4) is a metalloglycoprotein whose distinctive ATPase activity is transiently enhanced toward the end of the requisite cold period, leading to the suggestion that it functions as a cold-duration clock in the diapause eggs of the silkworm *Bombyx mori* (Hiraki et al. 2008). Though its time-measuring mechanism is entirely unknown, this EA4 timer property is known to begin with the dissociation of an inhibitory peptide (referred to as the "peptidyl inhibitory needle") under cold temperatures. The crystal structures and functional characteristics of EA4 with and without glycosylation are studied.

References

Asakura T (2021) Structure of silk I (*Bombyx mori* silk fibroin before spinning)-type II β-turn, not α-helix. Molecules 26(12):3706. https://doi.org/10.3390/molecules26123706

Chen Q, Ma Z, Wang X, Li Z, Zhang Y, Ma S, Zhao P, Xia Q (2015) Comparative proteomic analysis of silkworm fat body after knocking out fibroin heavy chain gene: a novel insight into cross-talk between tissues. Funct Integr Genomics 5(5):611–637. https://doi.org/10.1007/s10142-015-0461-0

Dong Z, Zhao P, Zhang Y et al (2016) Analysis of proteome dynamics inside the silk gland lumen of *Bombyx mori*. Sci Rep 6:21158. https://doi.org/10.1038/srep21158

Fujimoto Z, Suzuki R, Shiotsuki T, Tsuchiya W, Tase A, Momma M, Yamazaki T (2013) Crystal structure of silkworm *Bombyx mori* JHBP in complex with 2-methyl-2,4-pentanediol: plasticity of JH-binding pocket and ligand-induced conformational change of the second cavity in JHBP. PLoS One 8(2):e56261

Gamo T, Inokuchi T, Laufer H (1977) Polypeptides of fibroin and sericin secreted from the different sections of the silk gland in *Bombyx mori*. Insect Biochem 7:285–295

Hiraki T, Shibayama N, Akashi S, Park S (2008) Crystal structures of the clock protein EA4 from the silkworm *Bombyx mori*. J Mol Biol 377(3):630–635

Hou Y, Zhao P, Liu HL, Zou Y, Guan J, Xia QY (2007) Proteomics analysis of fat body from silkworm (*Bombyx mori*). Sheng Wu Gong Cheng Xue Bao 23(5):867–872

Hou Y, Guan J, Zhao P, et al. (2007) Proteomics analysis of midgut from silkworm, Bombyx mori. Science of Sericulture 33(2):216–222

Hou Y, Zou Y, Wang F, et al. (2010) Comparative analysis of proteome maps of silkworm hemolymph during different developmental stages. Proteome Sci. 8;8:45.

Hou Y, Li J, Li Y, Dong Z, Xia Q, Yuan YA (2014) Crystal structure of Bombyx mori arylphorins reveals a 3:3 heterohexamer with multiple papain cleavage sites. Protein Sci 23(6):735–746. https://doi.org/10.1002/pro.2457

Inoue S, Tanaka K, Arisaka F, Kimura S, Ohtomo K, Mizuno S (2000) Silk fibroin of *Bombyx mori* is secreted, assembling a high molecular mass elementary unit consisting of h-chain, l-chain, and p25, with a 6:6:1 molar ratio. J Biol Chem 275:40517–40528

Inoue S, Tanaka K, Tanaka H, Ohtomo K, Kanda T, Imamura M, Quan GX et al (2004) Assembly of the silk fibroin elementary unit in endoplasmic reticulum and a role of l-chain for protection of a1,2-mannose residues in N-linked oligosaccharide chains of fibrohexamerin/p25. Eur J Biochem 271:356–366

Kajiwara H, Imamaki A, Nakamura M, Mita K, Xia Q, IshizakaM. (2009) Proteome analysis of silkworm 2. Hemolymph. J Electrophor 53:27–31

Ma S, Shi R, Wang X, Liu Y, Chang J, Gao J, Lu W, Zhang J et al (2014) Genome editing of *bmfib-h* gene provides an empty *Bombyx mori* silk gland for a highly efficient bioreactor. Sci Rep 4:6867

Marsh RE, Corey RB, Pauling L (1955) An investigation of the structure of silk fibroin. Biochim Biopyhys Acta 16:1–33

Moghaddam H, Du S, Li X et al (2008) Proteome analysis on differentially expressed proteins of the fat body of two silkworm breeds, *Bombyx mori*, exposed to heat shock exposure. Biotechnol Bioproc E 13:624–631

Nagaraju J, Goldsmith M (2002) Silkworm genomics—Progress and prospects. Curr Sci 83(4):415–425

Pietrzyk AJ, Bujacz A, Mueller-Dieckmann J, Lochynska M, Jaskolski M, Bujacz G (2013) Two crystal structures of *Bombyx mori* lipoprotein 3 - structural characterization of a new 30-kDa lipoprotein family member. PLoS One 8(4):e61303. https://doi.org/10.1371/journal.pone.0061303

Suzuki R, Fujimoto Z, Shiotsuki T, Tsuchiya W, Momma M, Tase A et al (2011) Structural mechanism of JH delivery in hemolymph by JHBP of silkworm, *Bombyx mori*. Sci Reps 1:133. https://doi.org/10.1038/srep00133

Takahashi Y, Gehoh M, Yuzuriha K (1999) Structure refinement and diffuse streak scattering of silk (*Bombyx mori*). Int J Biol Macromol 24:127–138

Tanaka K, Inoue S, Mizuno S (1999) Hydrophobic interaction of P25, containing Asn-linked oligosaccharide chains, with the H–L complex of silk fibroin produced by *Bombyx mori*. Insect Biochem Mol 29:269–276

Wilson D, Valluzzi R, Kaplan D (2000) Conformational transitions in model silk peptides. Biophys J 78:2690–2701

Wu Z, Liu Y, Chen G, Wang J, Tan J (2014) Differential expression of sex-related fat body proteins during the larval–pupal developmental stages of the silkworm (*Bombyx mori*). J Asia-Pac Entomol 17(1):19–26

Xia Q, Li S, Feng Q (2014) Advances in silkworm studies accelerated by the genome sequencing of *Bombyx mori*. Annu Rev Entomol 2014(59):513–536

Yang H, Zhou Z, Zhang H, Chen M, Li J et al (2010) Shotgun proteomic analysis of the fat body during metamorphosis of domesticated silkworm (*Bombyx mori*). Amino Acids 38:1333–1342
Zhang P, Aso Y, Yamamoto K, Banno Y, Wang Y, Tsuchida K, Kawaguchi Y, Fujii H (2006) Proteome analysis of silk gland proteins from the silkworm, *Bombyx mori*. Proteomics 6(8):2586–2599. https://doi.org/10.1002/pmic.200500348
Zhang Y, Dong Z, Wang D, et al (2014) Proteomics of larval hemolymph in Bombyx mori reveals various nutrientstorage and immunity-related proteins. Amino Acids 46(4):1021–31
Zhang Y, Zhao P, Dong Z, Wang D, Guo P, Guo X et al (2015) Comparative proteome analysis of multi-layer cocoon of the silkworm, *Bombyx mori*. PLoS One 10(4):e0123403. https://doi.org/10.1371/journal.pone.0123403
Zubair M, Wang J, Yu Y, Faisal M, Qi M, Shah AU, Feng Z, Shao G et al (2022) Proteomics approaches: a review regarding an importance of proteome analyses in understanding the pathogens and diseases. Front Vet Sci 9:1079359. https://doi.org/10.3389/fvets.2022.1079359

Epigenomics: A Way Forward from Classical Approach

Ranjini, Raviraj, S. Manthira Moorthy, and S. Gandhi Doss

Abstract

Epigenomics is developing expeditiously in insect research wherein, without involvement of changes in DNA sequence, the adaptations to environmental changes are addressed. Besides, when there are endogenous and environmental stimuli, epigenetics is involved in the reprogramming of gene expression. Epigenetic changes occur through the presence of pathogens, parasites, harmful chemicals, varied temperature, humidity, and other stress factors. Epigenetic modifications are strongly evident in causing reprogramming of transcriptional process and known to be passed from parents to offspring. DNA methylation, histone acetylation, and histone methylation are the major epigenomic mechanisms added to which there are posttranscriptional gene regulation heritable mechanisms, i.e., the synthesis of noncoding microRNAs (miRNAs) known to bind corresponding messenger RNAs (mRNAs) and degrade its activity, thereby inhibiting translation. In addition to miRNAs, long noncoding RNAs (lncRNAs; >200 nucleotides in length) play an important role in regulating complex biological processes and also during stress through epigenetic control. Emerging evidences have revealed a key role of lncRNAs in regulating immunological functions and autoimmunity. Recent discoveries have shown that different epigenetic pathways along with microRNAs (miRNAs) form regulatory circuit

Ranjini (✉) · S. Gandhi Doss
Silkworm Bivoltine Breeding Laboratory, Central Sericultural Research & Training Institute, Central Silk Board, Mysuru, India
e-mail: ranjinims.csb@nic.in

Raviraj
Silkworm Breeding & Genetics Laboratory, Central Sericultural Research & Training Institute, Central Silk Board, Berhampore, West Bengal, India

S. Manthira Moorthy
Central Silk Board, Ministry of Textiles, Government of India, Bengaluru, India

R. V Suresh et al. (eds.), *Biotechnology for Silkworm Crop Enhancement*,
https://doi.org/10.1007/978-981-97-5061-0_5

involving epigenetics and miRNA which provides strong biological insights into gene-regulatory mechanism(s) underlying a variety of diseases. Collectively, progressing inputs by many studies on epigenetics, questioning whether the occurrence of phenotypic plasticity is through the role of epigenetics, are dealt, and the overall advancement in the field of epigenomics as a new avenue to unravel the silkworm biological system complexity under stressful environmental conditions, genetic diversity, heterosis, and impact on productivity information will be purveyed in this chapter.

Keywords
Epigenetics · Methylation · Transoposable elements · Transgeneic · RNAi

Introduction

Glimpses on the Journey of Genetics

The identity of every living organism that connects to one's existence, being individually different, with specific genetic architecture, right from the single celled organisms to *Homo sapiens*, is the reflection of genetics, a fascinating subject of one's origin. The cascade of events in genetics including recent advancement in silkworm is depicted in Fig. 1. The journey of genetics originated through Gregor Mendel's experiment on plant hybridization during 1866 which provided a new route. Since then, many scientists carried out research on this field and new avenue was generated. Gyon (2016) has given detailed information on history of genetics from Mendel (1865) to epigenetics. During 1900, Mendel's law was rediscovered independently by Hugo de Vries, Carl Correns, and Erich Von Tschermak. Bateson (1902) reported that the Mendelian laws of hybridization are not specifically for plants but also indeed applicable to animals. Sutton (1903) proposed chromosomes as the Mendelian factor bearers in support of explaining the meiotic division highlighting the laws of segregation and reassortment. Bateson coined the term genetics in the year 1905. The term pangene originated from gene (Johannsen 1909). The Mendelian genetics and the chromosomal theory of inheritance together gave rise to classical genetics which literally happened through Thomas Hunt Morgan experiments (Morgan et al. 1915). Sturtevant (1913, 1965) established the first genetic maps that locate genes along with chromosomes while working on *Drosophila*, until which it was known through calculation of crossing overs. Muller (1925) explained the effect of X-rays on *Drosophila* chromosome where specific allele was transformed into the another one and termed as mutant gene. Until DNA discovery, gene was thought to be a hereditary unit, unit of recombination, and unit of mutation. Creighton and McClintock (1931) explained the cytological evidence for genetic theory that linked genes on paired chromosomes and exchanged places from one homolog to the other. Later, based on the experiment conducted on *Neurospora* about the biochemical reactions controlled by genetics, one gene-one enzyme hypothesis was explained (Beadle and Tatum 1941).

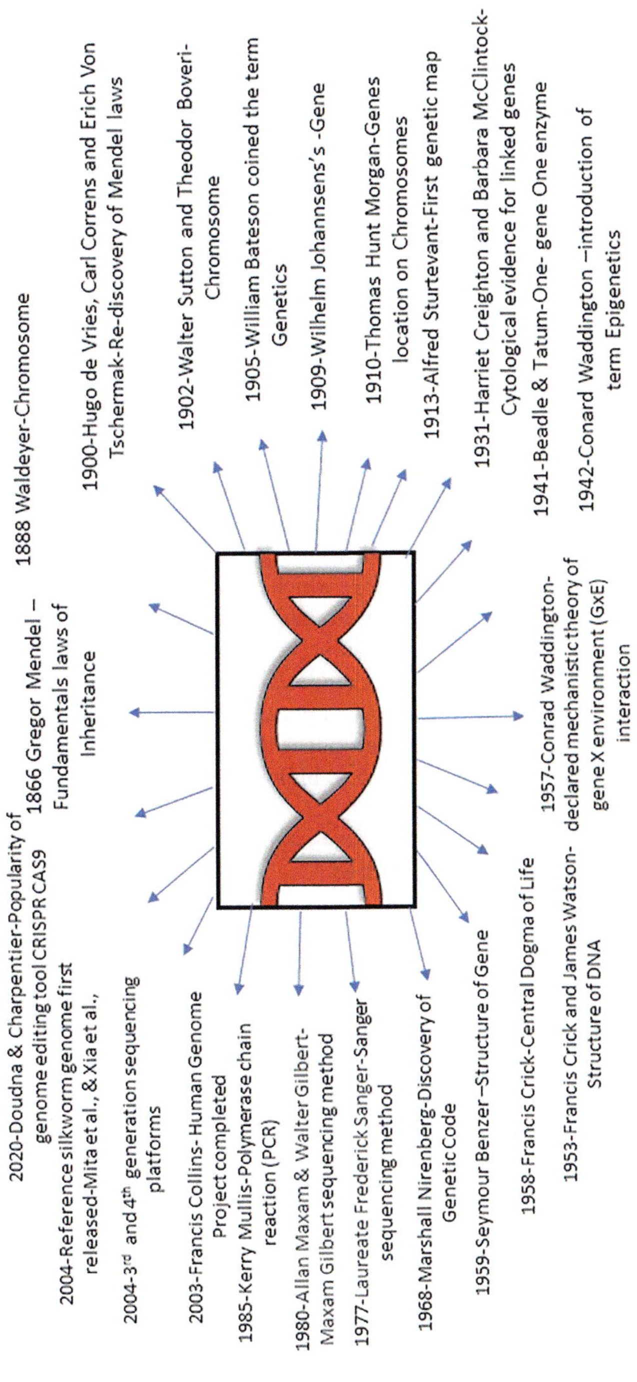
1888 Waldeyer-Chromosome
1900-Hugo de Vries, Carl Correns and Erich Von Tschermak-Re-discovery of Mendel laws
1902-Walter Sutton and Theodor Boveri-Chromosome
1905-William Bateson coined the term Genetics
1909-Wilhelm Johannsens's -Gene
1910-Thomas Hunt Morgan-Genes location on Chromosomes
1913-Alfred Sturtevant-First genetic map
1931-Harriet Creighton and Barbara McClintock-Cytological evidence for linked genes
1941-Beadle & Tatum-One- gene One enzyme
1942-Conard Waddington –introduction of term Epigenetics
1866 Gregor Mendel – Fundamentals laws of Inheritance
1957-Conrad Waddington-declared mechanistic theory of gene X environment (GxE) interaction
2020-Doudna & Charpentier-Popularity of genome editing tool CRISPR CAS9
2004-Reference silkworm genome first released-Mita et al., & Xia et al.,
2004-3rd and 4th generation sequencing platforms
2003-Francis Collins-Human Genome Project completed
1985-Kerry Mullis-Polymerase chain reaction (PCR)
1980-Allan Maxam & Walter Gilbert-Maxam Gilbert sequencing method
1977-Laureate Frederick Sanger-Sanger sequencing method
1968-Marshall Nirenberg-Discovery of Genetic Code
1959-Seymour Benzer–Structure of Gene
1958-Francis Crick-Central Dogma of Life
1953-Francis Crick and James Watson-Structure of DNA

Fig. 1 Journey of classical to modern genetics

Baseline Research in Molecular Genetics

Avery et al. (1944) carried out experiment on pneumococcus and explained that DNA can transform. However, only after the discovery of structure of DNA (Watson and Crick 1953), it was considered as a molecule that carries the hereditary properties. The process by which the instructions in DNA are converted into a functional product was proposed as central dogma by Francis Crick (1970). Later, the molecular structure of the gene was discovered by Seymour Benzer (1959). The discovery of the genetic code and the regulation of gene expression were depicted by Jacob and Monod (1961) through lactose operon concept. He explained the cis-trans test in which he distinguished the mutations affecting different genes and the different mutations affecting same gene. The genetic code was discovered by Nirenberg and Leder (1964). After genetic code was discovered, the first DNA sequencing was carried through Sanger sequencing method by Sanger et al. (1977). Maxam and Gilbert (1980) method of sequencing was proposed. Polymerase chain reaction was discovered by Mullis et al. (1986) which was very much essential for exonic and intronic sequencing by utilizing molecular markers, region specific primers, cloning, recombination, and hybridization process. The most exciting achievement of science in the modern genetics happened through the completion of the Human Genome Project by Collins et al. (2003) which was initiated through Crick and Watson. Since then, genome sequencing in many model organisms' approach was undertaken. Silkworm first reference genome was sequenced by Mita et al. (2004) and Xia et al. (2004). Doudna and Charpentier were awarded with Nobel prize in 2020 for the discovery of CRISPR/Cas9—a popular genome editing tool (Ledford & Callaway 2020) through which the new hopes in modern science have evolved to address many complex questions in modern biology.

Importance of Epigenetics Studies

Environmental cues are very important in organisms' survival. Since recent times, the climate change/variation is creating challenges in organisms' existence, being exposed to uncertainty created due to climate change/variation. What happens during such changes in one's cell level is an important concern, as all nucleated cells in an organism share the same genetic material, but these cells are differentiated based upon the specific genes expressed in each cell at specific time. During aging process, transition occurs from one stage to another by facing different challenges during development which is controlled or monitored by gene expression and regulatory mechanisms at transcriptional and translational levels in all organisms. The questions which have to be addressed are whether the organisms which face abiotic or biotic stresses can combat it or overcome it? I the stress response created heritable? If so, is it through Mendelian or non-Mendelian pattern? Is the gene regulation restricted only to transcriptional or translational levels, or are there key players in posttranslational levels? All these questions need to be studied in order to protect the havoc which is getting created by the climate variations and toxicity load.

The answers for all the abovementioned questions could be possible through the epigenetics/epigenomics study/approach. It is an emerging research field which is a need of the hour through which modulations could be possible to overcome the disturbances in the environmental cues. Way back in the 1940s, term epigenetics was introduced by Waddington (1940, 1942) that explains the interactions of environment and genes which reflects on the phenotypic/behavioral changes of an organism. In Epigenetics, the term "Epi" means over which presumes that operating over the genes (Tronick and Hunter 2016). The theory on how the gene and environment interaction works mechanistically was popularly known as GxE interaction (Waddington 1957). Homeostasis of an organism is a dynamic equilibrium that breaks down when the body cannot cope with stressful environment and attain disease state. Epigenetics is a rapidly expanding field of research in insects, which is involved in hereditary mechanism of an organism that does not involve changes to the DNA sequence; instead, reprogramming of gene expression is attained in response to endogenous and environmental stimuli, viz., pathogens, parasites, harmful chemicals, and other stress factors (Jaenisch and Bird 2003).

Whenever the organisms are exposed to environmental stresses, there will be stimulation of epigenetic mechanism through DNA methylation, histone acetylation, and histone methylation, and noncoding microRNAs (miRNAs) (Fig. 2) will be synthesized which bind to the corresponding mRNAs and interrupt or inhibit the translations processes which are considered to be as the epigenetic machineries. Epigenetic mechanisms act under both pre- and posttranscriptional gene regulations. During pre-transcriptional process, DNA methylation and histone acetylation/deacetylation occur, whereas the posttranscriptional gene regulation occurs with small noncoding RNAs/microRNAs (miRNAs) and long noncoding microRNAs (lncRNAs).

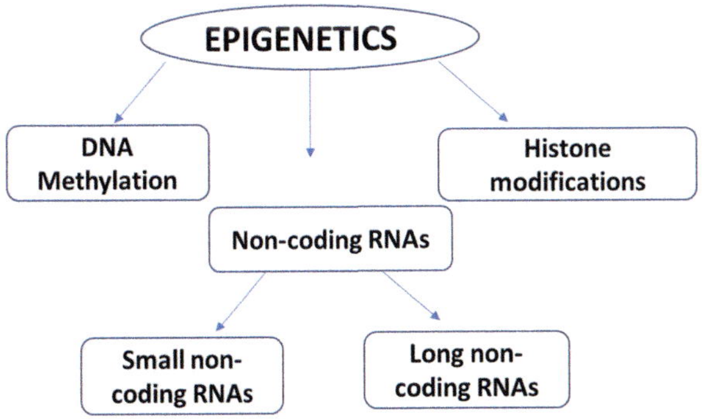

Fig. 2 Three major key processes involved in epigenetic mechanism

DNA Methylation

During this process, there will be addition of methyl group to cytidine residues in the CpG sequence, and changes take place in the interaction of DNA with proteins, and gene regulation is disturbed. The methylation occurrence is at the sites with twofold rotational axis of symmetry. Methylation mark is created which passes on to daughter cells during DNA replication, thus explaining inheritance mode of epigenetic state. DNA methyltransferases (DNMTs) are the enzymes which transfer methyl groups to DNA (Mukherjee et al. 2015).

Histone Acetylation

The DNA complex with proteins comprises chromatin and nucleosome that are fundamental structural unit of chromatin. The nucleosome consists of an octamer with core histones, viz., H2A, H2B, H3 and H4, in two numbers to which DNA is wrapped around. Each nucleosome is linked together through DNA paired with histones H1 and H5. The histones when acetylated promote the gene expression by getting coiled loosely with DNA and giving accessibility. However, during deacetylation process, the histones wrap DNA tightly so that gene expression is attenuated. Histone acetyltransferases (HATs) and histone deacetylases (HDACs) control acetylation processes (Mukherjee et al. 2015).

Acetylation, phosphorylation, methylation, ubiquitination, and ubiquitin-like modifier SUMOylation act through posttranslational modification of N- and C-terminal of the histone proteins (Dong et al. 2021).

Noncoding RNA

The noncoding RNAs (18–22-nucleotide length) that regulate the gene expression at posttranscriptional level by binding to the complementary mRNAs are known as microRNAs (miRNAs) (Bushati and Cohen 2007). miRNAs are known to transcribe from DNA sequences into primary miRNAs (pri-miRNAs) and get processed into pre-miRNAs which are the precursor miRNA and then finally to mature miRNAs. miRNAs are found to interact with 3'UTR (untranslated region) (Bandala et al. 2022) of target mRNAs to suppress expression and also interact with 5'UTR, coding region, and gene promoters by altering gene expression at translational and transcriptional level (Vasudevan 2012; Ha and Kim 2014; Broughton et al. 2016; Makarova et al. 2016). miRNAs are found as potential biomarkers for various diseases and also act as signaling molecules which mediate the cell-cell communications (Huang 2017; Kim and Kim 2007; Wang et al. 2016).

Studies made on epigenetics and its importance in development, aging, longevity, neurodegenerative disorders, cancer, inflammation, and sepsis have been well explained and reviewed by Mukherjee et al. (2015). Advanced discoveries include epigenetics-miRNA regulatory circuit which provides biological insights into gene-regulatory mechanism(s) underlying a variety of diseases (Sato et al. 2011). The first discovery of miRNA was in *C. elegans* (Lee et al. 1993; Reinhart et al. 2000); after that, many miRNAs are identified from several multicellular organisms. miRNAs play a vital role in the pathogenesis of human diseases including cancer and metabolic disorders (Esquela-Kerscher and Slack 2006; Hausen 2007; Kato et al. 2009; Park et al. 2006). In addition to miRNAs, long noncoding RNAs (lncRNAs; >200 nucleotides in length) have also attracted much attention (Kurokawa 2015), as its action in epigenetic control. lncRNAs have been reported in *D. melanogaster* (Brown et al. 2014), *Anopheles gambiae* (Jenkins et al. 2015), *Apis mellifera* (Jayakodi et al. 2015; Kiya et al. 2012), and *Bombyx mori* (Wu et al. 2016; Zhang et al. 2020). Emerging evidence reveals key role of lncRNAs in regulating immunological functions and autoimmunity (Satyavathi et al. 2016). These insights open up avenues to understand epigenetic mechanism regulating biological systems under disease/stress conditions and also to determine therapeutic measures to allay stress/diseases. Studies have been made on the generational toxicology which reports that exposure to environmental toxicants can increase disease rates in subsequent generations, which are devoid of direct exposure (Nilsson et al. 2018, 2022). The gene expression plays a major role during every organism life in order to regulate during transcriptional and translational levels.

Molecular and Epigenetics Studies in Silkworm

Xia et al. (2014) have provided comprehensive detailed review on the advances in silkworm studies accelerated by the genome sequencing of *Bombyx mori* starting from the initiation of silkworm genome project, genetic variation map, single-base-resolution methylome, genome databases, genome size, gene number, transposable elements, chromosomal synteny, origin of domesticated silkworm, transcriptomics, silk gland development and silk protein synthesis, metamorphic development, immune and disease resistance, genetic manipulation through systemic RNAi, transgenic silkworm techniques, and genome editing. A single-base-resolution silkworm genetic variation map has been constructed through the complete resequencing of 40 genomes of domesticated and wild silkworms (Xia et al. 2009). Construction of silkworm integrated linkage maps by single nucleotide polymorphism (SNP) markers (Yamamoto et al. 2006) and unique site-specific recombination sites (Zhan et al. 2009), and a single-base-resolution epigenetic methylome has been analyzed (Xiang et al. 2010). Integrated silkworm genome databases containing the two silkworm whole-genome shotgun (WGS) sequencing datasets have been established (Duan et al. 2010; Shimomura et al. 2009). Genetic manipulation tools and approaches are developed (Tamura et al. 2011; Zhang et al. 2012). Genome-based positional cloning studies identified a number of important genes related to mutations in body color, cocoon color, development, disease resistance, and feeding habits (Atsumi et al. 2012; Futahashi et al. 2008; Ito et al. 2008; Komoto et al. 2009; Liu et al. 2010; Meng et al. 2009; Sakudoh et al. 2007). These major scientific advances support *B. mori* as a model organism for studying and understanding biological phenomena and complexities.

Silkworm is a poikilothermic insect, and changes in the external environment (temperature and humidity) have profound effect on its growth and metabolism. The incidence of diseases during silkworm rearing is one of the major constraints in silk production. The major diseases affecting silkworm are flacherie, muscardine, grasserie, and pebrine caused by bacteria, fungi, viruses, and microsporidians (Alam et al. 2021). Silkworm diseases cause an estimated crop loss of 27–35% amounting to 11–15 kg/100 dfls cocoon yield (Selvakumar et al. 2002). Temperature and humidity are two important predisposing factors for silkworm diseases (Steinhaus 1958; Inoue 1972; Miyajima 1978; Pillai and Krishnaswamy 1980; Kato et al. 1989; Vijayakumari et al. 2001). Selvakumar et al. (2013) recorded higher susceptibility of silkworms to flacherie under fluctuating temperature and humidity. Low temperature and high humidity influence the incidence of muscardine (Nirupama 2014). Higher susceptibility to grasserie was observed in third instar larvae of silkworm due to increased temperature and relative humidity (Deb et al. 2015).

Based on dominance, overdominance, and epistasis, hypothesis classical quantitative genetics look forward to explain heterosis (Davenport 1908; East 1936; Li et al. 2001; Yu et al. 1997). The complex phenomenon of heterosis has to be dealt with advanced research (Birchler et al. 2003, 2006). Advances in molecular biology and genomics have provided the opportunity to explore important molecular mechanisms. Several molecular approaches, such as transcriptomics (Klosinska et al. 2016; Kong et al. 2020; Luo et al. 2021; Springer and Stupar 2007; Wang et al. 2014), proteomics (Ge et al. 2020), and epigenomics of parents and hybrids (Lauss et al. 2018; Li et al. 2018; Zhou et al. 2021), have done remarkable efforts to explore the classical genetic hypotheses.

Gao et al. (2020) studied animal domestication and explained the role of epigenetic modifications and investigated EMEs in the whole genome of silkworm. Two BmEMEs, namely, BmSW4–20 and BmDNMT2, which are strongly selected during silkworm domestication and expressed higher in testes and ovaries of domesticated silkworm when compared to wild silkworm are identified. They reported that BmEMEs might be involved in incurring changes in reproductive characters during domestication. Xu et al. (2022) have demonstrated that many heterosis related genes get regulated by DNA methylation which affects hybrid vigor.

Only few studies (Qian et al. 2016) have focused on miRNAs and lncRNAs in silkworm. Liu et al. (2009) reported expression profiles of miRNAs throughout the life cycle of *B. mori*. Four BmNPV-encoded miRNAs were identified and suggested key role of viral miRNAs in insect-pathogen interactions (Singh et al. 2010). Singh et al. (2012) have studied how baculovirus-encoded miRNA suppresses host (*B. mori*) miRNA biogenesis regulating Exportin-5 cofactor Ran. Involvement of miRNAs in the infection of silkworm with BmCPV was reported (Wu et al. 2013). Li et al. (2014) profiled 1229 microRNAs (miRNAs), including 728 novel miRNAs and 110 miRNA/miRNA* duplexes, of posterior silk gland at fifth larval instar. Differential expression of miRNAs (Lawrie 2013) has been studied in diapausing versus HCl-treated *B. mori* embryos (Fan et al. 2017).

The journey of genetics from classical to modern era has indeed revolutionized the arena of research in resolving many of the problems related to phenotypic, metabolic, physiological, neurological, biochemical, and molecular levels in many organisms. Through the advances in epigenetics/epigenomics transgenerational inheritance study, we can understand the diversity of disease and phenotype which has been induced by different climatic variations, viz., temperature (Mogilicherla and Roy 2023), nutrition, and environmental toxins/pollutants, for the betterment of living organisms. The short-term and long-term effect/impact of environmental variations through climate change on poikilothermic organisms with reference to silkworms could be effectively dealt with through epigenomics approaches. For sericulture industry, it could be very much essential in order to have better crop improvement and face the challenges of climatic variations.

References

Alam K, Raviraj VS, Chowdhury T, Bhuimali A, Saha GP, Soumen. (2021) Applications of biotechnology in sericulture: Progress, scope and prospect. Nucleus 65:129–150

Atsumi S, Miyamoto K, Yamamoto K, Narukawa J, Kawai S et al (2012) Single amino acid mutation in an ATP-binding cassette transporter gene causes resistance to Bt toxin Cry1Ab in the silkworm, Bombyx mori. Proc Natl Acad Sci USA 109:1591–1598

Avery O, MacLeod C, McCarty M (1944) Studies on the chemical nature of the substance inducing transformation of pneumococcal types: induction of transformation by a desoxyribonucleic acid fraction isolated from pneumococcus type III. J Exp Med 79:137–158

Bandala BG, Nolasco MT, Nova PC, Lara LA, Garcia BO (2022) Dendrimers as targeted systems for selective gene and drug delivery. In: Pharmaceutical Nanobiotechnology for targeted therapy. Springer

Bateson W (1902) A defence of Mendel's principles of heredity. Cambridge University Press, Cambridge

Beadle GW, Tatum EL (1941) Genetic control of biochemical reactions in Neurospora. Proc Natl Acad Sci 27:499–506

Benzer S (1959) On the topology of the genetic fine structure. Proc Natl Acad Sci USA 45:1607–1620

Birchler JA, Auger DL, Riddle NC (2003) In search of the molecular basis of heterosis. Plant Cell 15(10):2236–2239

Birchler JA, Yao H, Chudalayandi S (2006) Unraveling the genetic basis of hybrid vigor. Proc Natl Acad Sci USA 103(35):12957–12958

Broughton JP, Lovci MT, Huang JL, Yeo GW, Pasquinelli AE (2016) Pairing beyond the seed supports microRNA targeting specificity. Mol Cell 64:320–333

Brown JB, Boley N, Eisman R, May GE, Stoiber MH, Duff MO, Booth BW, Wen J, Park S, Suzuki AM et al (2014) Diversity and dynamics of the Drosophila transcriptome. Nature 512:393–399

Bushati N, Cohen SM (2007) MicroRNA functions. Annu Rev Cell Dev Biol 23:175–205

Collins FS, Morgan M, Patrinos A (2003) The human genome project: lessons from large-scale biology. Science 300(5617):289–290

Creighton HB, McClintock B (1931) A correlation of cytological and genetical crossing-over in Zea mays. Proc Natl Acad Sci 17:492–497

Crick F (1970) Central dogma of molecular biology. Nature 227:561–563

Davenport CB (1908) Degeneration, albinism and inbreeding. Science 28(718):454–455

Deb S, Chakraborty S, Hazra N, Saha AK, Chakraborty C (2015) Impact of Temperature and Humidity on Incidence of Grasserie in Silkworm, Bombyx mori L. in West Bengal. Journal of Plant Protection and Environment 12(1):19–26

Dong L, Yan Y, Youping L, Zeqin XZ (2021) Epigenetic regulation of gene expression in response to environmental exposures: from bench to model. Sci Total Environ 776:145998
Duan J, Li R, Cheng D, Fan W, Zha X et al (2010) SilkDB v2.0: a platform for silkworm (*Bombyx mori*) genome biology. Nucleic Acids Res 38:453–456
East EM (1936) Heterosis. Genetics 21(4):375–397
Esquela-Kerscher A, Slack FJ (2006) Oncomirs-MicroRNAs with a role in cancer. Nat Rev Cancer 6:259–269
Fan W, Zhong Y, Qin M, Lin B et al (2017) Differentially expressed microRNAs in diapausing versus HCl-treated Bombyx mori embryos. PLoS One 12(7):e0180085
Futahashi R, Sato J, Meng Y, Okamoto S, Daimon T et al (2008) Yellow and ebony are the responsible genes for the larval color mutants of the silkworm Bombyx mori. Genetics 180:1995–2005
Gao R, Li C-L, Tong X-L, Han M-J et al (2020) Identification, expression, and artificial selection of silkworm epigenetic modification enzymes. BMC Genomics 21:740
Ge Q, Xiao R, Yuan Y, He SQ, Chen L, Ma SS et al (2020) Transcriptome and proteomics-based analysis to investigate the regulatory mechanism of silk gland differences between reciprocal cross silkworm, Bombyx mori. J Asia Pac Entomol 23(4):1101–1113
Gyon J (2016) From Mendel to epigenetics: history of genetics. C R Biol 339:225–230
Ha M, Kim VN (2014) Regulation of microRNA biogenesis. Nat Rev Mol Cell Biol 15:509–524
Hausen HZ (2007) The role of microRNAs in human cancer. Int J Cancer 122:5
Huang W (2017) MicroRNAs: biomarkers, diagnostics, and therapeutics. Methods Mol Biol 1617:57–67
Inoue H (1972) Studies on the multiplication of infectious flacherie virus in the silkworm, *Bombyx mori*. J Seric Sci Jpn 41(6):437–444
Ito K, Kidokoro K, Sezutsu H, Nohata J, Yamamoto K et al (2008) Deletion of a gene encoding an amino acid transporter in the midgut membrane causes resistance to a Bombyx parvo-like virus. Proc Natl Acad Sci USA 105:7523–7527
Jacob F, Monod J (1961) Genetic regulatory mechanisms in the synthesis of protein. J Mol Biol 3:318–356
Jaenisch R, Bird A (2003) Epigenetic regulation of gene expression: how the genome integrates intrinsic and environmental signals. Nat Genet 33:245–254
Jayakodi M, Jung JW, Park D, Ahn YJ, Lee SC, Shin SY, Shin C, Yang TJ, Kwon HW (2015) Genomewide characterization of long intergenic non-coding RNAs (lincRNAs) provides new insight into viral diseases in honey bees, ApisceranaandApismellifera. BMC Genomics 16:680
Jenkins AM, Waterhouse RM, Muskavitch MA (2015) Long non-coding RNA discovery across the genus anopheles reveals conserved secondary structures within and beyond the Gambiae complex. BMC Genomics 16:337
Johannsen W (1909) Elements of the exact theory of heredity. Jena
Kato M et al (1989) Study on resistance of the silkworm, Bombyx mori, to high temperature. Proc. 61h /111. Congr. SABRAO 2:953–956
Kato M, Paranjape T, Muller RU et al (2009) The mir-34 microRNA is required for the DNA damage response in vivo in C. elegans and in vitro in human breast cancer cells. Oncogene 28(25):2419–2424
Kim YK, Kim VN (2007) Processing of intronic microRNAs. EMBO J 26:775–783
Kiya T, Ugajin A, Kunieda T, Kubo T (2012) Identification of kakusei, a nuclear non-coding RNA, as an immediate early gene from the honeybee, and its application for neuroethological study. Int J Mol Sci 13:15496–15509
Klosinska M, Picard CL, Gehring M (2016) Conserved imprinting associated with unique epigenetic signatures in the Arabidopsis genus. Nat Plants 2(10):16145
Komoto N, Quan GX, Sezutsu H, Tamura T (2009) A single-base deletion in an ABC transporter gene causes white eyes, white eggs, and translucent larval skin in the silkworm w-3oe mutant. Insect Biochem Mol Biol 39:152–156
Kong XP, Chen L, Wei TZ, Zhou HW, Bai CF, Yan XP et al (2020) Transcriptome analysis of biological pathways associated with heterosis in Chinese cabbage. Genomics 112(6):4732–4741
Kurokawa R (2015) Long noncoding RNAs structures and functions. Springer

Lauss K, Wardenaar R, Oka R, Van Hulten MHA, Guryev V, Keurentjes JJB et al (2018) Parental DNA methylation states are associated with heterosis in epigenetic hybrids. Plant Physiol 176(2):1627–1645
Lawrie C (2013) MicroRNAs in medicine. Wiley. https://doi.org/10.1002/9781118300312
Lee RC, Feinbaum RL, Ambros V (1993) The C. elegans heterochronic gene lin-4 encodes small RNAs with antisense complementarity to lin-14. Cell 75:843–854
Ledford H, Callaway E (2020) Pioneers of revolutionary CRISPR gene editing win chemistry Nobel. Nature. 586:346–347
Li ZK, Luo LJ, Mei HW, Wang DL, Shu QY, Tabien R et al (2001) Overdominant epistatic loci are the primary genetic basis of inbreeding depression and heterosis in rice. I. Biomass and grain yield. Genetics 158(4):1737–1753
Li J, Cai Y, Ye L, Wang S, Che J, You Z, Yu J (2014) MicroRNA expression profiling of the fifth-instar posterior silk gland of Bombyx mori. BMC Genomics 15(1):410
Li H, Yuan JY, Wu M, Han ZP, Li LH, Jiang HM et al (2018) Transcriptome and DNA methylome reveal insights into yield heterosis in the curds of broccoli (*Brassica oleracea* L var. italic). BMC Plant Biol 18(1):168
Liu S, Zhang L, Li Q, Zhao P et al (2009) MicroRNA Expression profiling during the life cycle of the silkworm (Bombyx mori). BMC Genomics. 10:455
Liu S, Li D, Li Q, Zhao P, Xiang Z, Xia Q (2010) MicroRNAs of Bombyx mori identified by Solexa sequencing. BMC Genomics 11:148
Luo JH, Wang M, Jia GF, He Y (2021) Transcriptome-wide analysis of epitranscriptome and translational efficiency associated with heterosis in maize. J Exp Bot 72(8):2933–2946
Makarova JA, Shkurnikov MU, Wicklein D, Lange T, Samatov TR, Turchinovich AA et al (2016) Intracellular and extracellular microRNA: an update on localization and biological role. Prog Histochem Cytochem 51:33–49
Maxam AM, Gilbert W (1980) Sequencing end-labeled DNA with base-specific chemical cleavages. Methods Enzymol 65(1):499–560
Mendel G (1865) Experiments in plant hybridization. Mendel's paper in English
Meng Y, Katsuma S, Daimon T, Banno Y, Uchino K et al (2009) The silkworm mutant lemon (lemon lethal) is a potential insect model for human sepiapterin reductase deficiency. J Biol Chem 284:11698–11705
Mita K, Kasahara M, Sasaki S, Nagayasu Y, Yamada T et al (2004) The genome sequence of silkworm, Bombyx mori. DNA Res 11:27–35
Miyajima J (1978) Effect of high temperature on the incidence of viral flacherie of the silkworm, Bombyx mori (L) Res. Bull Alchiken. Agric. Res. Center Jpn. D9:45–49
Mogilicherla K, Roy A (2023) Epigenetic regulations as drivers of insecticide resistance and resilience to climate change in arthropod pests. Front Genet 13:1–17
Morgan TH, Sturtevant AH, Muller HJ, Bridges CB (1915) The mechanism of Mendelian heredity. Henry Holt, New York
Mukherjee K, Twyman RM, Vilinskas A (2015) Insects as models to study the epigenetic basis of disease. Prog Biophys Mol Biol 118:69–78
Muller HJ (1925) The regionally differential effect of X rays on crossing over in autosomes of Drosophila. Genetics 10:470–507
Mullis K, Faloona F, Scharf S, Saiki R, Horn G, Erlich H (1986) Specific enzymatic amplification of DNA in vitro: the polymerase chain reaction. Cold Spring Harb Symp Quant Biol 51(1):263–273
Nilsson E, Sadler-Riggleman I, Skinner MK (2018) Environmentally induced epigenetic transgenerational inheritance of disease. Environ Epigenet 4:1–13
Nilsson EE, Maamar MB, Skinner MK (2022) Role of epigenetic transgenerational inheritance in generational toxicology. Environ Epigenet 8(1):1–9
Nirenberg M, Leder P (1964) RNA codewords and protein synthesis: the effect of trinucleotides upon the binding of sRNA to ribosomes. Science 145:1399–1407
Nirupama R (2014) Fungal disease of white muscardine in silkworm, Bombyx mori L.Mun. Ent. Zool. 9(2):870–874

Park SK, O'Neill MS, Wright RO, Hu H, Vokonas PS, Sparrow D et al (2006) HFE genotype, particulate air pollution, and heart rate variability: a gene-environment interaction. Circulation 114(25):2798–2805

Pillai SV, Krishnaswami S (1980) Effect of high temperature o the survival rate, cocoon quality and fecundity of Bombyx mori L.Proc.Sericult. Seminar and Sympo. TNAU(Coimbatore):141–148

Qian P, Wang X, Jiang T, Song F, Chen C, Fan Y, Shen X (2016) Gene screening and differential expression analysis of microRNAs in the middle silk gland of wild-type and naked pupa mutant silkworms (Bombyx mori). J Asia-Pac Entomol 19(2):439–445

Reinhart BJ et al (2000) The 21-nucleotide let-7 RNA regulates developmental timing in Caenorhabditis elegans. Nature 403:901–906

Sakudoh T, Sezutsu H, Nakashima T, Kobayashi I, Fujimoto H et al (2007) Carotenoid silk coloration is controlled by a carotenoid-binding protein, a product of the yellow blood gene. Proc Natl Acad Sci USA 104:8941–8946

Sanger F, Nicklen S, Coulson AR (1977) DNA sequencing with chain-terminating inhibitors. Proc Natl Acad Sci USA 74(12):5463–5467

Sato F, Tsuchiya S, Meltzer SJ, Shimizu K (2011) MicroRNAs and epigenetics. FEBS 278:1598–1609

Satyavathi V, Ghosh R, Subramanian S (2016) Long non-coding RNAs regulating immunity in insects. Non-coding RNA 3(1):14

Selvakumar T, Nataraju B, Balaventasubbaiah M, Sivaprasad V, Baig M, Virendrakumar Sharma SD et al (2002) A report on the prevalence of silkworm diseases and estimated crop loss. In: Proceedings of National Conference on Strategies for Sericulture Research and Development. CSR&TI, Mysore, pp 354–337

Selvakumar T, Sharma SD, Shashidhara M (2013) Effect of temperature on incidence of grasserie disease in silkworm Bombyx mori L. Indian J Seric 52(2):163–166

Shimomura M, Minami H, Suetsugu Y, Ohyanagi H, Satoh C et al (2009) KAIKObase: an integrated silkworm genome database and data mining tool. BMC Genomics 10:486

Singh J, Singh CP, Bhavani A, Nagaraju J (2010) Discovering microRNAs from *Bombyx mori* nucleopolyhedrosis virus. Virology 407:120–128

Singh J, Singh CP, Bhavani A, Nagaraju J (2012) A baculovirus-encoded microRNA (miRNA) suppress its host miRNA biogenesis by regulating the exportin-5 cofactor ran. J Virol 86(15):7867–7879

Springer NM, Stupar RM (2007) Allele-specific expression patterns reveal biases and embryo-specific parent-of-origin effects in hybrid maize. Plant Cell 19(8):2391–2402

Steinhaus EA (1958) Stress as factor in insect diseases. Proc Tenth Int Congr Entomol 4:725–730

Sturtevant AH (1913) The linear arrangement of six sex-linked factors in drosophila, as shown by their mode of association. J ElCp Zool 14:43–59

Sturtevant AH (1965) A his/Qry of genetics. Harper and Row, New York, p 165

Sutton WS (1903) The chromosomes in heredity. Biol. Bull. 4:231–251

Tamura K, Peterson D, Peterson N, Stecher G (2011) MEGA5: Molecular evolutionary genetics analysis using maximum likelihood, evolutionary distance, and maximum parsimony methods. Molecular Biology and Evolution 28(10):2731–2739

Vijaya Kumari KM, Balavenkatasubbaiah M, Rajan RK, Himantharaj MT, Nataraju B, Rekha M (2001) Influence of temperature and relative humidity on the rearing performance and disease incidence in CSR hybrid silkworms, ombyx mori L. Int.J.Ind.Entomol. 3:113–116

Tronick E, Hunter RG (2016) Waddington, dynamic systems, and epigenetics. Front Behav Neurosci 10(107):1–6

Vasudevan S (2012) Posttranscriptional upregulation by microRNAs. Wiley Interdiscip Rev RNA 3:311–330

Waddington CH (1940) Organisers and Genes. Cambridge University Press, Cambridge

Waddington CH (1942) The epigenotype. Endeavour 1:18–20

Waddington CH (1957) The strategy of genes. In: Routledge library editions: 20th century science

Wang SH, You ZY, Ye LP, Che JQ, Qian QJ, Nanjo Y et al (2014) Quantitative proteomic and transcriptomic analyses of molecular mechanisms associated with low silk production in silkworm Bombyx mori. J Proteome Res 13(2):735–751

Wang J, Chen J, Sen S (2016) MicroRNA as biomarkers and diagnostics. J Cell Physiol 231:25–30
Watson JD, Crick FHC (1953) Molecular structure of nucleic acids. Nature 171:737–738
Wu Y, Cheng T, Liu C, Liu D, Zhang Q, Long R, Zhao P, Xia Q (2016) Systematic identification and characterization of long non-coding RNAs in the silkworm, *Bombyx mori*. PLoS One 11:e0147147
Wu P, Han S, Chen T, Qin G, Li L Guo X (2013) Involvement of microRNAs in infection of silkworm with Bombyx mori cytoplasmic polyhedrosis virus (BmCPV). PloS One 2;8(7):e68209
Xia Q, Zhou Z, Lu C, Cheng D, Dai F et al (2004) A draft sequence for the genome of the domesticated silkworm (Bombyx mori). Science 306:1937–1940
Xia Q, Guo Y, Zhang Z, Li D, Xuan Z et al (2009) Complete resequencing of 40 genomes reveals domestication events and genes in silkworm (Bombyx). Science 326:433–436
Xia Q, Li S, Feng Q (2014) Advances in silkworm studies accelerated by the genome sequencing of Bombyx mori. Annu Rev Entomol 59:513–536
Xiang H, Zhu J, Chen Q, Dai F, Li X et al (2010) Single base-resolution methylome of the silkworm reveals a sparse epigenomic map. Nat Biotechnol 28:516–520
Xu H, Chen L, Tong X-L, Hu H, Li-Y L et al (2022) Comprehensive silk gland multi-omics comparison illuminates two alternative mechanisms in silkworm heterosis. Zool Res 43(4):585–596
Yamamoto K, Narukawa J, Kadono-Okuda K, Nohata J, Sasanuma M et al (2006) Construction of a single nucleotide polymorphism linkage map for the silkworm, *Bombyx mori*, based on bacterial artificial chromosome end sequences. Genetics 173:151–161
Yu SB, Li JX, Xu CG, Tan YF, Gao YJ, Li XH et al (1997) Importance of epistasis as the genetic basis of heterosis in an elite rice hybrid. Proc Natl Acad Sci USA 94(17):9226–9231
Zhan S, Huang J, Guo Q, Zhao Y, Li W et al (2009) An integrated genetic linkage map for silkworms with three parental combinations and its application to the mapping of single genes and QTL. BMC Genomics 10:389
Zhang X, Xue R, Cao G, Hu X, Wang X et al (2012) Effects of EGT gene transfer on the development of Bombyx mori. Gene 491:272–277
Zhang S, Yin H, Shen M, Huang H, Hou Q, Zhang Z, Zhao W, Guo X, Wu P (2020) Analysis of lncRNA-mediated gene regulatory network of Bombyx mori in response to BmNPV infection. J Invertebr Pathol 170:107323
Zhou SR, Xing MQ, Zhao ZL, Gu YC, Xiao YP, Liu QQ et al (2021) DNA methylation modification in heterosis initiation through analyzing rice hybrid contemporary seeds. Crop J 9(5):1179–1190

Applications of Marker Assisted Selection in Silkworm Breeding for Abiotic Stress Tolerance

Nalavadi Chandrakanth, Raviraj V Suresh, Mallikarjuna Gadwala, and Shunmugam Manthira Moorthy

Abstract

Silk produced from mulberry silkworm, *Bombyx mori* L., has commercial applications in various fields of sciences. Improvement in silk quality and quantity can have profound benefits to the applied fields. Abiotic stresses including high temperature, high humidity, and fluoride pollution during silkworm rearing can affect the silk quality and quantity. Abiotic stress tolerant silkworm breeds developed through molecular approaches could be an answer to avoid losses in silk quality and quantity. Recent advances in the silkworm genomics reported whole genome sequencing, draft genome, 40 genome resequencing, and construction of linkage maps using molecular markers, and successful marker-trait associations can pave the way for applying marker assisted selection (MAS) in silkworm breeding. Molecular MAS involves screening of the breeding population for DNA markers associated with targeted genes which improves the accuracy and efficiency and reduces breeding period with minimal drag around the target gene to achieve the breeding target. In this chapter, several aspects of breeding for tolerance to abiotic (high temperature, high humidity, fluoride pollution) stress with special emphasis on molecular approaches being utilized for silkworm breed development across the world are reviewed. This chapter provides insights on abiotic stress tolerant silkworm breeds/hybrids and their commercial utilization in India.

N. Chandrakanth (✉) · M. Gadwala
Central Sericultural Research and Training Institute, Central Silk Board, Mysuru, India

R. V Suresh
Central Sericultural Research and Training Institute, Central Silk Board, Berhampore, India

S. M. Moorthy
National Silkworm Seed Organization, Central Silk Board, Bangalore, India

R. V Suresh et al. (eds.), *Biotechnology for Silkworm Crop Enhancement*,
https://doi.org/10.1007/978-981-97-5061-0_6

Keywords

Mulberry silkworm · Silkworm genomics · Abiotic stress · Marker assisted selection

Introduction

Silkworm, *Bombyx mori*, is a silk secreting insect that belongs to Lepidoptera. Owing to its economic and commercial importance, breeding silkworms has excelled in the development of improved silkworm breeds with specific traits. Currently more than 4000 breeds are accessible in the germplasm of *B. mori* (Nagaraju and Goldsmith 2002; Kumaresan et al. 2007) around the world of which 450 bivoltine and 150 multivoltine are maintained in India with broader differences in qualitative and quantitative traits. Most of them are evolved and developed through conventional breeding methods. The breeding programs in silkworm targeting specific trait improvement are ending up with other undesired traits. Most of the economical and stress tolerance traits in silkworm are quantitative in nature and governed by multiple genes. Therefore, breeding programs involving markers would be more fruitful in improving silkworm breeds.

The first silkworm genome assembly was reported in 2004 that provided the basic information essential for genetic analysis (Mita et al. 2004; Xia et al. 2004). Later, the merger of 6× and 3× draft genome sequences, respectively, generated by Chinese and Japanese group has resulted in a new silkworm genome assembly (Xia et al. 2004; Mita et al. 2004) with 432Mb size and 14,623 estimated genes, which is highest in the so far whole genome sequenced insects. Totally, 11,104 full-length cDNAs and 408,172 expressed sequence tags (ESTs) are deposited publicly in KAIKO base database (http://sgp.dna.affrc.go.jp/KAIKObase/) of the Silkworm Genome Database (Mita et al. 2004) (Table 1). PacBio long reads and BAC sequences were used in 2019 to build the p50T genome sequence, producing a very contiguous assembly (Kawamoto et al. 2019). By expanding the knowledge of the molecular mechanics of silk formation, such as the identification of the transcription factor that controls key silk proteins and the genes linked to cocoon yield, these assemblies helped genetic study (Takiya et al. 2016; Tsubota et al. 2016; Li et al. 2017; Li et al. 2020). With an estimated size of 452 Mbp, the most high-quality genome assembly of silkworms to date was recently accomplished using the integration of PacBio Sequel II long-read and ddRAD-seq-based high-density genetic linkage map. With 95 Gb of mRNA-seq collected from 10 distinct organs, a total of 18,397 proteins were predicted, covering 96.9% of the full orthologs of the genes belonging to the lepidopteran family (Waizumi et al. 2023). The genomic information are encouraging to conduct molecular breeding experiment to establish marker/gene-trait relationship and for the development of new improved silkworm breeds.

Table 1 Basic information on genetics and genomics of *Bombyx mori*

Insect type	Holometabolous
Larval duration (days)	23–24
Feeding stage	Larval
Diapause stage	Egg
Copulation type	End to end
Sex determination	ZW-ZZ
Crossing over	Male
No crossing over	Female
Chromosome number (n)	28 (including sex chromosomes)
Mapping population	Backcross and F_2
Linkage group	28
Genome size (bp)	454,710,009
Protein-coding transcripts	16,069
Total scaffolds	28
GC (%)	38.32
Expressed sequence tags (ESTs)	408,172
Full-length cDNAs	11,104
Genes predicted by BGI gene Finder	21,302
No. of RFLP markers	61
No. of RAPD markers	168
No. of RAPD primers	140
SSRs	601,225
Databases	KAIKObase, SilkDb, SilkPathDB, Bombyx Trap DataBase, BmncRNAdb

In India, about 90% of the raw silk produced is contributed by domesticated mulberry silkworm, *B. mori*, which is sensitive to abiotic stresses. Therefore, the raw silk production can be boosted by introducing the productive silkworm hybrids or abiotic stress tolerant hybrids that prevents the crop losses during the adverse seasons. Among the abiotic stresses, high temperature and high humidity are the major culprits. The silkworm rearing temperature varies from 25 to 28 °C—higher than 28 °C is reported to cause reduction in economic traits. Exposure of silkworms at different developmental stages will have different negative effects on economic traits of silkworm (Fig. 1). Among the development stages of *B. mori*, the egg stage has the lowest tolerance to high temperature. The incubation temperature of eggs not only affects voltinism character but also reduces the hatching percentage (Kobayashi et al. 1986). In the larval stage, increase in rearing temperature proportionally decreases the growth and cocoon characters of silkworm (Chandrakanth et al. 2015a). Exposure of silkworms to high temperature in pupal and adult (moth) stages negatively impact the reproductive traits by reducing their pairing abilities, fertility, and oviposition (Rahmathulla 2012). The other stress includes fluoride pollution. Fluoride is one of the materials in the environment in varying amounts, and it can accumulate in some agricultural products. At high concentrations, fluoride is harmful to humans and animals and also small insects like silkworms (Czerwinski et al. 1988; Dote et al. 2000). Reports on deleterious effect of fluoride pollution on

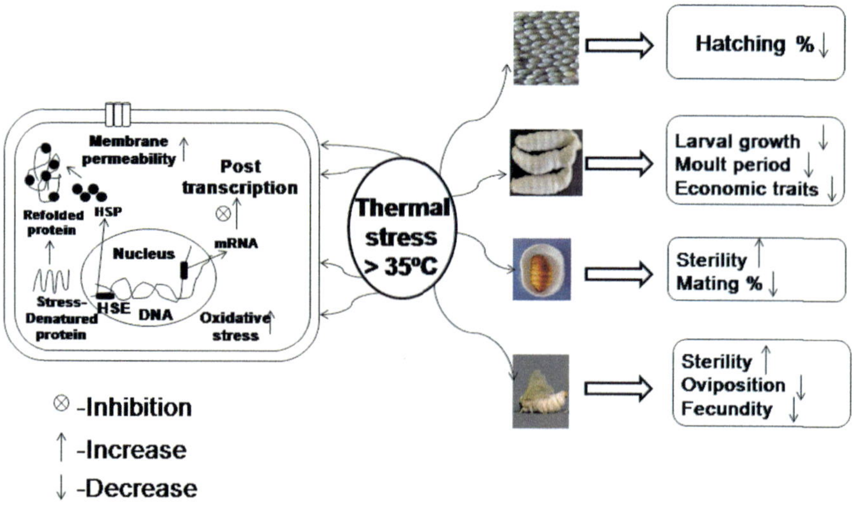

Fig. 1 Negative effects of high temperature on molecular mechanisms and economic traits of silkworm

silkworm and silk production in sericulture practiced countries like China, Japan, and India are well documented (Lin et al. 2006; Nath 1993; Fujii and Honda 1972; Kuribayashi 1977; Shong 1985; Wu 1990; Chen 1993; Chen and Wu 1995; Ahamed and Chandrakala 1999; Ahmed et al. 1999). Experiments showed that minute dose of fluoride (4 ppm) has no impact on the economic traits but higher concentrations significantly reduce the silkworm growth, cocoon weight, shell weight, silk index, fecundity, and filament length (Ramakrishna and Nath 2004).

One of the cost-effective ways to avoid cross losses due to abiotic stress is development of stress tolerant silkworm breeds and hybrids. So far, conventional breeding strategies are applied to develop such stress tolerant silkworm hybrids, but they failed to perform well in the fields. The abiotic stress tolerance traits are governed by genetic and environment factors and include the effects of multiple genes or quantitative trait loci. The conventional methods have focused on phenotypic traits not including the genetic factors that can be misleading and may yield undesired characters. Therefore, the application of genetic markers tightly linked to the abiotic stress tolerance in breeding programs could yield more effective results without undesired traits. Before that, it is important to understand the cellular and molecular mechanisms affected by abiotic stresses and the inheritance pattern of abiotic stress tolerance in silkworm.

Molecular Mechanisms Affected by Abiotic Stresses

All the abiotic stress tolerance in silkworm is measured in terms of pupation rate, but the lethal mechanisms underlying the different stress tolerance may not be same. Hence, it is important to study the molecular responses of the silkworm to understand the different mechanisms involved in stress induced mortality (Sivaprasad et al. 2022). One of the most popularly studied mechanisms is high temperature tolerance in relation to expression of heat shock proteins. *B. mori* domesticated for more than 5000 years lacks combating abilities against temperature and humidity stress. Exposure of silkworms at high temperature during different stages is reported to have negative effect on economic traits. The main reason for reduction in growth and development of silkworms at high temperature is loss of water from body surfaces and respiratory epithelium of tracheal system (Rahmathulla 2012). As water plays a crucial role in cellular reactions, reduction in water content will negatively affect the metabolic activities and optimum growth in silkworm. This problem is compounded by loss of water from mulberry leaves through evaporation causing early drying and poor feeding of leaves by silkworm (Suresh Kumar et al. 2012). Moreover, in a cell, higher temperature affects biologically important molecules like DNA, RNA, and lipids and even halts normal protein synthesis mechanism and nonfunctioning of the cellular proteins by unfolding them (Feder 1996; Feder et al. 1996). These processes jointly increase the vulnerability of other biological processes to heat stress resulting in the decline in the performance of the silkworm (Kumari et al. 2011).

In silkworms, the heat shock responses are well studied with respect to the synthesis of heat shock proteins (HSPs). HSPs are a group of proteins performing the function of molecular chaperons by folding proteins and refolding denatured proteins, forming structurally functional proteins (Tissiéres et al. 1974). There is ample scope for this type of research, as these proteins are involved in biochemical functions by assisting in the proper structure maintenance, regulation, and functioning of proteins by folding them into functional forms. They are also reported to have a significant role in protecting the organism during other environmental stresses (Feder and Hofmann 1999).

Sixteen different small hsp genes are identified in *B. mori* being highest to be characterized among the whole genome sequenced insects indicating their importance in the domesticated insect (Zi-Wen Li et al. 2009). The Hsps protect cell by performing chaperonin functions during stress, thereby offering resistance to thermotolerance, cytoskeletal modulation, apoptosis (Mehlen et al. 1997), protection against oxidative stress, and cell growth (Landry et al. 1989; van den Ijssel et al. 1994; Linder et al. 1996; Mehlen et al. 1996). Several reports confirm the role of sHsps in denaturation as well as renaturation of proteins (Lee et al. 1995; Kampinga et al. 1994). The ATP-independent chaperone like activity of these proteins is directed toward preventing protein aggregation early in denaturation, rather than the active refolding of compromised proteins (Wang and Spector 1994; Rajaraman et al. 1996; Wang et al. 2014).

Another mechanism affecting the survival of silkworm under high temperature condition is oxidative stress evidenced by increase in activity of catalase, superoxide dismutases, and ascorbate peroxidases at 35 and 40 °C in silkworm (Nabizadeh and Kumar 2011; Pooja Makwana et al. 2021) (Fig. 1). Higher membrane permeability was also seen in silkworm at high temperature and humidity conditions (Raviraj et al. 2022). Experiments conducted on the silkworm by feeding with fluoride-polluted mulberry leaves showed that the domesticated silkworm is the least tolerant compared to wild silkworms (Yuyin and Kumar 2003). Like thermal and high humidity stress, the tolerance to fluoride pollution also varies with the silkworm breeds (Chandrakanth et al. 2015a; Raviraj et al. 2022; Chen 2003). It is reported that fluoride induces SIRT1/autophagy through ROS-mediated JNK signaling in ameloblasts (Suzuki et al. 2015). Expression analysis of 36 potential membrane proteins of several pathways involved in the fluoride exposure confirmed that oxidative phosphorylation process was extremely inhibited by fluoride exposure resulting in the blockage of ATP synthesis. Fluoride stimulates MAPK signal pathways, induces cell stress response, and inhibits ATP synthesis by destroying the oxidative phosphorylation process in the mitochondrial membrane, finally inhibiting the growth and development of *B. mori* larva (Liu et al. 2021). With this, it is comprehended that the mechanisms involved in providing tolerance to abiotic stress are different and the gene-based studies are needed to further exploit the underlying mechanism for protecting *B. mori* under different abiotic stresses.

Inheritance

Any trait can be improved though breeding only if it is heritable from one generation to other. Understanding the pattern of inheritance of a trait helps the breeders' in selecting the breeding strategy to improve the trait. Among the abiotic stresses in sericulture, fluoride endurance and thermotolerance inheritance studies are done. Fortunately, a mutant, *Dtf* (dominant tolerance to fluoride), has been revealed in silkworms (Zhang and Qin 1991; Lin et al. 1996, 1997). The silkworms with this gene can tolerate the fluoride levels of up to 200 mg/kg when sprayed as NaF solution. Microsatellite markers linked to this gene are also mapped (Bai et al. 2008). While fluoride tolerance is a *Dtf* gene controlled, the thermotolerance is a complex trait controlled by multiple factors. A series of experiments conducted by exposing silkworm to high temperature has proven that the thermotolerance is genetically heritable (Kato et al. 1989). Like *D. melanogaster*, the heat sensitivity in silkworm also follows non-Mendelian inheritance (Stephanow and Alahiotis 1983; Chandrakanth et al. 2015b). To understand the inheritance pattern of thermotolerance in silkworm, the F_2 population derived from two different sets of genetic background consisting of crosses made between thermotolerant female and thermo-susceptible male was exposed to high temperature of 36 °C. Surprisingly, the segregation pattern did not fit the model of Mendelian inheritance pattern and deviated significantly from the expected phenotypic ratio of 3:1 as confirmed through chi-square test suggesting the role of multiple QTLs distributed on several chromosomes controlling the thermotolerance in silkworm as in *D. melanogaster* (Morgan and Mackay 2006; Lin et al. 2014).

Prerequisites of Marker Assisted Selection

Identification of Parents

For the identification of DNA markers linked to the trait of interest, it is imperative to select parents with contrasting types of responses to the trait of interest. For example, if thermotolerance is the trait of interest, then the parents should be thermotolerant and thermo-susceptible. The information on the traits of silkworm germplasm accessible at the Central Sericultural Germplasm Resource Centre (CSGRC)-Hosur could be useful in selecting the contrasting parents. Even though the information on characters of silkworm is accessible, it should be confirmed by phenotypic screening for the trait of interest before the development of mapping population. For example, for thermotolerance trait, the phenotypic screening is done from fifth instar third day to till spinning by rearing at 36 °C for 6 h daily. The silkworm breeds with highest and lowest pupation percent are considered as thermotolerant and thermo-susceptible, respectively, for developing a mapping population (Chandrakanth et al. 2015b).

Mapping Population

Owning to their silk producing ability, the silkworm breeds have undergone a high quantum of inbreeding and are comparatively homozygous. Backcross (BC) and F_2 populations obtained from a single pair are considered to map markers on silkworm chromosomes. While using BC population for linkage mapping is advantageous in generating information, using F_2 population would save eightfold in labor with loss of twofold in information. In silkworm, due to lack of crossing over in females, BC populations are popularly used for linkage mapping. If all the possible combinations of the initial cross, sex of the parent, and recurrent strain for the backcross are considered, then eight populations have to be screened for linkage mapping, which is time-consuming and expensive. Therefore, biphasic mapping strategy is adapted wherein a BC population is derived by crossing F_1 female with one of the male parent resulting in nonrecombinants as there is no crossing over in female used to identify the linkage groups right away. One more BC population, derived by crossing F_1 male with the female of the same parent resulting in recombinants, can be used to calculate recombination distance (Prasad et al. 2005a; Miao et al. 2005). The biphasic mapping strategy adapted in silkworm is depicted in Fig. 2.

Another method used for determining marker-trait association in silkworm is bulk segregant analysis (BSA). In this method, two pooled samples of DNA of individuals with contrasting characters from a segregating population developed from a single pair mating are compared. Two bulks are genetically different in the region of interest, but due to independent segregation and recombination, they are both equally heterozygous in all the other regions. Due to linkage and co-segregation, however, only markers that are genetically linked to the gene of interest are polymorphic between DNA pools. Thus, the DNA markers that are polymorphic between the bulks are likely to be linked to the target gene (Michelmore et al. 1991). This method is highly recommended to reduce the huge number of unrelated markers to a very few highly specific and tightly linked markers for the trait. This method has also been adapted for linking the DNA markers to the thermotolerance trait in Indian silkworm (Chandrakanth et al. 2015b).

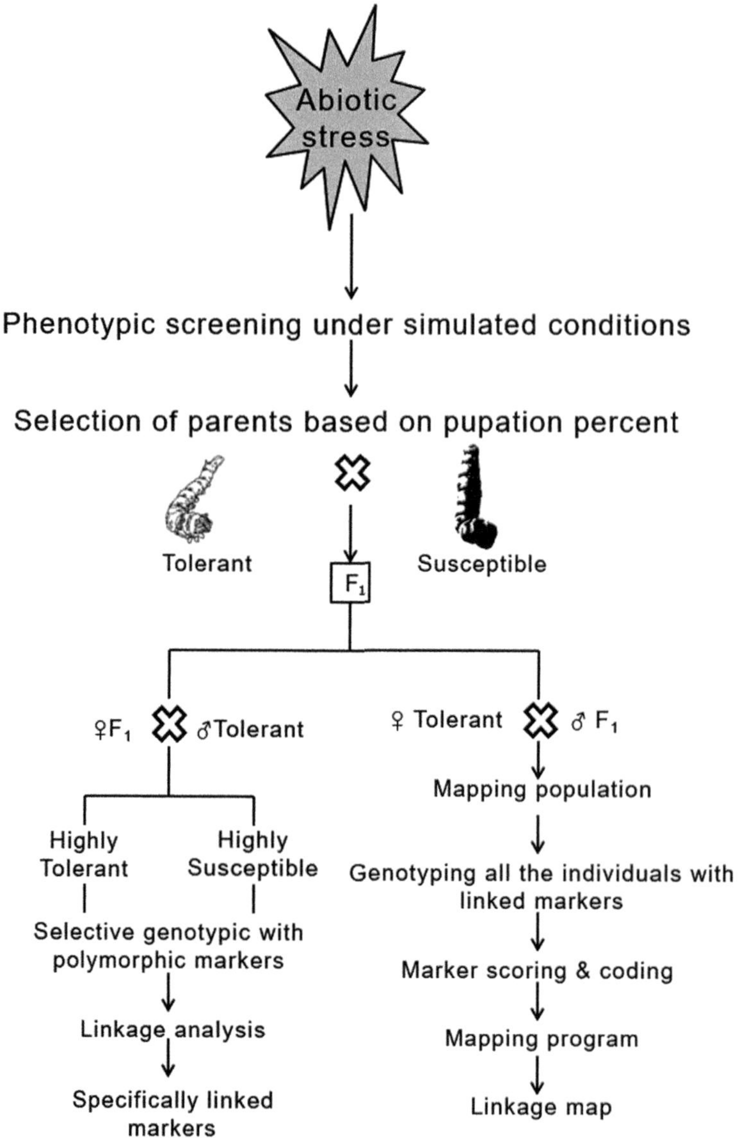

Fig. 2 Biphasic strategy adapted for construction of linkage map in silkworm

Markers

Generally marker represents the variation associated with the trait in morphological, biochemical, or molecular form that is inherited along with the target trait. Morphological marker is a tool by which any organism or group of organisms can be identified on the basis of the special bearing morphological character. The morphological characters in silkworm like body color, larval marking, cocoon color, and shape are generally used in selection and characterization process of silkworm breeds. Also, quantitative traits like fecundity, larval weight, yield/10,000 larvae by weight and number, shell weight, cocoon weight, filament length, and filament size have been traditionally used for diversity studies in silkworm (Jolly et al. 1989; Thangavelu et al. 2000; Kumaresan and Sinha 2002).

Genetic variation studies through biochemical approach by using isozymes are regular in many crop species and also in silkworm (Ashwath et al. 2001; Somasundaram et al. 2009; Etebari et al. 2005; Moorthy et al. 2007; El-Akkad et al. 2008; Anuradha et al. 2012). Isozymes are a group of structurally different molecular forms of an enzyme with the same catalytic function but coded by different genes. Isozymes originate through amino acid alterations, which cause changes in net charge, or the spatial structure (conformation) of the enzyme molecules (Staub et al. 1996). They are codominant markers with expression of both alleles in different form in an individual. Therefore, with codominant markers, heterozygotes can be distinguished from homozygotes, allowing the determination of genotypes and allele frequencies at loci (Kumar et al. 2009). Digestive amylase is reported as a marker in silkworm breeding for improving survival and productivity (Ashwath et al. 2001). Using heat stable esterase, thermotolerant silkworm has also been identified (Moorthy et al. 2007). However, the application of morpho-quantitative and isozyme markers in silkworm breeding programs is limited due to their lesser availability in number and varied expression with the environment conditions.

DNA or molecular markers are considered to be the most efficient and useful markers in tagging the variation with the traits and their application in molecular breeding programs. Unlike morpho-quantitative and biochemical markers, DNA markers are characterized by abundant polymorphism and neutral to environmental changes making them suitable markers for marker-trait associations (Jingade et al. 2011). DNA markers, particularly PCR based markers like randomly amplified polymorphic DNA (RAPD; Williams et al. 1990), inter simple sequence repeat polymorphism (ISSR; Zietkiewicz et al. 1994), amplified fragment length polymorphism (AFLP; Vos et al. 1995), simple sequence repeat (SSR; Tautz 1989), and single nucleotide polymorphism (SNPs), have emerged as the most powerful markers for high resolution genotyping in organisms. The hypothetical model of sequence variations in different DNA marker systems and their detection in gels after polymerase chain reaction based amplification are demonstrated in Fig. 3, and the basic information on DNA markers is provided in Table 2.

Most of the traits in silkworm are controlled by polygenes and quantitative in nature, governed by multiple loci called quantitative trait locus (QTL). QTL is a genomic region of DNA that is linked with a specific phenotypic trait (Beavis et al. 1991) distributed on different chromosomes. The genetic architecture of a trait can be understood by QTL numbers contributing to the phenotypic variation. A phenotypic trait can be controlled by few genes with large effect or vice versa. By using QTLs, candidate genes associated with a phenotypic trait can also be identified by sequencing the region and analyzing it by bioinformatics tools, thereby increasing the breeding accuracy. Therefore, information and knowledge on QTLs is essential

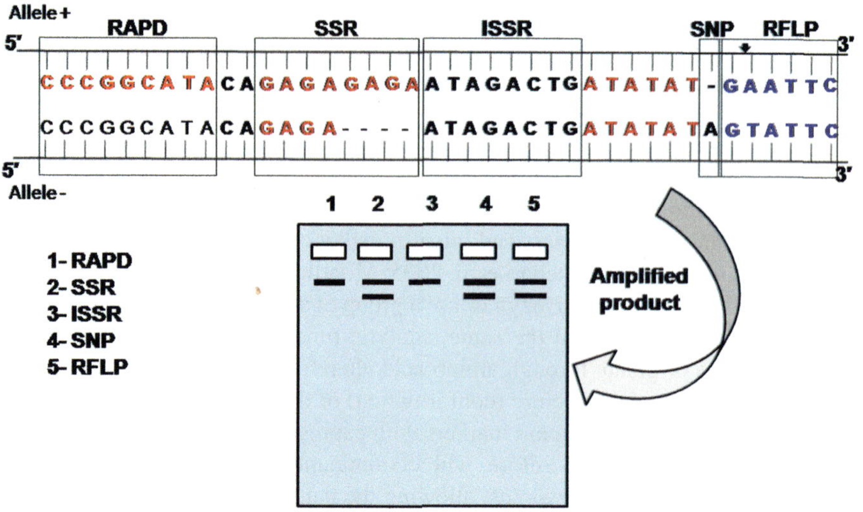

Fig. 3 Hypothetical model representing the sequence variations in different marker systems and their detection

for applying MAS in breeding programs (Mundkur and Muniraju 2018; Nagaraju and Goldsmith 2002).

Construction of Linkage Map

The first step in the construction of a linkage map is to identify DNA markers that are polymorphic between contrasting parents. A sufficient number of polymorphic markers between parents are required to construct a linkage map (Young 1994). Generally, the silkworms differing in voltinism have higher levels of DNA polymorphism when compared to the same voltinism breeds (Reddy et al. 1999). The choice of DNA markers for mapping depends on the availability of markers in the genome of the particular species (Collard et al. 2003, 2005). One of the most widely used markers in silkworm is microsatellites; microsatellites are codominantly inherited and multi-allelic employed for studies of genetic diversity, pedigree evaluation, and genetic mapping. The microsatellites would be a good choice for construction of linkage maps and marker-trait association in silkworm as they are abundant and cover about 0.31% of the silkworm genome (Prasad et al. 2005b). Several markers associated with the traits of commercial importance were reported in silkworm (Shi et al. 1995; Promboon et al. 1995; Yasukochi 1998; Tan et al. 2001; Miao et al. 2005; Yamamoto et al. 2006). The polymorphic markers identified need to be screened for across the F_1 hybrids and entire mapping/segregating population. This process is termed as "genotyping" of the population. In order to do so, the DNA must be isolated from each individual of the mapping population. Silkworm DNA is isolated by using standard phenol-chloroform method as mentioned by Sambrook

Table 2 Basic information on DNA markers used in silkworm

Methods/Markers	RAPD	AFLP	RFLP	SSR	ISSR	SNP
Type	Dominant	Dominant	Codominant	Codominant	Dominant	Codominant
Primer	Universal	Universal	Not required	Specific	Specific	Specific
Detection method	Gel	Gel	Hybridization	Gel	Gel	Gel/sequencing
Reproducibility	Low	Medium	High	High	Low-medium	High
DNA quantity	High	Medium	High	Low	Low	Low
DNA quality	High	High	High	Low	Low	High
Abundance	Very high	Very high	High	High	Medium	Very high
Genomic information	Not required	Not required	Not required	Required	Not required	Required
Software for mining	Not required	Not required	Not required	Necessary	Necessary	Necessary

and Russell (2001). Significant deviations from expected phenotypic ratios are analyzed through chi-square tests. Anyhow, the deviations from the phenotypic ratios will not affect the marker segregation in Mendelian heritance (Sayed et al. 2002).

The genotypic data generated by screening the individuals of the mapping population by polymorphic markers are coded and fed into the computer programs for linkage map construction. There are various software for linkage map construction, but Mapmaker/EXP (Lander et al. 1987; Lincoln et al. 1993) and MapManager QTX (Manly et al. 2001) are the majorly used in silkworm that are freely available. Linkage between markers is estimated by using odds ratio logarithm of the ratio of linkage against no linkage expressed as logarithm of odds (LOD) value (Risch 1992). Lower LOD values are applied (usually 3 or less) to detect a higher level of linkage per se to insert more markers within maps constructed with higher LOD values (Collard et al. 2003). Linked markers are positioned on the same linkage groups; the closer the markers are, the higher the likelihood that they would segregate together during meiosis, whereas distant markers have a higher likelihood of recombination between them.

Marker Validation

The marker-trait linkage has to be validated in the independent populations and different genetic backgrounds, as to test the reliability of markers to predict phenotype indicating the effectiveness or usage of marker in normal screening for MAS which is known as "marker validation" (Cakir et al. 2003; Collins et al. 2003; Sharp et al. 2001). Some reports have showed that assuming marker-QTL linkages unaltered in different genetic backgrounds and testing environments is not less than danger (Reyna and Sneller 2001). Even when a particular trait is controlled by a single gene, there is no assurance that the DNA markers identified in one population will remain same in different populations (Yu et al. 2000). Therefore, validation of markers is mandatory for using them in genotypes with different genetic background for MAS.

Marker Assisted Breeding

Generally, in silkworm, MAS is used to integrate two characters, one from each parent in the single breed developed through backcrossing assisted by markers linked to the trait of interest. Among the abiotic stresses, high temperature and fluoride pollution tolerant silkworm breeds are developed through MAS (Chandrakanth et al. 2021; Li et al. 2015). The first and crucial step in marker assisted breeding is selection of suitable parents. The selection of parents depends on the objective of the breeding program. For example, if a silkworm breeder wants to develop a breed with thermotolerance and productivity, then one of the parents should be thermotolerant and the other should be productive. The parents should possess the markers linked to the trait of interest in divergent form (tolerant and susceptible), preferably in homozygous condition for easing the marker assay throughout the breeding process. Initially, the selected parents are crossed to obtain F_1 progenies with heterozygous marker expression followed by six generations of backcrossing with the recurrent/recipient parent and two generations of selfing targeting specific region of the genome or genes from the donor parent. This method is used in nearly all the silkworm breeds developed through MAS. Likewise, a new silkworm breed developed would posses both the targeted characters from the parents. The breed obtained in such a fashion is termed as near isogenic line (NIL). NILs would be the ideal measure to identify loci which are fixed during the repetitive crossing and those segregated and not fixed through generations (Sivaprasad et al. 2022). The targeted region of the genome or genes from the donor parent is screened for heterologous pattern until BC_6F_1; in BC_6F_1 population, sib mating is done between the male and female with heterozygous banding pattern to obtain the segregating population with genotypic ratio of 1:2:1. From this generation, male and female individuals, which are found with donor type markers, are selected and self-mated to develop a new breed at BC_6F_3 generation (Fig. 4). Theoretically, six generations of backcrossing ensures 0.78% of DNA from the thermotolerant parent contributing to thermotolerance and remaining 99.22% for productivity from productive parent (Fig. 5). The

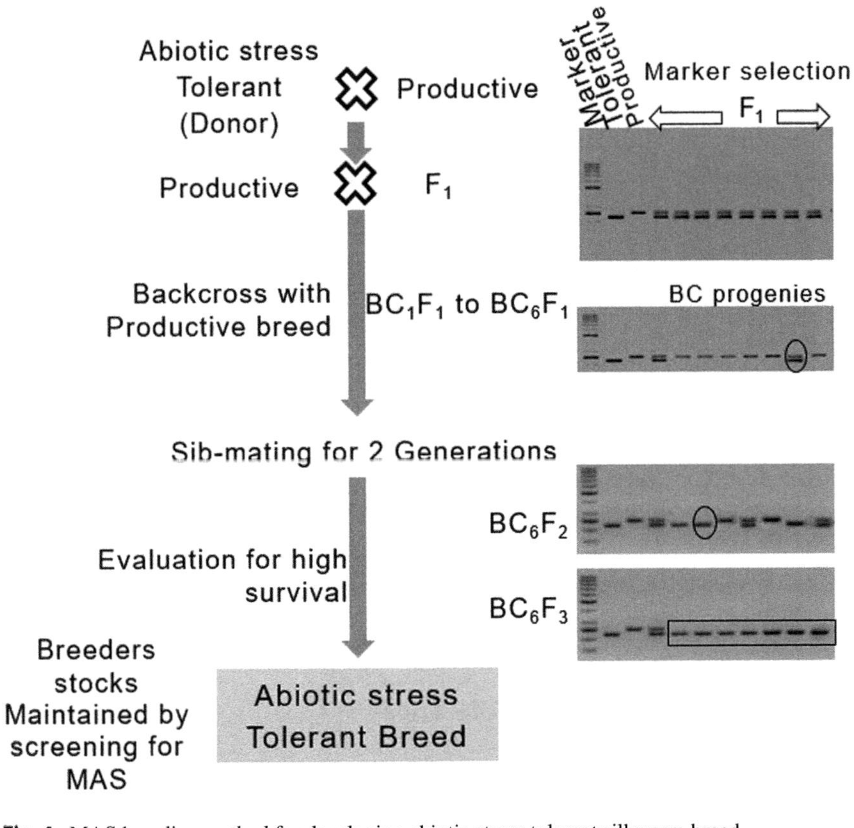

Fig. 4 MAS breeding method for developing abiotic stress tolerant silkworm breed

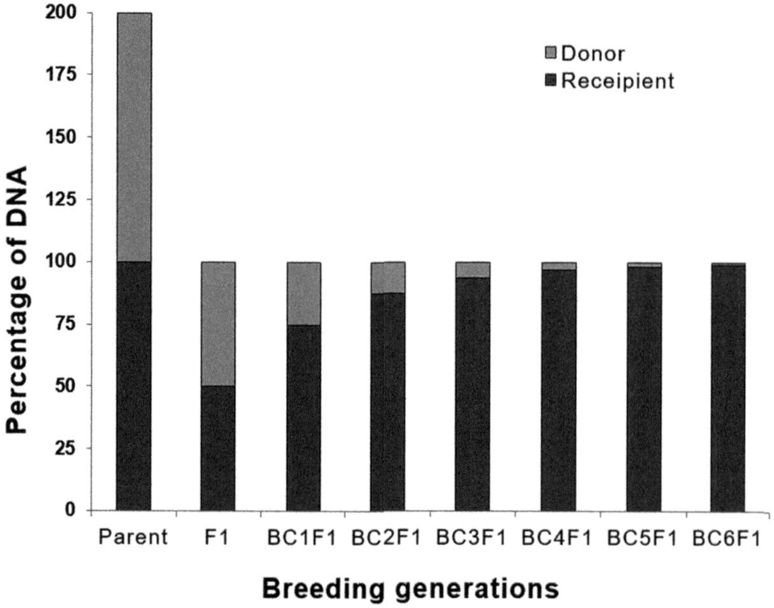

Fig. 5 Generation-wise DNA sharing of donor and recurrent parents in the breeding line

Table 3 Sharing of genome content and allele type in breeding generation

	Genome content		[a]Allele type		
Breeding generation	Recipient	Donor	Recipient type	Donor type	Hybrid
Parent	100	100	100	100	0
F_1	50	50	0	0	100
BC_1F_1	75	25	0	50	50
BC_2F_1	87.5	12.5	0	50	50
BC_3F_1	93.75	6.25	0	50	50
BC_4F_1	96.88	3.13	0	50	50
BC_5F_1	98.44	1.56	0	50	50
BC_6F_1	99.22	0.78	0	50	50
BC_6F_2	–	–	25	25	50
BC_6F_3	–	–	0	100	0

[a]Assuming the alleles are in homozygous condition and the marker is codominant

generation-wise DNA sharing of donor and recurrent parents along with allele types is presented in Table 3.

Breeding for Abiotic Stress Tolerance

To a great extent, silkworm breeders have focused for improving the productivity of silkworm with limited focus to develop abiotic stress tolerant breeds. During adverse seasons, a quantum of defective cocoons are generated due to prevailing high temperature and humidity causing the economic loss to the sericulture farmer and silk industry. Highly productive silkworm hybrids are sensitive to abiotic factors establishing negative relation between tolerance and productivity in silkworm. Even though several bivoltine single hybrids and double hybrids have been developed, their survival at high temperatures could not be exploited commercially (Shirota 1992; Suresh Kumar et al. 2002, 2003, 2004; Krishna Rao et al. 2003; Rao et al. 2007a, b; Singh and Suresh Kumar 2010). The silkworm hybrids developed for tolerance to abiotic stresses by using different breeding methods are summarized in Table 4.

Silkworm Hybrids for Thermotolerance

The growth of tropical sericulture industry is majorly hindered by issues for production of superior quality bivoltine silk under adverse seasons during the high temperature summer. Several reports are evident for results of low pre- and post-cocoon parameters when silkworms are reared at hot tropical seasons. This has encouraged the farmers to rear more adaptive crossbreeds developed by crossing multivoltine females with bivoltine males, as multivoltines are more high temperature tolerant than bivoltines. However, crossbreeds are highly tolerant to thermal stress but lack inherent potential to produce superior quality silk. Silkworm breeders from different countries made attempts to develop silkworm hybrids tolerant to high temperature based on selection of tolerant individuals under simulated high temperature conditions. Because of the simulated high temperature conditions, the developed hybrids showed less economic traits in the field conditions compared to the popular hybrids in the field (Kato et al. 1989; Coulon-Bublex and Mathelin 1991; Suresh Kumar et al. 2002, 2003, 2004 and Krishna Rao et al. 2003). Even though several researchers suggested that expression of heat stable esterase, catalase, and digestive amylase and higher expression of hsps could provide thermotolerance in silkworm, their application to develop thermotolerant silkworm breed is very much limited (Chandrakanth et al. 2015b). In this context, marker assisted selection (MAS) is emerging as a very promising strategy for retaining the selection gain. If a tightly linked marker is identified, it is possible to screen bulky populations for rapid identification of progenies with desired characteristic. DNA based markers have several advantages over others, as they are not affected by environmental fluctuations and have high reproducibility (Sivaprasad et al. 2022).

A very first attempt made to develop high temperature tolerant silkworm breeds by employing MAS was through introgression of heat stable esterase, a biochemical marker. The heat stable esterase from the high temperature tolerant multivoltine parent (donor) was integrated in the productive bivoltine parent (recipient) by crossing followed by three backcrosses. The two bivoltine breeds developed (D6PN and-SK3C) were tolerant to high temperature with>70% survival against recipient parents during unfavorable season (Moorthy et al. 2007). Similarly, digestive amylase was used to develop near isogenic lines of CSR2 (GEN1 and 2C) and CSR4 (4D and 4S) by crossing multivoltines (Pure Mysore, C. Nichi, Daizo, and Sarupat) followed by backcrossing. This leads to the development of single hybrid, GEN2 × GEN3

Table 4 Abiotic stress tolerant silkworm hybrids developed through different breeding methods

Silkworm breeds	Type	Country	Breeding method	Reference
Tolerant to high temperature				
873 × 874	Single	China	Crossing and pedigree selection	Zhao et al. (2007)
CSR18 × CSR19		India	Stress induction and directional selection	Suresh Kumar et al. (2002)
CSR46 × CSR47				Suresh Kumar et al. (2006)
SR1 × SR4				Sudhakara Rao et al. (2006)
CSR50 × CSR51			Crossing, stress induction, and directional selection	Dandin et al. (2006)
HTPO5 × HTP5				Lakshmi et al. (2011)
HH1.HH3 × HH8.HH12	Double			Singh and Suresh Kumar (2010)
(CSR50 × CSR52) × (CSR51 × CSR53)				Bindroo et al. (2014)
TT21 × TT56			Marker assisted selection	Moorthy et al. (2016)
WB79 × WB13				Chandrakanth et al. (2021)
Tolerant to high humidity and high temperature				
J108 × NL	Bi × multi	Thailand	Line selection and improvement	Gosalvitra et al. (2015)
Dok Bua (UB1 × NN)				
B.Con4HH	Improved line	India	Induction and directional selection and MAS	Raviraj et al. (2022)
HTH10HH				
N5HH				
SK7HH				
WB1HH				
NistariHH				Sivaprasad et al. (2022)
Tolerant to fluoride pollution				
Huayuan × Dongsheng	Single	China	Crossing and pedigree selection	Xu et al. (2006)
Qiu Feng × Baiyu	Single		By evaluation	Chen (2003)
Wu·Xuan × 28·32	Double		Crossing and pedigree selection	Guo et al. (2007)
Lu·Ping × Qing·Guang	Double		Systematic selection method	Lin et al. (2001)

(Ashwath et al. 2011), and double hybrid, G11 × G19 (2C.4S × 2D.4N), for suboptimal conditions introgressing amylase gene following marker assisted selection. Though these thermotolerant productive hybrids performed well under controlled laboratory condition, they were not able to make a significant mark in the field. Many reasons like fluctuations in temperature, humidity, and aeration combined with poor leaf quality and non-hygiene maintenance could have impacted the performance of these hybrids in the field conditions (Sivaprasad et al. 2022). However, the attempts were also made by silkworm breeders to develop thermotolerant silkworm hybrids by using DNA based markers. In silkworm, Chandrakanth et al. (2015b) showed association between five SSR markers and thermotolerance by conducting experiments through bulked segregant analysis with two different genetic backgrounds of contrasting parents. By using these markers, a thermotolerant bivoltine double hybrid, TT21 × TT56, was developed by crossing followed by six generations of backcrossing which displayed 56.16% and 73.98% improvement in cocoon yield and pupation, respectively, over the most popular double hybrid (FC1 × FC2) in India under laboratory conditions (Chandrakanth et al. 2015a; Moorthy et al. 2016). The SSR markers found associated with thermotolerance in *B. mori* are LFL1123, LFL329, S0809, and S0801 that could be applied to develop thermotolerant bivoltine silkworm breeds (Chandrakanth et al. 2015b). By using a SilkMap tool, a sequence based map was constructed locating three SSR markers (S0803, S0809, and S0816) and three gene sequences (BGIBMGA005249, BGIBMGA005250, and BGIBMGA005272) on eighth chromosome. Similar breeding strategy was used to develop thermotolerant bivoltine silkworm breeds for eastern India. As a result, MAS strategy was successfully applied in the development of thermotolerant silkworm double hybrid, WB79 × WB13, that showed 7.35% improvement in pupation at 36 °C over the ruling foundation cross, SK6 × SK7 (Chandrakanth et al. 2021). The thermotolerant hybrids developed through MAS exhibited promising yields in adverse conditions in southern (TT21 × TT56; 75 kg/100 dfls with 21.67% shell at farmers' level) and eastern zone (WB79 × WB13; 65 kg/100 dfls with 19.67% shell at farmers' level) under adverse seasons. Thermotolerant double hybrids can play a major role in boosting the raw silk production in India especially during adverse seasons and help farmers to take up bivoltine crop during adverse seasons. The thermotolerant hybrids developed through MAS appear to be more effective and remarkable in field conditions with less breeding period in contrast to the conventional methods.

Silkworm Hybrids for High Humidity Tolerance

Maintaining optimum relative humidity is among the environment factors that play a significant role in the development and survival of silkworms. Fluctuations in relative humidity negatively impact the physiology of silkworm affecting molting, spinning behavior, cocoon yield, silk reeling parameters, and silk quality. The humidity fluctuations are major problems in eastern and northeastern India and coastal regions of India. The effect of humidity on silkworm is more pronounced in combination with high temperature (Sivaprasad et al. 2022). Especially, the seed crops taken up in the eastern and northeastern India during August/September are severely affected by the high humidity seasons incurring great loss to the seed farmers. While the seed farmers rely on rearing the traditional indigenous multivoltine race Nistari, there will be shortage for bivoltine seed cocoons of foundation crosses, SK6 × SK7 and BCon1 × BCon4, to produce crossbreed to meet the demand for commercial season in October/November.

To provide choice for the commercial and seed farmers to rear high humidity tolerant bivoltine silkworm breeds/foundation crosses during adverse humidity seasons, it is necessary to develop humidity tolerant silkworms. CSRTI-Berhampore has initiated a breeding program for development of high humidity tolerant silkworms by inducing high humidity and temperature and directional selection for tolerance aided by MAS through *pyrexia* gene expression. Research reports on humidity sensing genes particular dry- and moist-sensing receptors, namely Nanchung (*Nan*) and Waterwitch (*Wtwr*), found in the model organism, *D. melanogaster*, provide hints about the involvement of pyrexia genes. Ionotropic receptors like pyrexia play vital role in regulation of membrane potential, propagation of action potentials, neurotransmitter release, and intercellular communication (Liu et al. 2015; Wallach et al. 2017). However, the role and mode of mechanism that triggers the pyrexia expression in silkworm is not yet identified. The effects of humidity are majorly studied in combination with temperature—as such, the role and effects of humidity on silkworm with respect to molecular mechanisms have wider scope for research. Preliminary investigations were carried out by Madhusudhan et al. (2019), and Raviraj et al. (2022) identify these ionotropic and transient receptor in silkworms. The expression of *Pyrexia* gene in the survived population under high temperature and high humidity condition was found to be higher and associated with high humidity tolerance (Raviraj et al. 2022). Further, an integrated approach of MAS with conventional breeding method was used to identify improved lines of B.Con4HH, HTH10HH, N5HH, SK7HH, and WB1HH with high humidity tolerance by inducing stress and selection by survival and higher expression of pyrexia gene (Raviraj et al. 2022). The developed high humidity tolerant breeds need to be tested in the field condition during adverse seed crop seasons to ensure their viability for commercial exploitation.

Silkworm Hybrids for Fluoride Pollution Tolerance

Atmospheric fluoride originating from the widely spread urban industries like cement, brick, tile, etc. has severely caused economic losses in agriculture. The adverse effects of fluoride on the growth and development of plants and silkworms are well reported (Davies et al. 1992; Chen and Wu 1995; Ahmed et al. 1999). Since silkworm rearing is a cottage agricultural industry, it is directly related to fluoride-rich water and atmospheric fluoride pollution affecting the cocoon production in sericulture (Nath 1993). Breeding silkworm breeds tolerant to fluoride pollution now is the central challenge being focused for the future environmental changes. In China, fluoride is the key component in air pollution and is particularly evident in autumn (Li et al. 2015). It is a serious hazard in some areas of China. Similarly, wing scales produced in silkworm egg production units are also harmful for the workers. The *Dtf* (dominant tolerance to fluoride) allele controls tolerance to fluoride in silkworms, and the *nlw* (recessive scaleless wings) allele controls the formation of scaleless wings in silkworms (Zhang and Qin 1991; Lin et al. 1996, 1997). These two mutants are mapped by using simple sequence repeat (SSR) and sequence tag site (STS) markers (Wang et al. 2010; Zhang and Qin 1991; Bai et al. 2008). Therefore, MAS is applied to develop a strain with fewer scales during egg production with fluoride tolerance which will benefit the health of the workers. Li et al. (2015) applied MAS as a method to integrate two characteristics in one silkworm by using two donor parents. Two donors are T6 silkworm strain (*Dtf/DtfDtf*) with high fluoride tolerance that can survive with extremely few effects after they are fed with NaF solution treated (200 mg/kg) mulberry leaves and U13 silkworm strain (*nlw/nlw*) with few scales on the wings. The Jingsong strain (*+Dtf/+Dtf*; *+nlw/+nlw*), which is susceptible to fluoride that cannot survive even when fed with low amount of NaF solution treated (50 mg/kg) mulberry leaves and has wild-type wings, is a productive and popular strain in China and is selected as recipient parent (Li et al. 2015). For the breeding program, a three way cross was made between a female of the Jingsong with a male of the T6 strain, and then (Jingsongx T6) females were crossed to the U13 strain to derive (Jingsongx T6) × U13, which was backcrossed to the Jingsong to obtain BC_1—the backcrossing was continued for six generations in total with T6 males. The SSR and STS markers located on the flanking regions of *Dtf*, S1214 and S121201 (Bai et al. 2008), and *nlw*, S1305 and cash2p (Wang et al. 2010), were screened to detect the genotype of male moths after mating with Jingsong females. Heterozygous progenies were selected in backcrossing generations for the desired characters. Self-mating was done in BC_6 generation, and these progenies were screened for fluoride tolerance by feeding with 200 mg/kg NaF-polluted mulberry leaves. The larvae that survived and emerged as scaleless wing moths with superior economic characteristics were selected and self-mated to generate the BC_6F_3. Simultaneously, the molecular markers were used to select the individuals that were homozygous at both *Dtf* and *nlw* sites. Thus, a new silkworm strain with economic characteristics of Jingsong with concurrent tolerance to fluoride and scaleless wings was developed (Li et al. 2015). These results prove that MAS methods are highly efficient in breeding silkworm strains with two or more desired characteristics. In comparison with the conventional breeding methods, it took shorter breeding time (4 years) to develop the new silkworm strain through MAS. MAS has several advantages over conventional breeding as fluoride tolerance can be screened by feeding every generation with fluoride-polluted/sprayed mulberry leaves, which is more laborious. Besides, wet leaves are not good for the larvae as they may encourage disease causing pathogens although the surviving ones show tolerance to fluoride. By using MAS methods, we can focus only on the economic characteristics desired, and allele can be selected by molecular markers later (Li et al. 2015).

Commercial Exploitation of Silkworm Hybrids

In India, newly developed silkworm hybrids with promising performance can be exploited for commercial utilization only after its successful trials in lab and field conditions. It has three tier system for evaluation of the new silkworm hybrids. A newly developed silkworm hybrid firstly undergoes laboratory trials in the developer institute and the selected stations in very small quantity (10–15 dfls/crop) along with the ruling hybrid. On successful trials at laboratory level, the silkworm hybrid will be subjected to on farm trials with the progressive farmers in medium quantity (20,000–30,000 dfls). On successfully conducting on farm trials, after thorough evaluation of the hybrid performance by the committee constituted as hybrid authorization committee by the central silk board, the hybrid will be recommended for final testing with the farmers in large scale termed as authorization trials. In authorization trials, the silkworm hybrid will be tested with the farmers along with the ruling hybrid in large scale (two to five lakh dfls depending on the region). Approximately, it can take 10 years for a silkworm hybrid to pass through this three tier system from initiation of breeding program. In India, the application of MAS in silkworm breeding has been taken up lately during the 2010s—so far, only two silkworm hybrids for high temperature tolerance and five silkworm breeds for humidity tolerance were developed. TT21 × TT56 and WB13 × WB75 are the two silkworm hybrids tolerant to high temperature developed through MAS. The TT21 × TT56 was developed by CSRTI-Mysuru and is under authorization trials with four lakh dfls tested all over India with 70 kg/100 dfls yield (Sivaprasad et al. 2022). The WB13 × WB75 was developed by CSRTI-Berhampore by following the same breeding strategy, and the small scale laboratory testing with 65 kg/100 dfls yield was completed (Chandrakanth et al. 2021). The development of high humidity tolerant silkworm breeds is under breeding process, and as to date, there are five different improved lines that are designated as B.Con4HH, HTH10HH, N5HH, SK7HH, and WB1HH (Raviraj et al. 2022). In India, as to our knowledge, no breeding program has been initiated through MAS for the development of silkworms tolerant to fluoride pollution.

Perspectives

In order to increase the raw silk, it is imperative to break the barrier of limited rearings of bivoltine silkworm. It can be achieved by introducing more number of silkworm hybrids tolerant to abiotic stresses, especially high temperature and high humidity in the fields. Integrated approaches in silkworm breeding are going to be vital for development of climate resilient silkworm hybrids, which could perform better under adverse conditions and keep up the silk productivity. In this line, we have to look for the development of multi-stress tolerant silkworm hybrids, which not only resist climatic fluctuation but are also productive. As several stress tolerance traits in silkworm are quantitative and polygenic, construction of QTL based linkage maps with different parental genetic backgrounds would reveal the minor and major QTLs and genes involved in providing tolerance to stresses. Such molecular information generated would assist the silkworm breeders to integrate molecular approaches with conventional methods to yield promising hybrids with remarkable stress tolerance and productivity. In the future, genome editing tools would greatly facilitate the molecular approaches that have the potential to provide solution against the various challenges in sericulture. Apart from it, deciphering the genetic information on *Bombyx mori* would help the scientific community for better understanding of biological processes of insects.

References

Ahamed CA, Chandrakala MV (1999) Effect of oral administration of sodium fluoride on food and water utilisation in silkworm, Bombyx mori L. Int J Trop Insect Sci 19(2-3):193–198

Ahmed CAA, Chandrakala MV, Maribashetty VG (1999) Influence of sodium fluoride on nutritional efficiency of a multivoltine race of the silkworm. *Bombyx mori* 2:127–133

Anuradha JH, Somasundaram P, Vishnupriya S, Manjula A (2012) Storage protein-2 as a dependable biochemical index for screening germplasm stocks of the silkworm Bombyx mori (L.). Albanian J Agric Sci 11(3):141

Ashwath SK, Morrison MN, Datta RK (2001) Development of near isogenic Lines of productive silkworm breeds by isozyme marker based selection. Proc Natl Acad Sci India Sec. A 71B(3&4):207–222

Ashwath SK, Sharmila KK, Mahalingappa KC, Nirmal, Kumar S, Qadri SMH (2011). Evolution and evaluation of single and double hybrids developed through Amylase marker assisted selection. In: Abstracts of Golden Jubilee National Conference on Sericulture Innovations: Before and Beyond, CSRTI, Mysore, India, pp 75–76

Bai HC, Xu AY, Li MW, Zhao YP (2008) SSR based linkage and mapping analysis of dominant endurance to fluoride gene (def) in the silkworm *Bombyx mori*. ACTA Seric Sin 34:191–196

Beavis WD, Grant D, Albertsen M, Fincher R (1991) Quantitative trait loci for plant height in four maize populations and their associations with qualitative genetic loci. Theor Appl Genet 83:141–145

Bindroo BB, Begum AN, Mal R (2014) New bivoltine silkworm double hybrid – Jayachamaraja – (CSR50 × CSR52) X (CSR51 × CSR53) for high egg recovery and crop stability and silk productivity

Cakir M, Gupta S, Platz GJ, Ablett GA, Loughman R, Emebiri LC (2003) Mapping and validation of the genes for resistance to Pyrenophora teres f. teres in barley (*Hordeum vulgare* L.). Aust J Agric Res 54(12):1369–1377

Chandrakanth N, Moorthy SM, Ponnuvel KM, Sivaprasad V (2015a) Screening and classification of mulberry silkworm, *Bombyx mori* based on thermotolerance. Int J Indust Entomol 31(2):115–126

Chandrakanth N, Moorthy SM, Ponnuvel KM, Sivaprasad V (2015b) Identification of microsatellite markers linked to thermo-tolerance in silkworm by bulk segregant analysis and *in silico* mapping. Genetika 47(3):1063–1078

Chandrakanth N, Lakshmanan V, Verma AK, Raviraj VS, Mitra G (2021) Development of thermo-tolerant bivoltine hybrids of silkworm through MAS. In: CSRTI-Berhampore Annual Report-2021, pp 45–50

Chen Y (2003) Variable tolerance of the silkworm *Bombyx mori* toatmospheric fluoride pollution. Fluoride Res Rep 36(3):157–162

Chen YY (1993) Changes in fluoride and calcium in the blood of the fluorosis-affected silkworm, *Bombyx mori* L. Sericologia 33:153–157

Chen YY, Wu YC (1995) Fluoride loading and kinetics in different tissues of larvae of fluorosis silkworm (*Bombyx mori* L.). Sericologia 34:1–10

Collard BC, Jahufer MZZ, Brouwer JB, Pang ECK (2005) An introduction to markers, quantitative trait loci (QTL) mapping and marker-assisted selection for crop improvement: the basic concepts. Euphytica 142:169–196

Collard BCY, Pang ECK, Taylor PWJ (2003) Selection of wild Cicer accessions for the generation of mapping populations segregating for resistance to ascochyta blight. Euphytica 130:1–9

Coulon-Bublex M, Mathelin J (1991) Variations in the rate of synthesis of heat shock proteins HSP70, between laying and neurula, the diapausing embryo of the silkworm *Bombyx mori*. Sericologia 3:295–300

Czerwinski E, Nowak J, Dabrowska D, Skolarczyk A, Kita B, Ksiezyk M (1988) Bone and joint pathology in fluoride-exposed workers. Arch Environ Health Int J 43(5):340–343

Dandin SB, Kumar NS, Basavaraja HK, Reddy NM, Kalpana GV, Joge PG (2006) Development of new bivoltine silkworm hybrid, Chamaraja (CSR50× CSR51) of *Bombyx mori* L. for tropics. Indian J Sericult 45(1):35–44

Davies MT, Davison AW, Port GR (1992) Fluoride loading of larvae of pine sawfly from a polluted site. J Appl Ecol:63–69

Dote T, Kono K, Usuda K, Nishiura H, Tagawa T, Miyata K (2000) Toxicokinetics of intravenous fluoride in rats with renal damage caused by high-dose fluoride exposure. Int Arch Occupat Environ Health 73:S90–S92

El-Akkad SS, Hassan EM, Abdel-Nabi IM, AI, A. H. (2008) The effect of mulberry varieties and nutritional additives on the protein patterns of the silkworm *Bombyx mori*. Egyptian J Biol 10:11–19

Etebari K, Mirhoseini SZ, Matindoost L (2005) A study on interaspecific biodiversity of eight groups of silkworm (*Bombyx mori*) by biochemical markers. Insect Sci 12(2):87–94

Feder ME (1996) Ecological and evolutionary physiology of stress proteins and the stress response: the *Drosophila melanogaster* model. In: Johnston IA, Bennett AF (eds) Animals and temperature: phenotypic and evolutionary adaptation to temperature. Cambridge University Press, Cambridge, pp 79–102

Feder ME, Hofmann GE (1999) Heat-shock proteins, molecular chaperones, and the stress response: evolutionary and ecological physiology. Annu Rev Physiol 61(1):243–282

Feder ME, Cartaño NV, Milos L, Krebs RA, Lindquist SL (1996) Effect of engineering Hsp70 copy number on Hsp70 expression and tolerance of ecologically relevant heat shock in larvae and pupae of *Drosophila melanogaster*. J Exp Biol 199(8):1837–1844

Fujii M, Honda S (1972) The relative oral toxicity of some fluorine compounds for silkworm larvae. J Sericult Sci Jpn 41(2):104–110

Gosalvitra P, Silapanapaporn O, Chuprayoon S (2015). Breeding line improvement of Thai silkworm (*Bombyx mori* L.) towards high quality products. In: 7th BACSA Conference, Romania

Guo DG, Lin JR, Huang P, Lin ZF, Zhong SY, Qiu GX (2007) Breeding of the new silkworm variety Yuefeng No.3 (Wu.Xuan x 28.32). Canye Kexue 33(3):466–469(Chinese)

Jingade AH, Vijayan K, Somasundaram P, Srinivasababu GK, Kamble CK (2011) A review of the implications of heterozygosity and inbreeding on germplasm biodiversity and its conservation in the silkworm, *Bombyx mori*. J Insect Sci 11(1):8

Jolly MS, Datta RK, Noamani MKR, Iyengar MNS, Nagaraj CS, Basavaraj HK (1989) Studies on the genetic divergence in mulberry silkworm *Bombyx mori L.* Sericologia 29(4):545–553

Kampinga HH, Brunsting JF, Stege GJJ, Konings AWT, Landry J (1994) Cells overexpressing Hsp27 show accelerated recovery from heat-induced nuclear-protein aggregation. Biochem Biophys Res Commun 204(3):1170–1177

Kato M, Nagayasu K, Ninagi O, Hara W, Watanabe A (1989). Studies on resistance of the silkworm, *Bombyx mori L.* for high temperature. In: Proceedings of the 6th International Congress of SABRAO (II), pp 953–956

Kawamoto M, Jouraku A, Toyoda A, Yokoi K, Minakuchi Y, Katsuma S (2019) High-quality genome assembly of the silkworm, *Bombyx mori*. Insect Biochem Mol Biol 107:53–62

Kobayashi J, Ebinuma H, Kobayashi M (1986) Effect of temperature on the diapause egg production in the tropical race of the silkworm, *Bombyx mori*. J Sericult Sci Jpn 55(4):343–348

Krishna Rao S, Raghuraman R, Bongale UD (2003) Silkworm breeding for sub-optimal conditions. In: Concept papers, mulberry silkworm breeders summit, APSSRDI, Hindupur, India, pp 52–59

Kumar NS, Basavaraja HK, Kumar CK, Reddy NM, Datta RK (2002) On the breeding of "CSR18 x CSR19"-A robust bivoltine hybrid of silkworm, *Bombyx mori* L. for the tropics. Int J Ind Entomol 5:155–162

Kumar NS, Basavaraja HK, Kalpana GV, Reddy NM, Dandin SB (2003) Effect of high temperature and high humidity on the cocoon shape and size of parents, foundation crosses, single and double hybrids of bivoltine silkworm, *Bombyx mori L.* Indian J Sericult 42(1):35–40

Kumar NS, Basavaraja HK, Dandin SB (2004) Breeding of robust bivoltine silkworm, *Bombyx mori* L. for temperature tolerance-a review. Indian J Sericult 43(2):11–124

Kumar NS, Basavaraja HK, Joge PG, Reddy NM, Kalpana GV, Dandin SB (2006) Development of a new robust bivoltine hybrid (CSR46× CSR47) of *Bombyx mori* L. for the tropics. Indian J Sericult 45(1):21–29

Kumar P, Gupta VK, Misra AK, Modi DR, Pandey BK (2009) Potential of molecular markers in plant biotechnology. Plant Omics 2(4):141–162

Kumaresan P, Sinha RK (2002) Genetic divergence in multivoltine silkworm germplasm in relation to cocoon characters. Indian J Gen Plant Breed 62:183–184

Kumaresan P, Sinha RK, Raje US (2007) An analysis of genetic variation and divergence in Indian tropical polyvoltine silkworm (*Bombyx mori* L.) genotypes. Caspian J Env Sci 5(1):11–17

Kumari S, Venkata Subbarao S, Misra S, Suryanarayana Murtyd U (2011) Screening strains of the mulberry silkworm, *Bombyx mori*, for thermotolerance. J Insect Sci 11(1):116

Kuribayashi S (1977) Effects of atmospheric pollution by hydrogen fluoride on mulberry tree and silkworm. J Sericult Sci Jpn 46(6):536–544

Lakshmi H, Chandrashekharaiah, Babu MR, Raju PJ, Saha AK, Bajpai AK (2011) HTO5 x HTP5, The new bivoltine silkworm (*Bombyx mori* L.) hybrid with thermo-tolerance for tropical areas. Int J Plant Anim Environ Sci 1:88–104

Lander ES, Green P, Abrahamson J, Barlow A, Daly MJ, Lincoln SE (1987) MAPMAKER: an interactive computer package for constructing primary genetic linkage maps of experimental and natural populations. Genomics 1(2):174–181

Landry J, Chrétien P, Lambert H, Hickey E, Weber LA (1989) Heat shock resistance conferred by expression of the human HSP27 gene in rodent cells. J Cell Biol 109(1):7–15

Lee GJ, Pokala N, Vierling E (1995) Structure and in vitro molecular chaperone activity of cytosolic small heat shock proteins from pea. J Biol Chem 270:10432–10438

Li C, Tong X, Zuo W, Luan Y, Gao R, Han M, Xiong G, Gai T, Hu H, Dai F, Lu C (2017) QTL analysis of cocoon shell weight identifies *Bm*RPL18 associated with silk protein synthesis in silkworm by pooling sequencing. Sci Rep 7(1):17985. https://doi.org/10.1038/s41598-017-18277-y

Li C, Tong X, Zuo W, Hu H, Xiong G, Han M (2020) The beta-1, 4-N-acetylglucosaminidase 1 gene, selected by domestication and breeding, is involved in cocoon construction of Bombyx mori. PLoS Genet 16(7):e1008907. https://doi.org/10.1371/journal.pgen.1008907

Li MW, Yu HJ, Yi XL, Li J, Dai FY, Hou CX (2015) Marker-assisted selection in breeding silkworm strains with high tolerance to fluoride, scaleless wings, and high silk production. Genet Mol Res 14(3):11162–11170

Li ZW, Li X, Yu QY, Xiang ZH, Kishino H, Zhang Z (2009) The small heat shock protein (sHSP) genes in the silkworm, *Bombyx mori,* and comparative analysis with other insect sHSP genes. BMC Evolut Biol 9(1):1–14

Lin CQ, Yao Q, Wu DX (1996) Investigation and analysis on endurance fluoride of silkworm strain resources. Acta Seric Sin 2:253–255

Lin CQ, Mi YD, Yao Q, Wu DX (1997) The discovery of dominant endurance to fluoride gene in silkworm, *Bombyx mori* L. Acta Seric Sin 23:237–239

Lin CQ, Yao Q, Fang QQ, Chen KP, Hou CX (2001) Breeding of summer-autumn rearing fluoride enduring Bombyx mori variety Lu Pingx Qing Guang and improvement of Guang's endurance to fluoride. Seric Sinic 27(1):24–28. (Chinese)

Lin CQ, Xu AY, Hou CX, Zhang YH (2006) Breeding of highly endurance fluoride silkworm strains Huayuan X Dongsheng for spring season and its characters. China Seric 27:30–32

Lin HJ, Li Z, Dang XQ, Su WJ, Zhou ZY, Wang LL (2014) Short-term increased expression of the heat shock protein 70 family in the midgut of three strains of the silkworm, *Bombyx mori*. Afr Entomol 22(1):4–29

Lincoln S, Daly M, Lander E (1993) Constructing genetic linkage maps with MAPMAKER/EXP Version 3.0. Whitehead Institute for Biomedical Research Technical Report, 3

Linder B, Jin Z, Freedman JH, Rubin CS (1996) Molecular characterization of a novel, developmentally regulated small embryonic chaperone from *Caenorhabditis elegans*. J Biol Chem 271(47):30158–30166

Liu WW, Mazor O, Wilson RI (2015) Thermo-sensory processing in the *Drosophila* brain. Nature 519:353–357

Liu Y, Liang Y, Yang C, Shi R, Lu W, Wang X (2021) A deep insight into the transcriptome of midgutand fat body reveals the toxic mechanism of fluoride exposure insilkworm. Chemosphere 262:127891. https://doi.org/10.1016/j.chemosphere.2020.127891

Madhusudhan KN, Ranjini MS, Moorthy SM (2019) Development of Hygro-tolerant bivoltine breeds/hybrids through molecular marker assisted selection. CSRTI-Mysore Annual Report (2018-19). pp 27–29

Makwana P, Kamidi R, Chattopadhyay S, Vankadara S (2021) Effect of thermal stress on antioxidant responses in *Bombyx mori*. Chem Sci Rev Lett 10(38):288–294

Manly KF, Cudmore RH Jr, Meer JM (2001) Map Manager QTX, cross-platform software for genetic mapping. Mammalian Genome 12:930–932

Mehlen P, Kretz-Remy C, Preville X, Arrigo AP (1996) Human hsp27, *Drosophila* hsp27 and human alphaB-crystallin expression-mediated increase in glutathione is essential for the protective activity of these proteins against TNFalpha-induced cell death. EMBO J 15(11):2695–2706

Mehlen P, Mehlen A, Godet J, Arrigo AP (1997) Hsp27 as a switch between differentiation and apoptosis in murine embryonic stem cells. J Biol Chem 272(50):31657–31665

Miao XX, Xub SJ, Li MH, Li MW, Huang JH, Dai FY (2005) Simple sequence repeat-based consensus linkage map of *Bombyx mori*. Proc Natl Acad Sci 102(45):16303–16308

Michelmore RW, Paran I, Kesseli R (1991) Identification of markers linked to disease-resistance genes by bulked segregant analysis: a rapid method to detect markers in specific genomic regions by using segregating populations. Proc Natl Acad Sci 88(21):9828–9832

Mita K, Kasahara M, Sasaki S, Nagayasu Y, Yamada T, Kanamori H (2004) The genome sequence of silkworm, *Bombyx mori*. DNA Res 11(1):27–35

Moorthy SM, Das SK, Rao PRT, Raje Urs S, Sarkar A (2007) Evaluation and selection of potential parents based on selection indices and isozyme variability in silkworm, *Bombyx mori*, L. Int J Ind Entomol 14:1–7

Moorthy SM, Ashwath SK, Kariyappa, Chandrakanth N (2016) Development of robust bivoltine hybrids of silkworm, *Bombyx mori* L. tolerant to high temperature environment of the tropics through DNA marker assisted selection. CSRTI-Mysore Annual Report (2016–2017), pp 14–17

Morgan TJ, Mackay TFC (2006) Quantitative trait loci for thermotolerance phenotypes in *Drosophila melanogaster*. Heredity 96:232–242

Mundkur R, Muniraju E (2018) Molecular marker-assisted selection breeding in silkworm, *Bombyx mori*. In: Kumar D, Gong C (eds) Trends in insect molecular biology and biotechnology. pp 3–24. https://doi.org/10.1007/978-3-319-61343-7_1

Nabizadeh P, Kumar TJ (2011) Fat body catalase activity as a biochemical index for the recognition of thermotolerant breeds of mulberry silkworm, *Bombyx mori* L. J Therm Biol 36(1):1–6

Nagaraju J, Goldsmith J (2002) Silkworm genomics-progress and prospects. Curr Sci 83:415–425

Nath BS (1993) Studies on organophosphorous insecticide toxicity during 5th instar of silkworm, *Bombyx mori* L. Ph.D. thesis. Sri Krishnadevaraya University, Anantapur India

Prasad MD, Muthulakshmi M, Arunkumar KP, Madhu M, Sreenu VB, Pavithra V (2005a) SilkSatDb: a microsatellite database of the silkworm, *Bombyx mori*. Nucl Acids Res 33(suppl_1):D403–D406

Prasad MD, Muthulakshmi M, Madhu M, Archak S, Mita K, Nagaraju J (2005b) Survey and analysis of microsatellites in the silkworm, *Bombyx mori*: frequency, distribution, mutations, marker potential and their conservation in heterologous species. Genetics 169(1):197–214

Promboon A, Shimada T, Fujiwara H, Kobayashi M (1995) Linkage map of random amplified DNAs (RAPDs) in the silkworm, *Bombyx mori*. Genet Res 66:1–7

Rahmathulla VK (2012) Management of climatic factors for successful silkworm (*Bombyx mori* L.) crop and higher silk production: a review. Psyche J Entomol. https://doi.org/10.1155/2012/121234

Rajaraman K, Raman B, Rao CM (1996) Molten-globule state of carbonic anhydrase binds to the chaperone-like α-crystallin. J Biol Chem 271(44):27595–27600

Ramakrishna S, Nath BS (2004) Evaluation of relative fluoride toxicity and its impact on growth, economic characters and fecundity of the silkworm, *Bombyx mori* L. Int J Ind Entomol 8(2):151–159

Rao PRM, Ravindra S, Premalatha V, Basavaraja HK (2007a) Identification of polyvoltine breeds of the silkworm *Bombyx mori* L. through evaluation index method. Indian J Sericult 46(2):163–168

Rao PS, Nayaka AR, Mamatha M, Sowmyashree TS, Bashir I, Ilahi I (2007b) Development of new robust Bivoltine silkworm hybrid SR2 x SR5 for rearing throughout the year. Int J Ind Entomol 14(2):93–97

Raviraj VS, Chandrakanth N, Lakshmanan V, Sivaprasad V (2022) Identification of markers for high humidity tolerance in silkworm breeds. In: CSRTI-Berhampore Annual Report-2022, pp 45–46

Reddy KD, Nagaraju J, Abraham EG (1999) Genetic characterization of the silkworm *Bombyx mori* by simple sequence repeat (SSR)-anchored PCR. Heredity 83:81–687

Reyna N, Sneller CH (2001) Evaluation of marker-assisted introgression of yield QTL alleles into adapted soybean. Crop Sci 41:1317–1321

Risch N (1992) Genetic linkage: Interpreting LOD scores. Science 255:803–804

Sambrook J, Russell DW (2001) Molecular cloning: a laboratory manual, vol 1, 3rd edn. Cold Spring Harbor Laboratory Press, New York

Sayed H, Kayyal H, Ramsey L, Ceccarelli S, Baum M (2002) Segregation distortion in doubled haploid lines of barley (*Hordeum vulgare* L.) detected by simple sequence repeat markers. Euphytica 225:265–272

Sharp PJ, Johnston S, Brown G, McIntosh RA, Pallotta M, Carter M (2001) Validation of molecularmarkers for wheat breeding. Aust J Agric Res 52:1357–1366

Shi J, Heckel DG, Goldsmith MR (1995) A genetic linkage map for the domesticated silkworm, *Bombyx mori*, based on restriction fragment length polymorphisms. Genet Res 66(2):109–126

Shirota T (1992) Selection of healthy silkworm strains through high temperature rearing of fifth instar larvae. Rep Silk Sci Res Inst 40:33–40

Shong LF (1985) A report on the effect of fluoride pollution on sericulture production. Jianshu Seri 3:16–19

Singh H, Suresh Kumar N (2010) On the breeding of bivoltine breeds of the silkworm, *Bombyx mori* L. (Lepidoptera: Bombycidae), tolerant to high temperature and high humidity conditions of the tropics. Psyche, https://doi.org/10.1155/2010/892452

Sivaprasad V, Chandrakanth N, Moorthy SM (2022) Genetics and genomics of *Bombyx mori* L. In: Chakravarthy AK (ed) Genetic methods and tools for managing crop pests. pp 127–210. https://doi.org/10.1007/978-981-19-0264-2_6

Somasundaram P, Ashok Kumar K, Babu GKS, Kamble CK (2009) Heat stable esterase – a biochemical marker for evolution of thermotolerant breeds of *Bombyx mori* (L). J Adv Biotechnol 1:20–21

Staub JE, Serquen FC, Gupta M (1996) Genetic markers, map construction, and their application in plant breeding. Hort Sci 31(5):729–739

Stephanow G, Alahiotis SN (1983) Non-Mendelian inheritance of "heat-sensitivity" in *Drosophila melanogaster*. Genetics 103:93–107

Sudhakara Rao P, Datta RK, Basavaraja HK (2006) Evolution of a new thermo-tolerant bivoltine hybrid of the silkworm (Bombyx mori L.) for tropical climate. Ind J Sericult 45(1):15–20

Suresh Kumar N, Lakshmi H, Saha AK, Bindroo BB, Longkumer N (2012) Evaluation of bivoltine silkworm breeds of bombyx mori L. under West Bengal conditions. Universal J Environ Res Technol 2(5):393–401

Suzuki M, Bandoski C, Bartlett JD (2015) Fluoride induces oxidative damage and SIRT1/autophagy through ROS-mediated JNK signaling. Free Radical Biol Med 89:369–378

Takiya S, Tsubota T, Kimoto M (2016) Regulation of silk genes by hox and homeodomain proteins in the terminal differentiated silk gland of the silkworm *Bombyx mori*. J Dev Biol 4(2):19. https://doi.org/10.3390/jdb4020019

Tan YD, Wan C, Zhu Y, Lu C, Xiang Z, Deng HW (2001) An amplified fragment length polymorphism map of the silkworm. Genetics 157(3):1277–1284

Tautz D (1989) Hypervariability of simple sequences as a general source for polymorphic DNA markers. Nucl Acids Res 17(16):6463–6471

Thangavelu K, Sinha RK, Mahadevamurthy TS, Radhakrishnan S, Kumaresan P, Mohan B, Rayaradder FR and Sekar S (2000) Catalogue on Silkworm Bombyx mori L. Germplasm, Central Sericultural Germplasm Resources Centre, Hosur, 2: 138

Tissiéres A, Mitchell HK, Tracy UM (1974) Protein synthesis in salivary glands of *Drosophila melanogaster*: relation to chromosome puffs. J Mol Biol 84(3):389–398

Tsubota T, Tomita S, Uchino K, Kimoto M, Takiya S, Kajiwara H (2016) A Hox gene, Antennapedia, regulates expression of multiple major silk protein genes in the silkworm *Bombyx mori*. J Biol Chem 291(13):7087–7096. https://doi.org/10.1074/jbc.M115.699819

van den Ijssel PR, Overkamp P, Knauf U, Gaestel M, de Jong WW (1994) αA-crystallin confers cellular thermoresistance. FEBS Lett 355(1):54–56

Vos P, Hogers R, Bleeker M, Reijans M, Lee TVD, Hornes M (1995) AFLP: a new technique for DNA fingerprinting. Nucl Acids Res 23(21):4407–4414

Waizumi R, Tsubota T, Jouraku A, Kuwazaki S, Yokoi K, Tetsuya I, Yamamoto K, Sezutsu H (2023) Highly accurate genome assembly of an improved high-yielding silkworm strain, Nichi01. G3 (Bethesda, Md.) 13(4):jkad044. https://doi.org/10.1093/g3journal/jkad044

Wallach J, Colestock T, Adejare A (2017) Receptor targets in Alzheimer's disease drug discovery. In: Drug discovery approaches for the treatment of neurodegenerative disorders. Academic Press, pp 83–107

Wang H, Fang Y, Bao Z, Jin X, Zhu W, Wang L (2014) Identification of a *Bombyx mori* gene encoding small heat shock protein BmHsp27. 4 expressed in response to high-temperature stress. Gene 538(1):56–62

Wang K and Spector A (1994) The chaperone activity of bovine alpha crystallin. Interaction with other lens crystallins in native and denatured states. J Biol Chem 269:13601–13608

Wang XY, Li MW, Zhao YP, Xu AY, Guo QH, Huang YP (2010) Mapping of non-lepis wing gene nlw in silkworm (Bombyx mori) using SSR and STS markers. Hereditas 32(1):54–58

Williams JG, Kubelik AR, Livak KJ, Rafalski JA, Tingey SV (1990) DNA polymorphisms amplified by arbitrary primers are useful as genetic markers. Nucl Acids Res. 18(22):6531–6535

Wu FZ (1990) Some problems on fluoride pollution of field by exhausting from brickklin. Acta Agric Universitatis Zhejiangensis 4:48–52

Xia Q, Zhou Z, Lu C, Cheng D, Dai F, Yang H (2004) A draft sequence for the genome of the domesticated silkworm (*Bombyx mori*). Science 306:1937–1940

Xu A, Changqi L, Chengxiang H, Yuehua Z, Muwang L, Pingjiang S (2006) A biovoltine silkworm variety, Huayuan × Dongshen, that is resistant to fluoride contamination. Int J Indust Entomol 13(1):1–5

Yamamoto K, Narukawa J, Kadono-Okuda K, Nohata J, Sasanuma M, Suetsugu Y (2006) Construction of a single nucleotide polymorphism linkage map for the silkworm, *Bombyx mori*, based on bacterial artificial chromosome end sequences. Genetics 173(1):151–161

Yasukochi Y (1998) A dense genetic linkage map of the silkworm, *Bombyx mori*, covering all chromosomes based on 1018 molecular markers. Genetics 150:1513–1525

Young ND (1994) Constructing a plant genetic linkage map with DNA markers. In: DNA-based markers in plants. Springer Netherlands, Dordrecht, pp 39–57

Yu K, Park SJ, Poysa V (2000) Marker-assisted selection of common beans for resistance to common bacterial blight: efficacy and economics. Plant Breed 119(5):411–415

Yuyin C, Kumar S (2003) Fluoride loading and distribution in insect-mulberry systems in polluted sites in China. Int J Trop Insect Sci 23(4):281–286

Zhang ZF, Qin J (1991) Genetics study on the transparent wing silkworm moth (a primary report). Acta Seric Sin 17:186–187

Zhao Y, Chen K, He S (2007) Key principles for breeding spring-and-autumn using silkworm varieties: from our experience of breeding 873×874. Caspian J Env Sci 5(1):57–61

Zietkiewicz E, Rafalski A, Labuda D (1994) Genome fingerprinting by simple sequence repeat (SSR)-anchored polymerase chain reaction amplification. Genomics 20:176–183

Mutation Breeding in *Bombyx mori*: Current Trends and Future Avenues

Nalavadi Chandrakanth, Khasru Alam, M. S. Ranjini, and Shunmugam Manthira Moorthy

Abstract

Genetic improvement of silkworm has been a prime focus since decades through conventional and molecular approaches, and heritable variation is an important prerequisite. Silkworm breeding for crop enhancement necessitates genetic variation in beneficial properties. In order to create new impulses in silkworm breeding, mutation breeding is rather effective in generating genetic variation targeting the desired trait. Owing to the availability of more than 400 visible mutations, the silkworm, *Bombyx mori*, is the model insect for genetic studies next only to *Drosophila*. Of which, 200 visible mutations have been well established to conventional linkage groups of *B. mori* spanning 900.2 CM. These mutations have been reported to affect various basic characteristics of silkworm that comprises different qualitative and quantitative traits, viz., larval weight, fecundity, disease tolerance/resistance, and thermotolerance, which are responsible for development and distinction of silkworm races and inbred lines. Mutation breeding estimates higher genetic variability which could be sparse through the conventional methods in achieving the targeted traits within short breeding time. In silkworm, it has been observed that the naturally occurred mutagenesis is not successful in generating commercially viable breeding lines. However, it could be successful through artificially induced mutagenesis by employing chemical mutagens and radiation which could be quick enough to develop new alleles and phenotypes. Through induction of X-ray irradiation, sex-limited breeds had been developed

N. Chandrakanth (✉) · M. S. Ranjini
Central Sericultural Research and Training Institute, Central Silk Board, Mysuru, India

K. Alam
Central Sericultural Research and Training Institute, Central Silk Board, Berhampore, India

S. M. Moorthy
National Silkworm Seed Organization, Central Silk Board, Bangalore, India

R. V Suresh et al. (eds.), *Biotechnology for Silkworm Crop Enhancement*,
https://doi.org/10.1007/978-981-97-5061-0_7

by researchers which gained commercial importance before its downfall which will be discussed in detail in this chapter. In recent years, silkworm mutants had been developed by utilizing various genome editing tools like zinc finger nucleases (ZFNs), transcription activator-like effector nucleases (TALENs), and the clustered regularly interspaced palindromic repeats/CRISPR-associated (CRISPR/Cas) system by enabling efficient gene knockouts. In this chapter, the silkworm mutants developed through different mutagens and gene editing tools will be reviewed, and the future prospects of mutation breeding in sericulture improvement will be discussed.

Keywords

Silkworm · Quantitative traits · Mutation breeding · Gene editing tools

Background

Mutation is an unexpected irreversible and rare change in the number or sequence of nucleotides that is heritable in the phenotypic form of an individual. Mutation is the foundation of all genetic variations of an organism, which is necessary for any kind of evolution (Aparupa et al. 2022). In 1791, mutation was first discovered by Wright in male lamb—later on it was reported by Hugo de Vries in 1900 in the *Drosophila*. Hugo de Vries coined the term "mutation," which means "to change" in Latin (Sridhar Rao 2006). Silkworm is the best biological model for carrying out basic scientific research due to its short life cycle and easy handling (Tanaka 1953).

Mutations are classified according to the cells in which they occur. If they occur in non-reproductive cells, they are called somatic mutations; if they occur in reproductive cells, they are called germline mutations. Based on the size and quality, the mutations are classified as point mutation, transversion, and insertion or addition mutations.

Point Mutation Point mutations are those which show a single nucleotide or base pair variation. They are further subclassified in to two different types.

i. Base Substitution If a single base of a DNA is substituted by another one, then it is called as base substitution. If a purin like cytosine is replaced by thymine, then the mutation is called transition. If a purin is replaced by a pyrimidin, for example, cytosine is replaced by guanine, then the mutation is called transversion. It can occur in eight different possibilities. A single substitution in the base sequence of one codon can alter only one amino acid in the protein. This demonstrates the immense impact even a small genetic modification can have on the final protein product.

ii. Deletion Deletion is a type of mutation where one or more bases are deleted or lost from a gene.

iii. Insertion and Addition Mutation This type of mutation is a result of certain viruses like *Rous sarcoma* virus responsible for disruption of gene function due to addition of viral DNA into the host genome. Bacteria are not an exemption to it—like viruses, some bacteria like *Helicobacter pylori* produce reactive oxygen species damaging DNA and reduced DNA repair (Aparupa et al. 2022). Therefore, viruses and bacteria causing mutations are considered as mutagens.

Mutagens and their Mode of Action

Mutagens are harmful to DNA forming genotoxic. They impair the central molecular dogma process of replication, transcription, and translation. Some mutagens act directly and others indirectly on DNA. Mutagens can cause modification in the number or structure of chromosomes. Mutagens cause various chromosomal abnormalities such as monosomy, translocation, insertion, deletion, duplication, and non-duplication. The role of chemical and radiation-induced mutagens in the sericulture industry is discussed in a later part of this section.

Silkworm Mutants and Mutation Breeding

Accessibility of a large number of inbred and mutated lines is one major advantage of using domesticated silkworm in the experiments. With more than 400 Mendelian mutations described in silkworm, it forms a non-model interesting insect to study the mutation process. Most of the mutations are spontaneous and governed by one or a few alleles for the majority of the characters. Some important exceptions are the larval marking locus, *p* (plain, 2–0.0), with more than 10 alleles; the famous homeotic E-mutants (extra legs, 6–21.1), with more than 35 alleles; and oily mutants with more than 20 alleles (Fujii 1998; Goldsmith 1995). The classical linkage maps drawn for domesticated silkworm cover ~240 biochemical and visible markers spanning 28 linkage groups, with a recombination size of ~900 cM (Fujii 1998; Lu et al. 2003). Some of the morphological larval mutants of *B. mori* are depicted in Fig. 1.

The use of mutagens in exploring genetic variations in silkworms for both quantitative and qualitative characters in mutation breeding is an essential tool (Lokesh and Ananthanarayana, 2008). It has been proven that several attempts have led to the induction of beneficial mutations in silkworms, which have been useful for the sericulture sector (Narayanaswamy et al. 1990). Polygenic inheritance, along with the gene interactions with the environment, collectively influences the development stages of silkworms. In mutation breeding, chemical mutagenesis is highly efficient, relatively specific, and easy to handle, and chemicals are readily available, making it more significant (Chaturvedi 1981). Most chemical mutagens induce chromosomal aberrations (Tazima 1978), where an increase in polyploidy may indicate the chemical's potential to induce aberrations. The fact that chromosomes are the cytologically smallest in silkworms is interesting (Kawaguchi 1928, 1936). Notably,

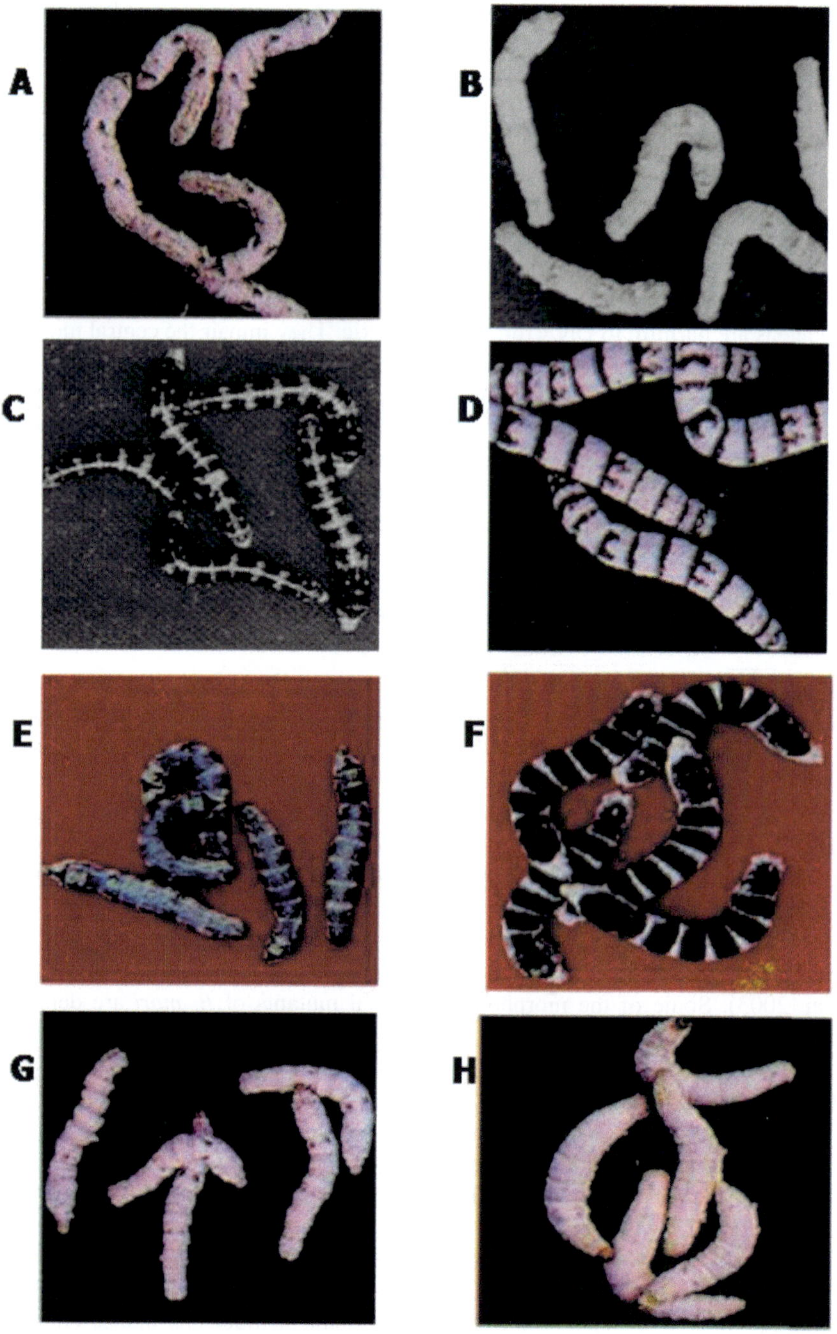

Fig. 1 Some of morphological larval mutants: (**A**) Quail, (**B**) Knobbed, (**C**) Ursa, (**D**) Zebra, (**E**) Moricaud, (**F**) Striped, (**G**) Stony, (**H**) Narrow breast

during spermatogenesis and oogenesis, chromosomal aberrations like clumping, translocation, stickiness, fragmentation, ring formation, etc. can occur as a result of effect of chemical mutagens (Tazima and Onimaru 1968; Datta et al. 1978; Sinha et al. 1993).

Diethyl sulfate (DES) is a very potent alkylating agent that has been found to cause severe physiological aberrations in silkworms. In a study, it was observed that DES caused chromosome clumping and stickiness under different meiotic stages, which was the result of depolymerization and cross-linking of DNA of the chromosomes (Lokesh and Ananthanarayana 2008). The study revealed that alkylation-induced damage during DNA replication was responsible for the majority of physiological chromosomal aberrations (Lewis and John 1966). In addition, the appearance of dicentric bridges was caused as a result of the spontaneous breakage in the meiotic chromosome. Some studies also reported that chemical mutagens can induce stage-specific mutations in silkworms during spermatogenesis. It has been suggested that the stage-specific induction of mutations by chemical agents could be attributed to their unique pathways and diverse effects on the structure and macromolecular processes during germ cell development (Sinha et al. 1993).

Mutagenesis through irradiation and chemical mutagens are well-studied in *B. mori*. Irradiation of silkworm has resulted in recessive lethal, in addition to variants in morphological characters and feeding behavior (Bt, Beet feeder, 1–40.8; Np, Non-preference, 11–32.7; and Nps, Non-preference Shokei, 3–2.2). Irradiations of silkworms are also known to cause "mottled" silkworms with extra fragment of chromosome with visible markers that have been used to study epidermal cell migration patterns (Fujiwara and Maekawa 1994). Irradiation of silkworms has also led to stable chromosomal translocations of the autosomal segment carrying a visible marker to the W chromosome (which determines the female), resulting in the evolution of "autosexing" breeds. Since the visible marker is only expressed in females, females can be visibly distinguished from males at different stages of development - egg, larva or cocoon color (Tazima 1964). These sex-limited breeds were the one that reached the level of commercial utilization in the field, which will be discussed in detail.

Sex-Limited Silkworm Strains

Sex determination in silkworm can be done visibly in larval and cocoon stages. In larval stage, it is based on the imaginal disc presence as tiny spot, namely, "Herolds gland," situated on the ventral side between 11th and 12th segment for male—otherwise, a pair of small spots called "Ishiwata glands" for female. The most popular stage of sex determination is pupal stage easing the crossing process in the adult stage. In pupal stage, the male has a dot mark on the ventral side between eighth and nineth segment, whereas female possesses an "X" mark on the ventral side of eighth segment. Sex separation during commercial preparation of silkworm hybrid eggs is laborious, time-consuming, and expensive. Therefore, this process can be eased by utilization of mutant sex-limited breeds, whose characters are limited to sex with visible differences between the males and females.

Chromosomal basis of sex determination is ZW for females and ZZ for males in *B. mori* (Tanaka 1916). However, male produces only one kind of gamete and the female produces different kinds of gametes—they are referred to as homogametic and heterogametic sex, respectively. The series of experiments conducted by Hashimoto and Tazima (1941, 1944) showed the near monopoly power of W-chromosome in determining femaleness and proved that the Z-chromosome has no role in maleness with ZZW in triploid and with ZZZW in tetraploid chromosome constitution. While many genes accountable for morphological characters have been mapped on the Z-chromosome, no such gene was found on W-chromosome. Irradiation of silkworm with X-rays caused translocation from second chromosome to one end of the W-chromosome harboring two genes (+P and pSa) for larval characters. The W-chromosome carrying translocated dominant gene for larval marking was present; the individuals were always females with larval markings. This discovery not only determined the independent segregation of Z- and W-chromosomes but also led to the idea of developing sex-limited breeds. All the larvae without marking turned out to be males allowing easy determination of sex at larval stage. As the larval markings were confined to one sex, they were termed as sex-limited breeds. But the females containing the translocated W-chromosome were weak due to the hyperploidy caused by the translocated part (Nagaraju 1996). Therefore, Tazima (1954) through a series of irradiation experiments was able to translocate the region responsible only for the larval marking (+P) eliminating the superfluous part of the second chromosome on to the W-chromosome. Further improving the survival of the females through selection and breeding procedures makes sex-limited breeds a reality in Japan. Similarly, Hashimoto (1948) developed a sex-limited breed with zebra marking on female larvae. Tazima et al. (1951) developed sex-limited black egg color females by translocation of dominant allele from tenth to W-chromosome.

In 1971, Kimura et al. were successful in developing a new silkworm breed with females spinning yellow color cocoons and males with white color cocoons by irradiating the γ-rays with a dose rate of 18,000R per hour for 20 min with a dose of 6000R. The silkworm strain was the result of translocation of yellow blood genes on to the W-chromosome of the female pupae of yellow color race. The translocated Y genes coexist with the genes responsible for cocoon color (C) on the 12th chromosome, thus producing yellow blood and yellow cocoons. In this case, a very large portion on the second chromosome was translocated to W-chromosome. Therefore, it was impossible to remove the nonessential chromosome fragment in such a large volume resulting in physiological instability of the silkworm strain. The females from these cocoons have inferior egg producing ability, weaker larvae, and low cocoon quality compared to normal races. Sometimes, it is also observed the female cocoon and shell weight are lesser when compared to the male, which is not in line with the silkworm breeding principles. Many attempts made by silkworm breeders' are successful in breeding new sex-limited breeds that performed on par with their normal breeds. Yamamoto (1989) developed a sex-limited double hybrid [(NSY3 × NSY7) × (CSY6 × CSY7)] by utilizing four sex-limited breeds that excelled in most of the commercial traits than the control hybrid. N131 × C131,

N136 × C131, N140 × C140, N142.N134 × C142.C143, N140 × C145, and N147 × C145 (Mano et al. 1969; Mano 1984; Mano et al. 1991) are the other silkworm sex-limited hybrids authorized for commercial utilization in Japan.

Sex-Limited Strains in India and Their Importance

In India, sex-limited strains are developed by introducing the sex-limited character in the desirable silkworm strain. Usually, sex-limited female will be crossed with the recurrent male with desired characters; the F_1 sex-limited female is backcrossed for up to four generations with the recurrent parent to obtain the silkworm strain with a fixed sex-limited character (Fig. 2). A Russian sex-limited strain, Saniish 18, was utilized for the development of sex-limited strains J112 and C110 (Sengupta 1968). Nistari (SL) and AP1 (SL) were developed at the Central Sericultural Research and Training Institutes (CSRTIs) in Berhampore and Mysuru, India. Karnataka State Sericulture Research and Development Institute, Bengaluru, India, has also developed a sex-limited larvae marking hybrid HDO × HND. Later the trend was more biased toward developing cocoon color sex-limited breeds that are popular in the field. Sex-limited Pure Mysore race was developed by using AP1 (SL) sex-limited multivoltine race. At CSRTI, Mysuru had picked up with several CSR races with sex-limited characters like CSR3 (SL), CSR8 (SL), CSR12 (SL), CSR18 (SL), and CSR19 (SL). Among them, CSR8 (SL) was the most popular with cocoon color as sex-limited which spins dumbbell-shaped cocoons. The performance of CSR8 (SL) was on par with the NB4D2 than the male component of ruling crossbreed (PM × NB4D2). Therefore, a new silkworm breed, CSR2 (SL) with

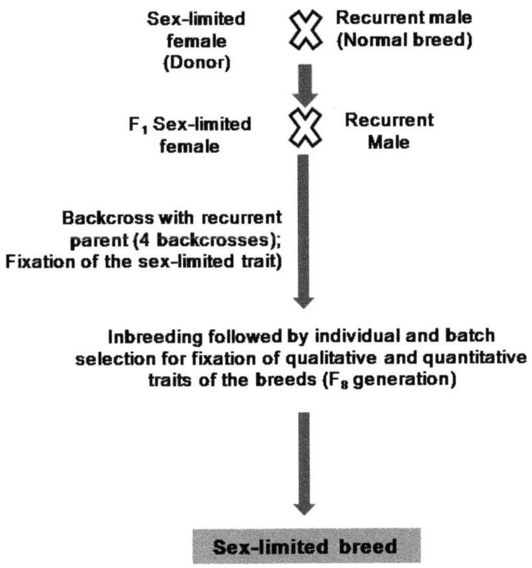

Fig. 2 Breeding plan for development of sex-limited breeds

sex-limited character for cocoon color, that spun oval-shaped cocoons was developed. Recently, PM × CSR2(SL) is on par with PM × CSR2 and superior to PM × CSR8 and PM × NB4D2 in terms of yield and reeling characters. The egg number is slightly less in CSR2 (SL) (450–475) when compared to CSR2 (Normal). Similarly, in place of normal CSR2 as a male constituent with PM, the new breed CSR2 (SL) with sex-limited character for cocoon color was developed to facilitate sex separation based on cocoon color at grainages (Basavaraja et al. 2005). SK6 (SL)Y, a popular bivoltine silkworm race with sex-limited character for cocoon color was developed for West Bengal conditions (Fig. 3).

India produces raw silk of 27654 MT (2022–2023) of which 70% is contributed by crossbreed obtained by crossing indigenous multivoltine females with productive bivoltine males. In order to achieve it, separation of males and females needs to be done efficiently; otherwise, self-pairing of the parent will take place losing in hybrid vigor. Therefore, separation of sex during cocoon stage is very important. It involves cutting of cocoons and manually separating male and female pupa, which is laborious and expensive, to meet the requirement of the dfls. In this direction, introduction of sex-limited breeds in Indian sericulture can have a great economic impact. Utilizing of such breeds in the hybrid dfls production can accelerate the process of sex separation without cutting the cocoon shell saving the labor cost and time; besides, the uncut cocoons of bivoltine females and multivoltine males can be sold for reeling to achieve higher returns. It should be noted that the economic benefits of the sex-limited breed can be harvested only when the sex-limited breed performs on par with the normal silkworm breed.

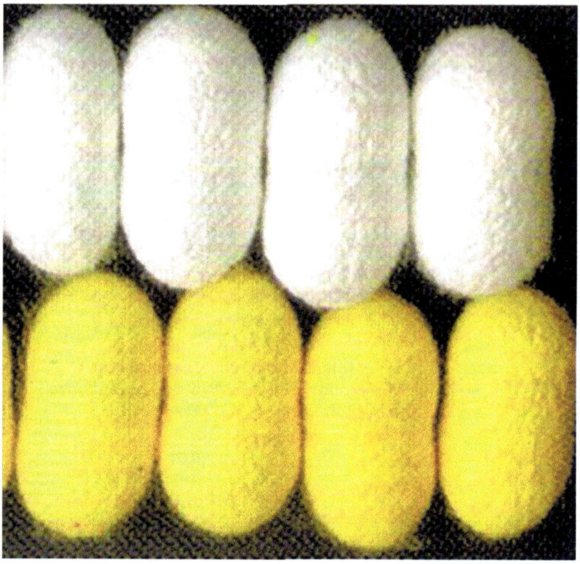

Fig. 3 SK6 (SL)Y- A popular bivoltine silkworm race of West Bengal with sex-limited character for cocoon color

Genome Editing Tools

Owing to the simplicity and efficiency in using genome editing tools, their application has rapidly shifted in economically important organisms like *B. mori* (Ma et al. 2017). The development of zinc finger nuclease (ZFN) in 2002 marked the emergence of gene editing technology, which has since made significant strides. *B. mori* has fortunately become the first choice for the researchers for genome editing experiments due to its economic importance. Takasu et al. (2010) targeted two *Bombyx* genes (BmBLOS2 and Bmwh3) responsible for urate granule formation in the larval epidermis (Quan et al. 2002; Komoto et al. 2009; Fujii et al. 2010). Loss of the urate granules in the epidermis leads to the visible "oily" phenotype with translucent skin, and disruption of BmBLOS2 and Bmwh3 genes results in an oily phenotype on the skin in G0 animals. This result suggests that interruption of the targeted gene can be achieved in *B. mori* using ZFNs, but very low mutation rate was observed when compared to the *Drosophila*. Therefore, ZFN-mediated gene knockout experiments in *B. mori* still require an immense effort, especially when it comes to genes with unknown phenotypes (Daimon et al. 2014). In 2010, transcription activator-like effector nuclease (TALEN) was developed as the second-generation gene scissor, a more efficient approach for genome editing (Christian et al. 2010; Joung and Sander 2013). Ma et al. (2012) showed that large chromosomal deletions can be induced by the simultaneous expression of two pairs of TALENs in entire organism (Joung and Sander 2013). Nakade et al. (2014) developed an innovative strategy for gene knock-in using TALEN and CRISPR/Cas9 and microhomology-mediated end joining, which they termed the TAL-PITCh (precise integration into target chromosome) system and CRIS-PITCh, respectively. The PITCh strategies have been shown to be efficient for various applications in cultured cells and different organisms (Ma et al. 2019). In addition, Park et al. (2019) and Wei et al. (2014) have demonstrated the molecular breeding potential using CRISPR/Cas9 systems for the white egg 2 gene (w-2 gene) and heritable genome editing with CRISPR/Cas9 in *B. mori*.

Conclusion

New silkworm breeds developed through mutation breeding by using suitable mutagens should possess the normal rate of fecundity and fertility, to overcome the negative effect on the commercial traits in silkworm. Targeted gene editing approaches have now emerged as an easy and standard technique in *B. mori*. However, the efficiency of gene knock-in is still very low in *Bombyx*. Therefore, in the future, efficient gene knock-in systems should be developed to achieve success in sophisticated genome editing. In India, only the sex-restricted mutants of mutant silkworms have been able to pass through the approval channel for commercial use before perishing. Therefore, when breeding silkworm mutants, breeder needs to focus on other commercially important traits in addition to the targeted trait. Any silkworm hybrid developed through mutation breeding should be tested for commercial viability. Further research on mutation breeding and available silkworm mutants is very much essential for overall development of sericulture industry.

References

Aparupa B, Parishmiti S, Nanita B, Priyanka C (2022) Mutation and its significance in sericulture industry: a review. J Exp Zool India 25(2):1597–1600

Basavaraja HK, Aswath SK, Suresh Kumar N, Mal Reddy N, Kalpana GV (2005) Silkworm breeding and genetics published by Central Silk Board. Bangalore, Karnataka, India

Chaturvedi (1981) The chemical and physical mutations. In: The text book of genetics. IBH Publication, New Delhi

Christian M, Cermak T, Doyle EL, Schmidt C, Zhang F, Hummel A, Bogdanove AJ, Voytas DF (2010) Targeting DNA double-strand breaks with TAL effector nucleases. Genetics 186:757–761

Daimon T, Kiuchi T, Takasu Y (2014) Recent progress in genome engineering techniques in the silkworm, *Bombyx mori*. Dev Growth Differ 56:14–25

Datta RK, Sengupta K, Das SK (1978) Sensitivity of male germ cells in silkworm to EMS. Mutation Res 51:199–304

Fujii H (1998) Genetical stocks and mutations of *Bombyx mori*: important genetic resources. Kyushu University, Fukuoka Japan, p 54

Fujii T, Daimon T, Uchino K, Banno Y, Katsuma S, Sezutsu H, Tamura T, Shimada T (2010) Transgenic analysis of the BmBLOS2 gene that governs the translucency of the larval integument of the silkworm *Bombyx mori*. Insect Mol Biol 19:659–667

Fujiwara H, Maekawa H (1994) RFLP analysis of chromosomal fragments in genetic mosaic strains of *Bombyx mori*. Chromosoma 103:468–474

Goldsmith MR (1995) Genetics of the silkworm: revisiting an ancient model system. In: Goldsmith MR, Wilkins AS (eds) Molecular model systems in the Lepidoptera. Cambridge University Press, New York, pp 21–76

Hashimoto H (1948) Sex-limited Zebra, an X-ray mutation the silkworm, *Bombyx mori*. Bull Sericult Exp Station Jpn 16:62–64

Joung JK, Sander JD (2013) TALENs: a widely applicable technology for targeted genome editing. Nat Rev Mol Cell Biol 14:49–55

Kawaguchi E (1928) Zytologische unterscuhungen an Seiden spinner and seinen Verwanten I. Gametogenesis von *Bombyx mori* L. and Bombyx mandarina M Land ihrer bastarde. *Z Zelforsch* 7:519–552

Kawaguchi E (1936) Der Eingluss der Eierbenhandlung mit Zebtrifugierung auf die vererbung bei dem seiden spinner. I. Uber experimentalle Auslosung der polyploiden Mutation. J Fac Ar Hokkaido Imp Uni 38:111–133

Kimura K, Harada C, Akai H (1971) Studies on "W" chromosome translocation in yellow blood gene in silkworm. Jpn J Breed 21:199–203

Komoto N, Quan GX, Sezutsu H, Tamura T (2009) A single-base deletion in an ABC transporter gene causes white eyes, white eggs, and translucent larval skin in the silkworm w-3(oe) mutant. Insect Biochem Mol Biol 39:152–156

Lewis KR, John B (1966) The meiotic consequence of spontaneous chromosome breakage. Chromosoma 18:287

Lokesh G, Ananthanarayana SR (2008) Mutagenic effect of Diethyl Sulphate (DES) on the Chromosomes of Silkworm, *Bombyx mori* L (Lepidoptera: Bombycidae). J Appl Sci Environ Manage 12(3):45–50

Lu C, Li B, Zhao A, Xiang Z (2003) Construction of AFLP molecular linkage map and localization of green cocoon and economically important traits QTL in silkworm (*Bombyx mori*). In: Collected papers of key sericultural laboratory of agricultural ministry, 2002–2003. Southwest Agricultural University, Chongqing, China, pp 159–171

Ma SY, Zhang SL, Wang F, Liu Y, Liu YY, Xu HF et al (2012) Highly efficient and specific genome editing in silkworm using custom TALENs. PLoS ONE 7:e45035

Ma SY, Liu Y, Liu YY, Chang JS, Zhang T, Wang XG et al (2017) An integrated CRISPR Bombyx mori genome editing system with improved efficiency and expanded target sites. Insect Biochem Mol Biol 83:13–20
Ma S-Y, Smagghe G, Xia Q-Y (2019) Genome editing in *Bombyx mori*: New opportunities for silkworm functional genomics and the sericulture industry. Insect Sci 26:964–972
Mano Y (1984) Studies on the breeding of autosexing silkworm races in Japan. Sericologia 24(3):384–392
Mano Y, Nagasawa K, Yamamoto I (1969) on the breeding of auto sexing silkworm varieties, N131 x C131. Bull Sericult Exp Station 23(5):1–22
Mano Y, Ohyanagi M, Nagayasu K, Murakami A (1991) Breeding of sex-limited larval marking silkworm race, N147 x C145. Bull Sericult Exp Station 2:1–29
Nagaraju J (1996) Sex determination and sex-limited traits in the silkworm, *Bombyx mori*: Their application in sericulture. Indian J Sericult 35(2):83–89
Nakade S, Tsubota T, Sakane Y, Kume S, Sakamoto N, Obara M et al (2014) Microhomology-mediated end-joining-dependent integration of donor DNA in cells and animals using TALENs and CRISPR/Cas9. Nat Commun 5:5560
Narayanaswamy KC, Govindan R, Narayanaswamy TK (1990) Physical mutagens and their effects on silkworms. Indian Silk 1990:39–41
Park J-W, Jeong Hee Y, Kim S-B, Kim S-W, Kim S-R, Choi K-H, Kim JG, Kim KY (2019) Analysis of silkworm molecular breeding potential using CRISPR/Cas9 systems for white egg 2 gene. Int J Ind Entomol 39(1):14–21
Quan GX, Kanda T, Tamura T (2002) Induction of the white egg 3 mutant phenotype by injection of the double-stranded RNA of the silkworm white gene. Insect Mol Biol 11:217–222
Sengupta K (1968) Sex-limited characters in two races of B. mori. Indian J Seric 1:79–86
Sinha RK, Srivastava PK, Sinha SP (1993) Effect of mutagens on the structure and movement of chromosomes in the primary spermatocytes of silkworm *Bombyx mori* L. National Conference on Mulberry Sericulture Research, Bhagalpur, India
Sridhar Rao PN (2006) Bacterial genetics. http://www.microrao.com
Takasu Y, Kobayashi I, Beumer K, Uchino K, Sezutsu H, Sajwan S, Carroll D, Tamura T, Zurovec M (2010) Targeted mutagenesis in the silkworm *Bombyx mori* using zinc finger nuclease mRNA injection. Insect Biochem Mol Biol 40:759–765
Tanaka Y (1916) Genetic studies in the silkworm. J Coll Assoc Sapporo 7:129–155
Tanaka Y (1953) Genetics of the silkworm *Bombyx mori* L. Adv Genet 5:239–301
Tazima Y (1941) A simple method of sex discrimination by means of larval markings in *Bombyx mori*. J Sericult Sci Jpn 12:184–188
Tazima Y (1944) Studies on chromosome aberrations in the silkworm, II Translocation involving second and W-chromosome. Bull Sericult Exp Station Jpn 12:109–181
Tazima Y (1954) Mechanism of sex-determination in the silkworm, *Bombyx mori*. Proc Int Congr Genet Cytol (Supplementary) 6:858–960
Tazima Y (1964) The Genetics of the Silkworm. Logos Press, London/Englewood Cliffs, p 253
Tazima Y (1978) Radiation mutagenesis of the silkworm. In: Tazima Y (ed) The silkworm: an important laboratory tool. Kodansha Ltd., Tokyo, pp 213–245
Tazima Y, Onimaru K (1968) Mutagenic action of Mytimyci-C and EMS on pre-meiotic cells of male silkworms. Ann Rep Natt Inst Genet (Japan) 19:62–63
Tazima Y, Ohta N, Harada C (1951) On the discrimination method by colouring gene of silkworm eggs. Inductions of translocation between the W and the 19th chromosome. Jpn J Breed 1:47–50
Wei W, Xin HH, Roy B, Dai JB, Miao YG, Gao GJ (2014) Heritable Genome Editing with CRISPR/Cas9 in the Silkworm. Bombyx mori PLoS ONE 9:e101210
Yamamoto T (1989) Breeding of sex-limited yellow cocoon races of silkworms by chromosome manipulation. Farm Jpn 23(5):42–48

Application of Marker-Assisted Selection in Silkworm Breeding for Disease Resistance

L. Satish, L. Kusuma, Raviraj V Suresh, S. M. Moorthy, and V. Sivaprasad

Abstract

The development of pathogens that cause silkworm diseases and their outbreak is clearly visible as a result of climate change, particularly in tropical nations like India, where environmental humidity and temperature are frequently favorable for pathogen survival and infection, causing significant financial losses to the silk industry. The main factor in increasing the sustainability of sericulture is illness prevention in silkworms. Use of silkworm breeds or hybrids that are not disease-

L. Satish
Silkworm Breeding and Molecular Biology Laboratory II, Central Sericultural Research and Training Institute, Central Silk Board, Ministry of Textiles, Government of India, Mysuru, Karnataka, India

Central Sericultural Research and Training Institute, Central Silk Board, Ministry of Textiles, Government of India, Mysuru, Karnataka, India

Central Sericultural Research and Training Institute, Central Silk Board, Mysuru, Karnataka, India

Central Silk Board, Berhampore, West Bengal, India

L. Kusuma (✉)
Silkworm Breeding and Molecular Biology Laboratory II, Central Sericultural Research and Training Institute, Central Silk Board, Ministry of Textiles, Government of India, Mysuru, Karnataka, India

Central Sericultural Research and Training Institute, Central Silk Board, Ministry of Textiles, Government of India, Mysuru, Karnataka, India

R. V Suresh
Central Sericultural Research and Training Institute, Central Silk Board, Ministry of Textiles, Government of India, Mysuru, Karnataka, India

R. V Suresh et al. (eds.), *Biotechnology for Silkworm Crop Enhancement*,
https://doi.org/10.1007/978-981-97-5061-0_8

resistant or tolerant, as well as insufficient disinfection, are the main causes of the occurrence of diseases that cause partial or entire crop loss. Because of the enhanced survival rate, using disease-tolerant/resistant cultivars is the most obvious way to increase productivity in any crop system. The ability to isolate and use genetic resources that are disease-resistant or tolerant from the rich and diverse germplasm collections is made possible by advances in molecular biology. Recent biotechnological technologies have been effectively used to pinpoint several disease-tolerant genes and QTLs. Through marker-assisted selection (MAS), which can hasten the formation of tolerant/resistant silkworm hybrid in the fewest feasible generations with the highest degree of precision, it is possible to pyramid putative tolerance genes into a single genotype.

Keywords

Silkworm · Disease · Marker-assisted selection · Markers · Resistant · Tolerant · Susceptible · Genes · Genotype

Silkworm and Its Genome

Bombyx mori, the silkworm, is a crucial economic insect for the production of silk and is known as the "queen of fabrics." *Bombyx mori* was domesticated from its wild parent, *B. mandarina*, about 5000 years ago. The reference silkworm genome (450 Mb) was first made available in 2004 (Mita et al. 2004). This draft sequence of the silkworm *Bombyx mori* was established by whole-genome shotgun (WGS) sequencing and assembled into 49,345 scaffolds that span a total length of 514 mb, including gaps and 387 mb without gaps. Since the silkworm's genome is thought to be 530 megabytes in size, over 97% of it has been structured into scaffolds, of which 75% have been sequenced. Many websites, such as SilkDB 2017 (Silkworm Knowledgebase from China) (http://silkworm.genomics.org.cn), SilkSatDB

S. M. Moorthy
Silkworm Breeding and Molecular Biology Laboratory II, Central Sericultural Research and Training Institute, Central Silk Board, Ministry of Textiles, Government of India, Mysuru, Karnataka, India

Central Sericultural Research and Training Institute, Central Silk Board, Ministry of Textiles, Government of India, Mysuru, Karnataka, India

Central Silk Board, Bengaluru, Karnataka, India

V. Sivaprasad
Silkworm Breeding and Molecular Biology Laboratory II, Central Sericultural Research and Training Institute, Central Silk Board, Ministry of Textiles, Government of India, Mysuru, Karnataka, India

Central Sericultural Research and Training Institute, Central Silk Board, Ministry of Textiles, Government of India, Mysuru, Karnataka, India

Central Silk Board, Mysuru, Karnataka, India

Table 1 Genome composition of silkworm, *Bombyx mori* L.

Haploid genome size	~530 mb
Linkage groups	28
Linkage map distance	1000 cM
Number of cloned genes	>20
Silk gland genes	1874
Genetic markers	1018
Average interval between markers	~2 cM (=~500 kb)
No. of RFLP markers	61
No. of RAPD markers	168
Independent cDNA BAC libraries	>11,000
cDNA libraries	209
EST	64,038
Scaffolds	23,156
Contigs	66,482
SSRs	601,225
Genes predicted by BGI gene finder	21,302
Homologous genes from other lepidopterans	521

(SilkSatDB 2017) (a microsatellite database of silkworm from CDFD, India) (www.cdfd.org.in/silksatdb), and SilkBase (SilkBase 2017) (EST database and BAC library from Japan) (www.ab.a.u-tokyo.ac.jp/silkbase) (Mita et al. 2004), provide updated information about genome sequence assembly, cDNAs, ESTs, SNPs, and functional annotations of genes of silkworm (Table 1).

Silkworm Diseases and Pathogen Evolution

With the changing climate and incorrect temperature and humidity control in the rearing house, diseases can grow more easily under conditions of high temperature and high humidity (28 to 30 °C, 85 to 95%). In addition, since silkworm feces and litter provide a favorable environment for the growth of pathogens that cause disease in the insect, control measures, such as sanitary management, will always be crucial. The development of new silkworm diseases has made them resistant to the effects of bed disinfectants, which are utilized in sericulture, adding to the issue.

The many forms of silkworm diseases (Fig. 1) have various symptoms. Infectious and noninfectious silkworm disorders can be generically categorized as silkworm diseases. Based on the agents that cause them, the infectious diseases of silkworms can also be divided into four groups: diseases caused by viruses, bacteria, fungi, and protozoa. The viruses that cause viral illness include nucleopolyhedrosis, cytoplasmic polyhedrosis, viral flacherie, and densonucleosis. Bacteria are the cause of bacterial intestinal disease, Sotto disease, and septicemia. Fungus is the root cause of aspergillosis, white muscardine, green muscardine, yellow muscardine, gray

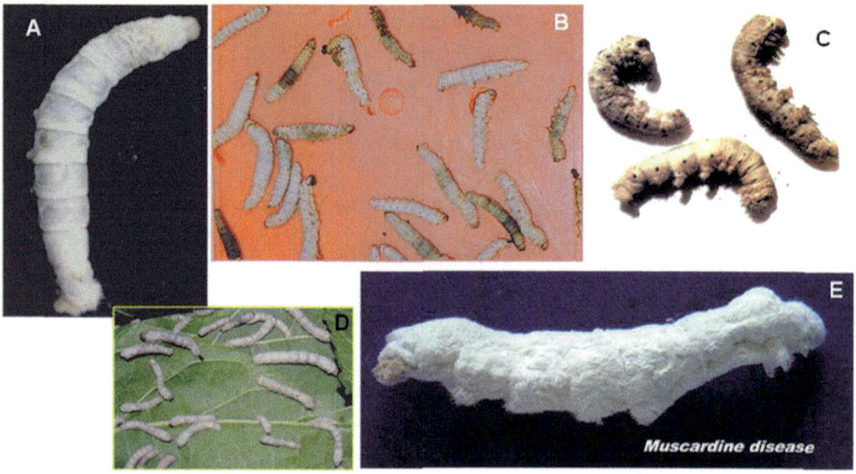

Fig. 1 Diseases of silkworm (A. Grasserie B. Flacherie C. Septicemia D. Pebrine E. Muscardine)

muscardine, and black muscardine. Pebrine is an illness caused by protozoa. Arthropods and poisonous substances are the main causes of noninfectious silkworm illnesses.

A great example of silkworm fungal illnesses can be used to show the evolutionary aspect of silkworm pathogens. Muscardine illness is caused by several fungi; however, the name varies depending on the color of the conidiospores those fungi generate and deposit on silkworm bodies. The skin is the primary route of skin-transmitted muscardine infections. However, it is also noted that oral transmission through spiracles and through consumption of contaminated leaves is also possible for muscardine infection. When fungal spores land on a silkworm's body surface, they immediately germinate and slowly work their way through the cuticle via enzymatic and mechanical forces. The host is eventually killed as the hyphae feed on its tissue and continue to develop. White muscardine, which is caused by the widespread entomopathogenic fungus *Beauveria bassiana*, is one of the most prevalent silkworm disorders. This is a major issue in areas where silkworms are produced all over the world, particularly in India's tropical rearing beds, where environmental humidity and temperature frequently favor *B. bassiana* survival and infection and sporadically favor epizootic outbreaks, causing significant financial losses to the silk industry (James and Li 2012). More than ten species of *Aspergillus*, i.e., *A. flavus, A. oryzae, A. taliutrii, A. ochraceus, A. soicte, A. oryzae* var. *fulvus, A. flavipes, A. terreus, A. melleus, A. clavatus, A. fumigatus, A. nidulans, A. elegans, A. parasiticus,* etc., have been reported to be parasitic to silkworm. Fifteen pathogenic strains of *A. flavus* of silkworm *(Bombyx mori)* produced aflatoxins in vitro, and aflatoxins have been extracted from infected larvae, though it is not clear whether in sufficient quantities to be the cause of death (Murakoshi et al. 1977).

Sericulture, where farmers produce silk, is a major source of subsistence income. Like other commercially significant insects, silkworms are susceptible to a wide range of pathogen-caused illnesses. The sericulture sector has suffered enormous financial losses as a result of these infections. Among them, *Nosema bombycis*, a microsporidian, causes pebrine disease in silkworms. This disease is spread not only horizontally through the silkworm larva, pupa, and moth but also vertically through the silkworm eggs to the following generation, negatively affecting the quantity and quality of silk produced and causing significant losses to sericulture.

Markers in Silkworm, *Bombyx mori*

The domesticated silkworm, *Bombyx mori*, is one of the most extensively researched insects and serves as a molecular model for lepidopterans. Geographic races, genetically enhanced silkworm strains with qualitative and quantitative features, and more than 400 identified mutations have been placed on linkage maps (Doira et al. 1992). According to DNA marker techniques used in insect studies, mtDNA, microsatellites or simple sequence repeats (SSR), random amplified polymorphic DNA (RAPD), inter simple sequence repeats (ISSR), expressed sequence tag (EST), amplified fragment length polymorphism (AFLP), restriction fragment length polymorphism (RFLP), microarray, and single nucleotide polymorphism (SNP) have all made significant contributions to understanding the genetic basis and in the evolution.

Numerous academics have embraced the silkworm genomic analysis programs on discovery of DNA markers for QTLs (Reddy et al. 1999). The use of microsatellite markers in DNA profiling of silkworms has also been conducted, and the results have revealed breed-specific profiles for 15 silkworms, demonstrating the potential of microsatellite markers for the creation of molecular Iientities for differentiating silkworm breeds (Sreekumar et al. 2011). Thirteen silkworm strains have been used for genetic characterization using simple sequence repeats (SSR) and inter-SSR (ISSR) (Reddy et al. 1999).

Marker-Assisted Breeding/Selection for Development of Disease-Resistant Silkworm Breeds/Hybrids

Using contemporary breeding methods with biotechnological components like marker-assisted selection, crop yield has been successfully increased in a number of crops. Positional cloning and marker-assisted breeding both use genetic linkage maps, which are crucial for mapping desirable features. Numerous genetic markers, including restriction fragment length polymorphisms (RFLPs), random amplified polymorphic DNA (RAPD), amplified fragment length polymorphisms (AFLPs), simple sequence repeats (SSRs), and single nucleotide polymorphisms (SNPs), have been used to create some genetic maps for the silkworm.

In practically every region of India where there is sericulture, viral infections are common. The greatest method to stop infectious diseases from spreading is to use breeds of silkworms that are relatively resistant. The main viral diseases causing significant crop loss and impeding the development of bivoltine sericulture in India are *Bombyx mori* densonucleosis virus (BmDV), *Bombyx mori* nuclear polyhedrosis virus (BmNPV), and *Bombyx mori* infectious flacherie virus (BmIFV) (Sivaprasad et al. 2021). In this regard, numerous scientists in India have created strains of silkworms that are resistant to particular viral diseases. Sivaprasad et al. (2003a) evaluated 145 germplasms from APSSRDI, Hindupur, against BmNPV illnesses and classified the breeds' responses as apparent tolerance, real tolerance, and susceptibility against BmNPV; they discovered that 18 bivoltines and 16 polyvoltines were really tolerant to the disease (Nataraju et al. 1998). The relative vulnerability of several silkworm races to BmNPV was investigated, and it was discovered that bivoltine breeds were, on average, more vulnerable than multivoltine races. According to reports, no race of silkworm is totally immune to this disease, and the susceptibility of different silkworm races to various diseases varies. Pure *Mysore*, *Nistari*, and *C. nichi* are found to be resistant to BmNPV among the breeds, and genetic divergence has been detected in silkworm stocks with regard to toleration to BmNPV (Sivaprasad et al. 2003b). Efforts made for the development of silkworm races tolerant to NPV have resulted in the evolution of the bivoltine breed, DR-1 polyvoltine breed NP1, and productive bivoltine breeds, 2N, 5N, and 61N (Sen et al. 1999). Inheritance analysis has indicated the possibility of a major dominant gene conferring resistance to NPV (Sen et al. 2004). Sowmyashree and Nataraju (2008) administered 24 bivoltine and 23 multivoltine silkworm breeds with BmNPV (6×10^6 PIBs/ml) and the highest survival per cent of 58.80 noticed in CSR19 compared to CSR2 39.70% in CSR2 and 39.61% in CSR19. Two races—APDR15 × APDR115 (crossbreed) and APDR105 × APDR126 (bivoltine hybrid)—were created to be BmNPV-tolerant and BmDNV1-resistant. In India, the prevalence of infectious flacherie alone is 22.37%. Numerous strains of silkworm are genetically resistant to BmDV or BmBDV. BmNPVand BmIFV tolerance in silkworms is regulated by many genes. It is still entirely unknown which major dominant genes and minor effector genes are responsible for a disease's tolerance to BmNPV and BmIFV (Chen et al. 2003). The non-susceptibility to BmDV and BmBDV is controlled by nsd-1 and nsd-2 recessive gene, respectively (Kadono-okuda et al. 2014). Productive breeds have been developed by transferring resistant gene from donor parent to BmDV susceptible breeds. Awasthi et al. (2007) identified SSR marker associated with NPV tolerance in silkworm, and same was used in the development of NPV-tolerant breeds employing MAS. Selot et al. (2007) found that the Nistari strain's stomach fluid contained the 26.5 kDa protein (soluble NADPH oxidoreductase) known as BmNOX, which was antiviral to BmNPV. NPV-tolerant breeds are being developed utilizing MAS and this protein marker. A marker-assisted selection (MAS) software was utilized to screen the resistant strains in segregating populations using SSR markers to select silkworms that were resistant to DNV-Z. Awasthi et al. (2007) have proven that the PCR approach may be utilized in breeding programs to determine if a silkworm strain has BmDV or not. The development of transgenic silkworms that can block multiple BmNPV critical genes resulted in the development of high-yielding cocoons, silk characteristics, and impaired occlusion body infectivity in transgenic lines (Subbaiah et al. 2013).

India is a tropical country, and the conditions favor the growth of viruses; it is highly expected that the silkworm can be infected by multi-virus.

The silkworm is prone to infectious illnesses brought on by many pathogens. Their susceptibility to certain diseases varies among different silkworm strains. BmIFV and BmDV together cause flacherie to a 43.05% level. Interstrain variations are inherited (Watanabe 1994). A new Chinese silkworm strain resistant to BmNPV and BmDNV-Z was developed by virus administration and molecular-assisted breeding technology. Extensive studies on inheritance and resistance to densonucleosis have been conducted by Seki (1984), Seki and Iwashita (1983), and Kobayashi et al. (1986). Resistance of silkworm to nuclear and cytoplasmic polyhedrosis virus is controlled by polygenes (Watanabe 1986). Breeding of silkworm breeds tolerant to BmNPV is possible by selection of survivors after viral exposure (Aratake 1973; Watanabe 1994). Aizawa (1991) has studied the defense reactions of the silkworm *Bombyx mori* against nuclear polyhedrosis. The susceptibility of Indian silkworm breeds to nuclear polyhedrosis virus has also been studied by Furata (1995). RAPD markers linked to the gene controlling non-susceptibility to BmDV virus have been mapped by Abe et al. (1998). Hara et al. (2008) developed a race "Taisei" which is non-susceptible to a densonucleosis virus type I. Recently, Fan et al. (2012) developed a new strain resistant to two virus, BmNPV and BmDNV-Z, which was obtained through virus administration and molecular marker breeding method.

Fifteen to twenty percent of the farmers' silkworm crops in India are lost due to viral infections. Although earlier efforts to generate silkworm breeds/hybrids exhibiting tolerance to all three viruses (BmDV, BmIFV, and BmNPV) were made, a number of silkworm breeds that are tolerant/resistant to either of the viruses (BmDV, BmIFV, and BmNPV) have already been developed. In this study, oral inoculations of certain viruses (BmNPV, BmIFV, and BmDV) were tested on 120 silkworm breeds from various geographic origins (tropical, subtropical, and temperate) throughout five generations. Populations that expressed a particular virus' tolerance were kept alive. The BmDV-tolerant broods were utilized to cross with specific virus-tolerant populations of BmIFV and/or BmNPV to pyramid the tolerant genes since inheritance of BmDV resistance is mediated by a single recessive gene (nsd1). SSR markers from Miao et al. (2005) were used to screen for the presence of tolerant genes. The 25 SSR markers with protein-coding sequences that were relevant to antibacterial activity were chosen. Out of the 25 SSR markers, 8 SSRs demonstrated PCR amplification and were used in the study to choose silkworm breeds that were resistant to all three viruses. The biological activity of each of the 8 SSR markers linked with virus tolerance was relevant to antibacterial action (Satish et al. 2022) (Fig. 2). The isocitrate dehydrogenase SSR marker was present in both parental strains, with the exception of HBM10N male, which was reported for the overexpression of IDH protein in *B. mori* that may be necessary to suppress BmNPV infection (Wang et al. 2017). *B. mori* serine lipase-1 extracted from the digestive juice of larval midgut has been reported to exhibit strong antiviral activity in vitro (Ponnuvel et al. 2003). It is possible for BmNPV to acquire the protein tyrosine phosphatase gene from the host in *B. mori*, and this gene is thought to play a significant role in the organism's heightened locomotor activity during BmNPV infection (Kamita

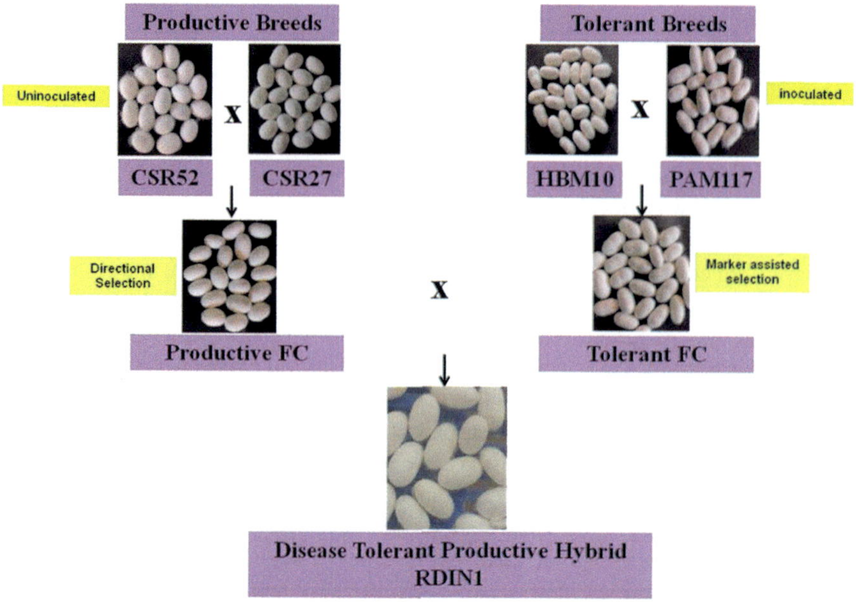

Fig. 2 Representation of the development of multi-viraltolerant bivoltime silkworm double hybrid (RDINI) Multi-viral-tolerant parental breeds HBM10 and PAM117 were inoculated and productive susceptible parental breeds, CSR52 and CSR27 were uninoclated. Productive foundation cross (FC) was established by directional selection and multiviral - tolerant foundation cross (FC) was established by markerassisted selected. Productive FC females were crossed with tolerant FC males to obtain RDINI

et al. 2004). LIP283, PTP284, and GDH306 amplified in all the silk moths, except for PAM117N females and HBM10N males. Attacin (ATT) SSR marker showed amplification in both the virus-tolerant parental strains and hybrids. Attacin is a well known antimicrobial peptide involved in the *B. mori* defense system (Nesa et al. 2020). The ankyrin repeats in the BmRelish gene are thought to play a role in the activation of antimicrobial peptides *in B. mori. Escherichia coli* was unable to activate antimicrobial peptide genes in transgenic silkworms when the BmRelish gene was knocked down (Kaneko et al. 2007), indicating that BmRelish is crucial for the expression of antimicrobial peptide genes ankyrin (ANK165). More BmTKs were identified by GO analysis to play roles in binding, catalysis, signal transduction, metabolism, biological regulation, and stimulus response; however, functional activities of BmTKs must still be investigated to determine the precise function of TKs in *B. mori* against viral infections (Kang et al. 2011). Glucose dehydrogenase expressed in wild and domesticated *B. mori* has oxidoreductase activity, which is involved in antioxidant activity. Alkaline tyrosine kinase (ATK285) (He et al. 2018) and dipeptidyl peptidase (DPP150), which are also reported to be expressed in wild and domesticated silkworms, have some relevance in silk protein synthesis, but its exact role in antiviral activity is not known so far (Fang et al. 2015). In this study, the genes, which are used as markers for viral resistance/tolerance, are associated with antiviral or antimicrobial activity. The presence of markers reflects the higher antimicrobial activity, thus conferring virus tolerance.

References

Abe H, Harada T, Kanehara M et al (1998) Genetic mapping of RAPD Markers linked to the Densonucleosis refractoriness gene, nsd-1 in the silkworm, *Bombyx mori*. Genes Genet Syst J 73:237–242. https://doi.org/10.1266/ggs.73.237

Aizawa K (1991) The nature of infections caused by nuclear polyhedrosis virus. In: Steinhaus EE (ed) Insect pathology, an advance treatise. Academic Press, New York, pp 381–412

Aratake Y (1973) Strain difference of the silkworm *Bombyx mori* L in the resistance to a nuclear polyhedrosis virus. J Seric Sci Jpn 42(3):230–238

Awasthi AK, Pradeep AR, Srivatava PP, Vijayan K, Kumar V, Urs SR (2007) PCR detection of densonucleosis virus isolates in silkworm (*Bombyx mori*) from India and their nucleotide variability. Indian J Biotechnol 7P:55–60

Chen K, Yao Q, Wang Y et al (2003) Genetic Basis of screening of molecular markers for nuclear Polyhedrosis virus Resistance in Bombyx mori L. Int J Ind Entomol 7(1):5–10

Doira H, Fujii H, Kawaguchi Y, Kihara H, Banno Y. (1992) Genetic stock and mutation of *Bombyx mori*. Institute of Genetic Resources, Kyushu University

Fan F, Haifeng S, Jingang F et al (2012) Double molecular markers-assisted breeding of silkworm, Bombyx Mori, resistant to the virus, BmNPV and BmDNV-Z. Indian J Animal Res 46

Fang SM, Bi-Li H, Zhou QZ et al (2015) Comparative analysis of the silk gland transcriptomes between the domestic and wild silkworms. BMC Genomics 16(1):60. https://doi.org/10.1186/s12864-015-1287-9

Furata Y (1995) Susceptibility status of the races of the silkworm *Bombyx mori* L. preserving in NISES to the NPV and Densonucleosis virus. Bull NISES 15:119–145

Hara W, Ann Y, Eguchi R et al (2008) Mapping of a novel virus resistant gene, Nid-1, in the silkworm, *Bombyx mori* based on the restriction fragment length polymorphism (RFLIP). J Insect Biotechnol Sericol 77:59–66. https://doi.org/10.11416/jibs.77.1_59

He S, Tong X, Han M et al (2018) Genome-wide identification and characterization of tyrosine kinases in the silkworm, *Bombyx mori*. Int J Mol Sci Article 19:934. https://doi.org/10.3390/ijms19040934

James RR, Li ZZ (2012) From silkworms to bees: diseases of beneficial insects. In: Insect pathology. Academic Press, London, Waltham, San Diego

Kamita SG, Nagasaka K, Chua JW, Shimada T, Mita K, Kobayashi M, Maeda S, Hammock BD (2004) A baculovirus-encoded protein tyrosine phosphatase gene induces enhanced locomotory activity in a lepidopteran host. Proc Natl Acad Sci 102(7):2584–2589. https://doi.org/10.1073/pnas.040945710

Kaneko Y, Furukawa S, Tanaka H, Yamakawa M (2007) Expression of antimicrobial peptide genes encoding Enbocin and Gloverin isoforms in the Silkworm, *Bombyx mori*. Biosci Biotechnol Biochem 71(9):2233–2241. https://doi.org/10.1271/bbb.70212

Kang L, Shi H, Liu X et al (2011) Arginine kinase is highly expressed in a resistant strain of silkworm (Bombyx mori, Lepidoptera): Implication of its role in resistance to *Bombyx mori* nucleopolyhedrovirus. Elsevier Comp Biochem Physiol Part B 158:230–234. https://doi.org/10.1016/j.cbpb.2010.12.001

Kobayashi J, Edimura HE, Kobayashi N (1986) The effect of temperature on the diapauses eggs production in the tropical silkworm *Bombyx mori* L. J Seric Sci Jpn 55(4):345–348

Miao XX, Xu SJ, Li MH et al (2005) Simple sequence repeat-based consensus linkage map of Bombyx mori. Proc Natl Acad Sci USA 102(45):16303–16308. https://doi.org/10.1073/pnas.050779410

Mita K et al (2004) The genome sequence of silkworm, *Bombyx mori*. DNA Res 11:27–35

Murakoshi S, Ichinoe M, Kumata H, Kurata H (1977) Productivity of aflatoxins and some biological effects in *Aspergillus jlavus* link isolated from cadavers of the silkworm *Bombyx mori L.* Appl Entomol Zool 12:255–259

Nataraju B, Datta RK, Baig M, Balavenkatasubbaiah M, Samson MV, Sivaprasad V (1998) Studies on the prevalence of nuclear polyhedrosis in sericultural areas of Karnataka. Indian J Seric 37(2):154–158

Nesa J, Sadat A, Buccini DF, Kati A, Mandal AK, Franco OL (2020) Antimicrobial peptides from *Bombyx mori*: a splendid immune defense response in silkworms. RSC Adv 10:512–523. https://doi.org/10.1039/C9RA06864C. No 2.107–113

Okuda KK, Ito K, Murthy GN, Sivaprasad V, Ponnuvel KM (2014) Molecular mechanism of densovirus resistance in silkworm, *Bombyx mori*. Sericologia 54(1):1–10

Ponnuvel KM, Nakazawa H et al (2003) A Lipase Isolated from the Silkworm *Bombyx mori* shows antiviral activity against nucleopolyhedrovirus. J Virol 77(19):10725–10729. https://doi.org/10.1128/jvi.77.19.10725-10729.2003

Reddy KD, Nagaraju J, Abraham EG (1999) Genetic characterization of the silkworm *Bombyx mori* by simple sequence repeat (SSR)-anchored PCR. Heredity 83:681–687. https://doi.org/10.1046/j.1365-2540.1999.00607.x

Satish L, Kusuma L, Mary Josepha Shery AV et al (2022) Development of productive multiviral disease tolerant bivoltine silkworm breeds of *Bombyx mori* (Lepidoptera: Bombycidae). Springer Nat Appl Entomol Zool 58:61–71. https://doi.org/10.1007/s13355-022-00803-8

Seki H (1984) Mode of inheritance of the resistance to the infection with the densonucleosis virus (Yamanashi isolate) in the silkworm *Bombyx mori*. J Seric Sci Jpn 53:472–475

Seki H, Iwashita Y (1983) Histopathological features and pathogenicity of a densonucleosis virus of the silkworm *Bombyx mori* isolated from sericultural farms in Yamnashi prefecture. J Seric Sci Jpn 52:400–405

Selot R, Kumar V, Shukla S, Chandrakuntal K, Brahmaraju M, Dandin SB, Laloraya M, Kumar PG (2007) Identification of a soluble NADPH oxidoreductase (BmNOX) with antiviral activities in the gut juice of *Bombyx mori*. Biosci Biotechnol Biochem 71(1):200–205

Sen R, Ahsan MM, Datta RK (1999) Induction of resistance to *Bombyx mori* nuclear polyhedrosis virus into a susceptible bivoltine silkworm breed. Indian J Seric 38(2):107–112

Sen R, Nataraju B, Balavenkatasubbaiah M, Premalatha V, Thiagarajan V, Datta RK (2004) Resistance to BmNPV type 1 and its inheritance in silkworm *Bombyx mori* L. Int J Indust Entomol 9(1):35–40

Sivaprasad V, Chandrasekharaiah M, Ramesh C et al (2003a) Screening of silkworm breeds for tolerance to bombyx mori Nuclear polyhedro Virus (BmNPV). Int J Ind Entomol 7(1):87–91

Sivaprasad V, Chandrasekharaiah M, Ramesh C, Misra S, Kumar K, Rao Y (2003b) Screening of silkworm breeds for tolerance to *Bombyx mori* nuclear polyhedron virus (BmNPV). Int J Indust Entomol 2(2):123–127

Sivaprasad V, Rahul K, Makwana P (2021) Immunodiagnosis of silkworm diseases. Methods Microbiol 49:27–46. https://doi.org/10.1016/bs.mim.2021.04.00

Sowmyashree TS, Nataraju B (2008) Identification of silkworm breeds resistant to nuclear polyhedrosis through BmNPV inoculation and induction. Indian J Seric 46(1):32–37

Sreekumar S, Ashwath SK, Slathia M et al (2011) Detection of a single nucleotide polymorphism (SNP) DNA marker linked to cocoon traits in the mulberry silkworm, *Bombyx mori* (Lepidoptera: Bombycidae). Eur J Entomol 108:347–354. https://doi.org/10.14411/eje.2011.043

Subbaiah EV, Royer C et al (2013) Engineering silkworms for resistance to Baculovirus through multigene RNA interference. Genetics 193:63–75. https://doi.org/10.1534/genetics.112.144402

Wang XY, Yu HZ, Xu JP et al (2017) Comparative subcellular proteomics analysis of susceptible and near-isogenic resistant *Bombyx mori* (Lepidoptera) Larval Midgut response to BmNPV infection. Sci Rep 7. https://doi.org/10.1038/srep45690

Watanabe H (1986) Resistance to the silkworm, *Bombyx mori* to viral infection. Agric Ecosyst Environ 15:131–139

Watanabe H (1994) Densonucleosis of the silkworm, *Bombyx mori*. Indian J Seric 33(2):114–117

Engineered Disease Resistance Silkworm Using Genome Editing

Katsuhiko Ito, Pooja Makwana, and Kamidi Rahul

Abstract

Silk cocoons obtained from silkworms are the primary source of commercial silk, making silkworm an economically important insect. However, the silk industry has experienced significant losses owing to various viral infections. *Bombyx mori* bidensovirus (BmBDV) is one of the causative pathogens of flacherie disease in silkworms. Although most silkworm strains die after BmBDV infection, certain strains have shown resistance to the virus, which is conferred by a single recessive gene, *nsd-2*. The $+^{nsd\text{-}2}$gene (the allele of *nsd-2*, the susceptibility gene) encodes a protein containing a putative 12-pass transmembrane domain, expressed only in the silkworm's midgut, where BmBDV can infect. In contrast, the *nsd-2* gene (the resistance gene) encodes a truncated membrane protein that contains only the first three-pass transmembrane domain. These results suggest that the complete membrane protein may function as a receptor for BmBDV, with viral resistance caused by a defective membrane protein. To take advantage of this relationship between membrane protein structure and viral infectivity, we are currently attempting to artificially generate BmBDV-resistant silkworms using genome editing methods. Theoretically, the deletion of the $+^{nsd\text{-}2}$ gene by genome editing could convert all silkworm strains from virus susceptible to virus resistant.

K. Ito (✉)
Department of Science of Biological Production, Graduate School of Agriculture, Tokyo University of Agriculture and Technology, Fuchu, Tokyo, Japan
e-mail: katsuito@cc.tuat.ac.jp

P. Makwana · K. Rahul
Central Sericultural Research and Training Institute, Central Silk Board, Ministry of Textiles, Government of India, Berhampore, Murshidabad, West Bengal, India

R. V Suresh et al. (eds.), *Biotechnology for Silkworm Crop Enhancement*,
https://doi.org/10.1007/978-981-97-5061-0_9

Keywords

Silkworm · Sericulture industry · Bidensovirus · Virus resistance · Genome editing

Introduction

Bombyx mori, which has been bred in captivity for approximately 5000 years, is now a completely domesticated species of silkworm (Goldsmith et al. 2005). The silkworm is an economically important insect that makes silk cocoons. Therefore, sericulture is a valuable industry that spread from China through the Silk Road and has been promoted in various Asian and European countries. The most severe threat to the sericulture industry is the lack of control of silkworm diseases. Many pathogens, including viruses, bacteria, fungi, and microsporidia, have been found to infect silkworms, making it challenging to control the spread of these diseases. The economic damage caused by viruses including nucleopolyhedrovirus (NPV), cypovirus (CPV), infectious flacherie virus (IFV), and densovirus (DNV), is so significant that the disease control for sericulture industry is critical to keep the silk production. Many studies have attempted to elucidate the interaction between the silkworm and viruses; a variety of molecules and pathways involved in the immune responses of the silkworm to viruses, such as antimicrobial peptides, prophenoloxidase-activating system, apoptosis, reactive oxygen species, small RNA, and related molecules, have been identified (Lü et al. 2018).

Considering this background, we focused on the mechanisms of DNV infection and host responses to the virus, and we achieved the first successful isolation of the important host factors, the putative virus receptors, in the insect virus research field (Ito et al. 2008, 2018). In this manuscript, we would like to discuss the following items: (i) the mechanism of silkworm resistance against DNV, (ii) the identification of the resistance gene through positional cloning approach using *B. mori* genome information, and (iii) the generation of the virus resistant silkworms by genome editing methods.

Bombyx mori Densovirus

Bombyx mori densovirus (BmDNV) is a pathogen that causes flacherie disease in the silkworm. BmDNV was previously divided into two types, type-1 (BmDNV-1) and type-2 (BmDNV-2 and BmDNV-Z), based on differences in their virulence toward silkworms, serological characteristics, and genomic structures (Watanabe and Maeda 1978; Seki 1984; Watanabe et al. 1986) (Table 1). BmDNV-1 was classified into the *Iteravirus* genus, while BmDNV-2 and BmDNV-Z were classified into the *Bidensovirus* genus, in the *Densovirinae* subfamily or parvovirus-like viruses. Currently, BmDNV-1 has been reassigned to the *Iteradensovirus* genus and has been renamed as BmDV1 (Adams and Carstens 2012) (Table 1). In contrast, BmDNV-2 and BmDNV-Z were excluded from the *Parvoviridae* family because of

Table 1 Salient features of BmDV and BmBDV

Virus type	*Bombyx mori densovirus* (BmDV1)	*Bombyx mori bidensovirus* (BmBDV)
Previous classification	*Bombyx mori* densovirus type-1 (BmDNV-1)	*Bombyx mori* densovirus type-2 (BmDNV-2)
Isolates	Ina isolate (Japan)	Yamanashi isolate (Japan), Saku isolate (Japan), China isolate (China), Zhenjiang isolate (China), Indian isolate (India)
Family/genus	*Parvoviridae/Iteradensovirus*	*Bidnaviridae/Bidensovirus*
Virion size	20 nm	24 nm
Genome topology	Linear	Linear
Genome type	ssDNA	Segmented ssDNA
Genome size	5.0 kb	6.0 kb, 6.5 kb[a]
Number of ORFs	3	6 or 7
Pathogenicity	Acute	Chronic
Virus affected tissue	Midgut columnar cell	Midgut columnar cell[b]
Resistance gene	*Nid-1* (Chr. 17)	*nsd-2* (Chr. 17)
	nsd-1 (Chr. 21)	*nsd-z* (Chr. 15)

[a]BmBDV has bipartite genomic structures {VD1 (6.0 kb) and VD2 (6.5 kb)}
[b]Chinese isolates of BmBDV infect the goblet cells of midgut epithelium during the late stage of infection

their bipartite genome structure and the presence of a DNA polymerase motif within their genomes. These viruses have been reassigned to the *Bidensovirus* genus, *Bidnaviridae* family, and are designated as *Bombyx mori* bidensovirus (BmBDV) (Adams and Carstens 2012) (Table 1). To date, one isolate of BmDV1 was reported, namely, the Ina isolate (Japan; BmDNV-1) (Shimizu 1975) (Table 1), whereas four BmBDV isolates have been reported, namely the Yamanashi and Saku isolates (Japan; BmDNV-2) (Seki and Iwashita 1983) and the China and Zhenjiang isolates (China; BmDNV-3, BmDNV-Z) (Wang et al. 2007) (Table 1). A new BmBDV isolate was recently identified in the Indian sericulture industry, named the Indian isolate (Gupta et al. 2018) (Table 1).

Pathogenicity of BmDV1 and BmBDV to Silkworms

BmDV1 and BmBDV multiply only in the nuclei of the columnar cells in the larval midgut epithelium; no pathological changes are observed in the other tissues (Shimizu 1975; Seki and Iwashita 1983; Watanabe et al. 1976) (Fig. 1A, Table 1). Histopathological studies of the midgut epithelium from the silkworm infected with BmDV1 revealed hypertrophic nuclei in the columnar cells that were markedly stained by the Feulgen reaction with methyl green (Watanabe et al. 1976). The degenerated columnar cells are eventually liberated into the lumen of midgut cells

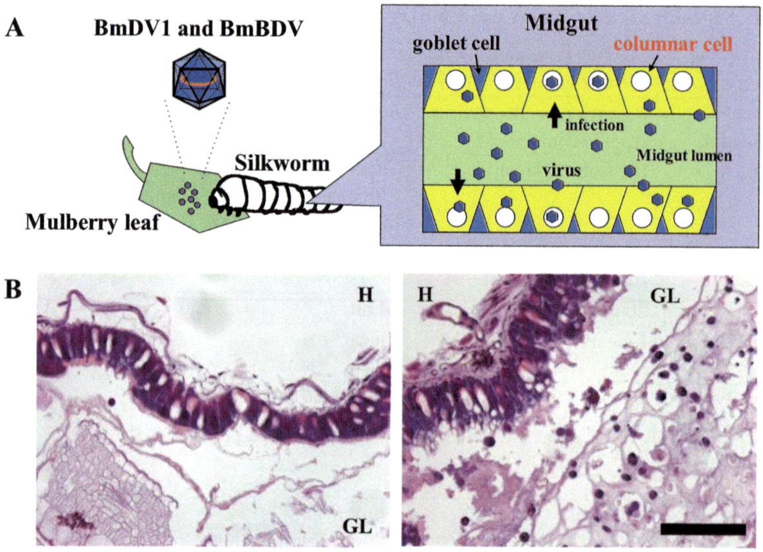

Fig. 1 Mechanisms of BmDV1 and BmBDV transmission to silkworms. (**a**) The silkworm larvae infected per os with BmDV1 and BmBDV usually die with body flaccidity as the major sign. BmDV1 and BmBDV multiply only in the nuclei of the columnar cells in the larval midgut epithelium, with no pathological changes observed in other tissues. (**b**) The histopathological analysis of BmDV1 in the anterior part of the midgut at 8 DPI. Sections were prepared from the larvae not exposed (left) and exposed (right) to BmDV1 and stained with hematoxylin and eosin. The degenerated columnar cells and nuclei are eventually liberated into the midgut cells' lumen. *GL* gut lumen, *H* hemolymph. Scale bar, 100 μm

(Watanabe et al. 1976; Ito et al. 2013) (Fig. 1B). In BmBDV infection, almost identical features were observed in the midgut (Seki and Iwashita 1983). However, the course of BmDV1 infection was significantly different to that of BmBDV infection. Larvae infected with BmDV1 exhibit flaccidity of the body as the major sign of infection and die approximately 7 days postinfection (DPI) (Ito et al. 2013; Watanabe and Kurihara 1988). Conversely, BmBDV infection is accompanied by no appreciable external symptoms until 6–7 DPI, and the time of death varies between 10 and 20 DPI (Watanabe and Kurihara 1988; Kobayashi et al. 1986; Ito et al. 2016). After becoming infected, a few infected larvae manage to pupate (Watanabe and Kurihara 1988). These results indicate that BmDV1 infection is acute, whereas BmBDV infection is chronic (Watanabe and Kurihara 1988) (Table 1).

Silkworm Resistance to BmDV1 and BmBDV

In general, there are two ways by which insects resist infection/disease. The first way is "infection resistance," in which it is difficult for the virus to establish infection in the insects, and the alternative way is "disease resistance," in which it is

Fig. 2 The effect of BmBDV infection on silkworm larvae. A comparison between resistant (left) and susceptible (right) silkworm strains at 9 DPI. Both are BmBDV-infected silkworms. Scale bar, 10 mm

difficult to propagate the infection and spread the infection even after initiation of infection (Watanabe and Maeda 1978). Through such a resistance mechanism, the host suppresses the progression and damage of virus infection. However, silkworms' resistance to BmDV1 and BmBDV differs from these mechanisms; they exhibit a unique phenomenon called "complete resistance" in which insects are completely unaffected by viruses, even when the amount and frequency of virus inoculation increases (Fig. 2). Interestingly, such resistances have been found in certain silkworm strains, and their offspring inherit the resistance trait (Watanabe and Maeda 1978; Seki 1984; Watanabe and Maeda 1981). These data suggest that the resistance gene is transferred from generation to generation.

Four unrelated resistance genes to BmDV1 or BmBDV infections, *non-susceptibility gene against BmDNV-1* (*nsd-1*) (Watanabe and Maeda 1981), *noninfectious gene against BmDNV-1* (*Nid-1*) (Eguchi et al. 2007), *non-susceptibility gene against BmDNV-2* (*nsd-2*) (Ogoyi et al. 2003), and *non-susceptibility gene against Zhenjiang (China) strain of BmDNV* (*nsd-Z*) (Qin and Yi 1996), have been reported thus far (Table 1). We have identified the *nsd-1* and *nsd-2* gene through a positional cloning approach (Ito et al. 2008, 2018). In this manuscript, we would like to discuss the *nsd-2* gene in detail.

Identification of the *nsd-2* Gene, Responsible for Resistance to BmBDV Infection

Based on the locus information of the *nsd-2* gene on chromosome 17 of *B. mori* (Ogoyi et al. 2003), we performed positional cloning using *B. mori* genome information (Mita et al. 2004; Xia et al. 2004) to identify the candidate region of *nsd-2* (Ito et al. 2008). Through positional cloning, one candidate gene was identified to be delimited to region on chromosome 17 of approximately 400 kb in length (Ito et al. 2008). The candidate gene encoded a putative transporter protein, and significant differences were noted in the sequence of the gene between resistant and susceptible strains (Fig. 3). The candidate gene derived from the $+^{nsd\text{-}2}$ gene (an allele of *nsd-2*; the susceptibility gene) encoded a protein containing a putative 12-pass transmembrane domain, whereas that from the *nsd-2* gene (the resistance gene) encoded a truncated membrane protein that contained only the first 3-pass transmembrane domain (Fig. 4). In addition, the expression analysis revealed that the candidate gene was expressed only in the midgut where BmBDV can infect and replicate (Table 1). These results suggested that this gene is the most likely candidate for *nsd-2* and that the complete membrane protein translated by $+^{nsd\text{-}2}$ may function as a putative receptor for BmBDV (Ito et al. 2008) (Fig. 4).

To verify whether this candidate gene was responsible for virus resistance in the *nsd-2* mutants, we generated a transgenic silkworm in which the $+^{nsd\text{-}2}$ candidate gene (the susceptibility gene) was introduced into a resistant strain (homozygous *nsd-2* gene genotype) using the GAL4-UAS system (Tamura et al. 2000). After virus inoculation, the transgenic silkworm displayed a remarkable susceptibility phenotype, strongly suggesting that the candidate gene is *nsd-2* itself, the virus resistance gene, and that the complete membrane protein expressed by the allele, $+^{nsd\text{-}2}$, is required for infection by BmBDV (Ito et al. 2008) (Figs. 3 and 4).

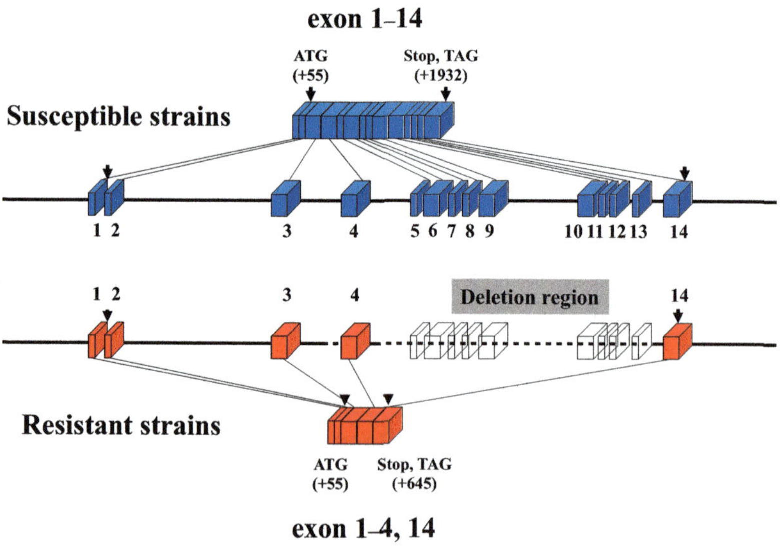

Fig. 3 Schematic genome structure of susceptible and resistant strains located on the *nsd-2* and $+^{nsd\text{-}2}$ gene. The relative position and size of the exon/intron in the genome of susceptible and resistant strains. The boxes indicate exons. The dotted line shows the deletion in the genome of resistant strains. The arrows indicate the start and stop codons

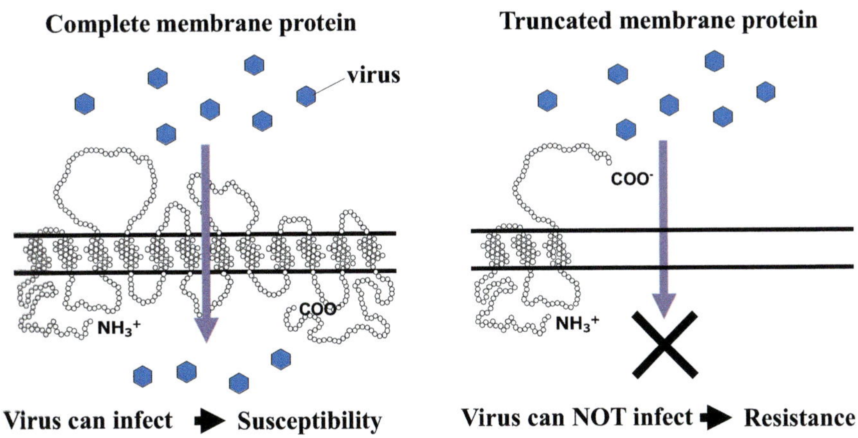

Fig. 4 Relationship between BmBDV infectivity and virus-susceptibility ($+^{nsd\text{-}2}$)/ resistance (*nsd-2*) gene products. The secondary structures of the NSD-2 protein derived from the $+^{nsd\text{-}2}$(left) and the *nsd-2* (right) gene are based on the topology prediction method, SOSUI (http://harrier.nagahama-i-bio.ac.jp/sosui/)

Putative Function of NSD-2

The analysis of the sequence through NCBI-BLAST homology searching (https://blast.ncbi.nlm.nih.gov/Blast.cgi) revealed that the NSD-2 protein expressed by the $+^{nsd\text{-}2}$ gene was highly homologous with two amino acid transporters of the tobacco hornworm, *Manduca sexta* (GenBank accession nos. AF006063 and AF013963 with 77% and 76% identities, respectively) (Castagna et al. 1998; Feldman et al. 2000). These amino acid transporters function as a K^+-coupled amino acid transporter (Castagna et al. 1998) and a Na^+- or K^+-activated nutrient amino acid transporter (Feldman et al. 2000). It remains unknown whether the NSD-2 protein functions as a transporter or not, but it may have a similar role to that of *M. sexta*-derived transporters because of its high sequence homology. Interestingly, the resistant strains possessing a large deletion of this transporter grow normally under appropriate conditions, which suggests that the NSD-2 protein is not essential for silkworm development.

During the positional cloning, at the genome locus near *nsd-2*, we found one other gene, *AK378309*, that encoded a putative transporter; however, this was considered to have no relationship to virus resistance because there was no difference in the sequence between the resistant and susceptible strains (Ito et al. 2008). The deduced amino acid sequence from this gene had a relatively high sequence identity (68.2%) with the NSD-2 protein expressed by the $+^{nsd\text{-}2}$ gene. Therefore, the protein expressed by this gene might function as an amino acid transporter instead of NSD-2.

Impact of BmBDV in India

India is currently the world's second largest producer of silk production, and the silk industry is being promoted throughout the country. Recently, the loss of cocoon production resulting from BmBDV has become a problem in India (Gupta et al. 2018; Kadono-Okuda et al. 2014; Gupta et al. 2015). We have been investigating the prevalence of BmBDV in India since 2014 and detected a high rate of BmBDV in dead silkworm individuals with flacherie disease from sericultural farmers across different parts of India (Gupta et al. 2018; Kadono-Okuda et al. 2014; Gupta et al. 2015). Consequently, we have investigated the genotypes of BmBDV resistance gene, *nsd-2*, in various local varieties of silkworm in India as a measure to prevent virus infection and have proposed an efficient method for selecting and breeding resistant silkworms by marker selection (Gupta et al. 2015). Our approaches have enabled the rapid and accurate diagnosis of virus resistance in silkworm varieties and strains that are maintained today in India (Gupta et al. 2015).

Isolation of the BmBDV Indian Strain

During the international cooperation between Japan and India, we isolated and identified a new Indian isolate of BmBDV from dead silkworm individuals with flacherie disease in a sericultural farm (Gupta et al. 2018; Kadono-Okuda et al. 2014; Gupta et al. 2015, 2022). Subsequently, the phylogenetic analysis revealed that BmBDV Indian isolate was most closely related to the BmBDV Yamanashi isolate found in Japan (Gupta et al. 2018; Kadono-Okuda et al. 2014; Gupta et al. 2015) (Table 1). Recent studies have revealed that the infectivity of the BmBDV Indian isolate to silkworms is also similar to that of the BmBDV Yamanashi strain (Gupta et al. 2022).

Two types of *Bombyx mori* densovirus, BmDV1 and BmBDV, have been reported so far; interestingly, BmDV1 has only been found in Japan (Table 1) (Ito et al. 2021). When considering the infectivity of BmBDV and BmDV1 against *Bombycidae*, *B. mandarina*, which is considered to have a common ancestor to *B. mori* (Goldsmith et al. 2005), is susceptible to BmBDV (Ito et al. 2008). Native (ancient) strains of *B. mori* also tend to be susceptible; however, these susceptible strains, including *B. mandarina*, have become resistant to BmDV1 (Ito et al. 2018; Furuta 1994). This suggests that BmBDV is likely to be the native disease causing agent of *Bombyx*, whereas BmDV1 is likely to originate from other insects. Interestingly, BmBDVs have been isolated in Japan (Yamanashi isolate) (Seki and Iwashita 1983), China (Chinese Zhenjiang isolate) (Wang et al. 2007), and India (Indian isolate) (Gupta et al. 2018), whereas BmDV1 has only been found in Japan (Kawase and Kurstak 1991). Therefore, the origin of BmDV1 may be an insect that inhabits Japan. Cotmore et al. (2014) reported that all viruses classified, to date, as *Iteradensovirus* have been isolated from lepidopteran insects, suggesting that a specific lepidopteran insect living in Japan may be the real host of BmDV1 (Cotmore et al. 2014). Conversely, BmBDV has been observed in sericultural regions around the world (Japan, China, and India), resulting in severe damage to farmers (Table 1). This suggests that other countries and regions where sericulture is flourishing may be affected by BmBDV in the future; therefore, it is extremely important to implement measures against this problem as soon as possible.

Utilization of Genome Editing Methods

In utilizing the BmBDV resistance gene, *nsd-2*, which we have successfully isolated and identified, the generation of homozygotes for resistance genes through crossbreeding can convert all silkworms to having viral resistance. However, crossbreeding causes the problem of the separation of useful traits possessed by each silkworm strains. Therefore, it is necessary to develop new approaches that can achieve breeding while retaining the preferred quality traits.

In recent years, genome editing methods have become widespread in various species, including silkworms. Genome editing methods allow the introduction of pinpointed mutations (insertions or deletions) into a target DNA region. Essentially, this method makes it possible to edit only the genome regions where genes involved in BmBDV infectivity are located. Currently, we are attempting the generation of new BmBDV-resistant silkworm strains by genome editing, specifically using the clustered regularly interspaced palindromic repeats/CRISPR-associated (CRISPR/Cas) system (Fig. 5).

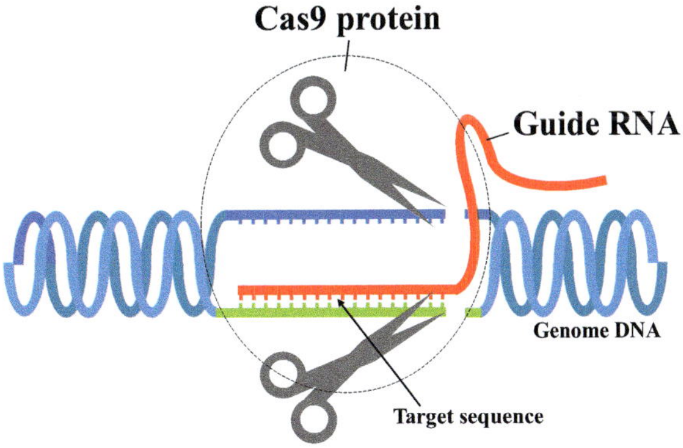

Fig. 5 Schematic of the CRISPR/Cas9 genome editing method. The action of two reagents, sgRNA and the Cas9 protein, induces the cleavage of genomic DNA corresponding to the target sequence

Clustered Regularly Interspaced Palindromic Repeats/ CRISPR-Associated (CRISPR/Cas) System

The CRISPR/Cas9 system is based on a bacterial immune system in which Cas9 is guided to its specific target DNA by CRISPR RNAs (crRNAs) and the trans-activating CRISPR RNA (tracrRNA) (Beumer et al. 2008; Jinek et al. 2012; Daimon et al. 2013). The experimental method mainly involves the introduction of mutations into the target DNA region through the injection of single-guide RNA (sgRNA), which is RNA fused with crRNA and tracrRNA, and the Cas9 protein into the eggs immediately after they are laid (Figs. 5 and 6). The injection of just two reagents, sgRNA and Cas9 protein, induces the cleavage of the genomic DNA corresponding to the target sequence (Figs. 5 and 6). Owing to the advantages and simplicity of these experimental techniques, many research papers using this method in silkworms have now been published (Daimon et al. 2013; Yuasa et al. 2016; Suzuki et al. 2022; Mang et al. 2023).

In this manuscript, we present an example of our experiments targeting the *BmBLOS2* gene. The *BmBLOS2* gene is essential for the formation of urate granules in the larval epidermis that confer an opaque white color to larval skin (Fujii et al. 2010). The loss of these urate granules in the epidermis results in a visible "oily" phenotype with translucent skin. We synthesized sgRNA targeting the internal

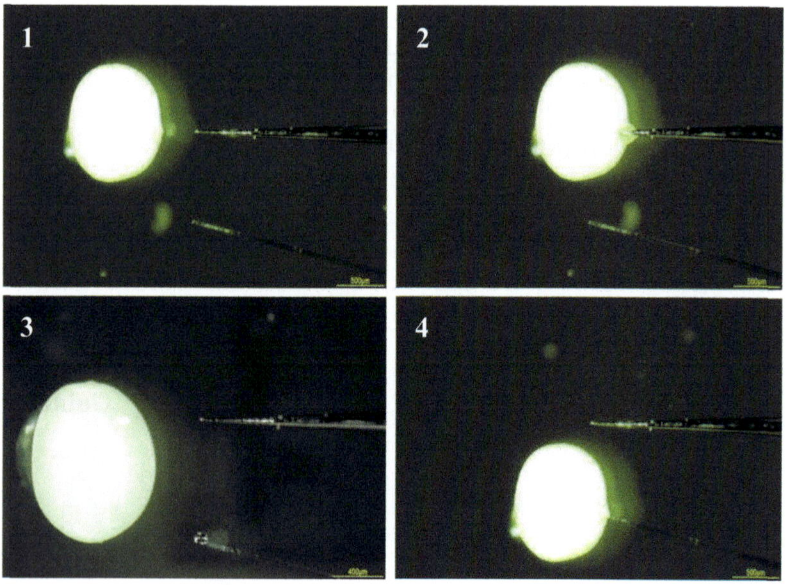

Fig. 6 Schematic of the injection of the CRISPR/Cas9 construct into silkworm eggs. The injection procedure consists of the following four steps (1–4). (1) Prepare a hard tungsten needle (upper) and a glass capillary needle (lower) containing the CRISPR/Cas9 construct (the mixture of sgRNA and the Cas9 protein). (2) Drill a hole in the egg with a hard tungsten needle. (3) Pull out the tungsten needle. (4) Insert a glass capillary needle into the hole drilled with a tungsten needle, and inject the CRISPR/Cas9 construct

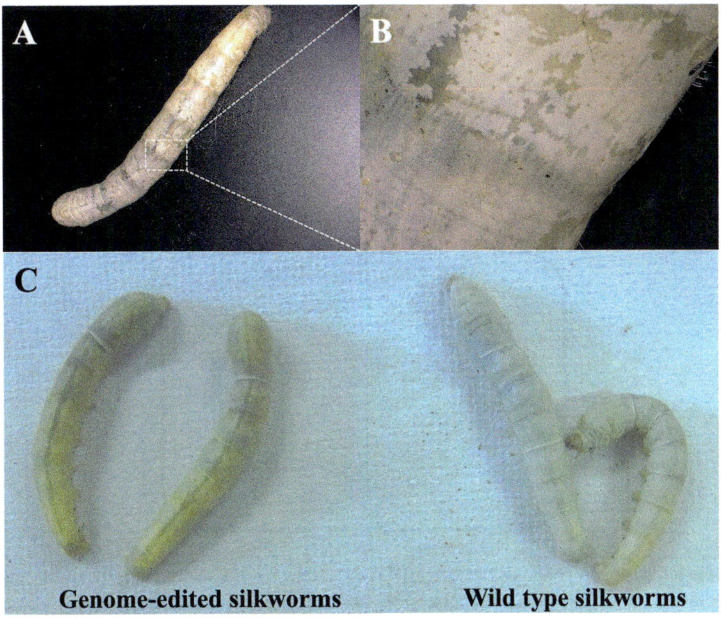

Fig. 7 Silkworms in which mutations have been introduced by genome editing. (**a**) Photo of the oily mosaic larva (Generation 0) induced by genome editing of the *BmBLOS2* gene. The *BmBLOS2* gene is essential for the formation of urate granules in the larval epidermis that confer an opaque white color to larval skin (Fujii et al. 2010). (**b**) Zoomed-in images corresponding to the dotted box in photo (**a**). (**c**) Photograph of the oily larva (Generation 1) induced by genome editing of the *BmBLOS2* gene. Left and right are the genome-edited silkworms and wild-type silkworms, respectively. The larval skin of the genome-edited silkworms has a completely oily phenotype

sequence of *BmBLOS2* gene as described by Daimon et al. (2013) and injected it together with the Cas9 protein into eggs immediately after oviposition (Figs. 5 and 6) (Daimon et al. 2013). Following the injection, mosaic larvae with normal white and oily mixed skin appeared (Fig. 7A, B), indicating that genome editing had occurred in the somatic cells. Furthermore, when mutations are introduced to the germline by mating genome-edited individuals, all the epidermis of the progeny displayed the oily phenotype (Fig. 7C). Thus, this method makes it easy to introduce mutations artificially.

Future Challenges and Countermeasures for BmBDV Research

The resistance by *nsd-2* against BmBDV is acquired by defects in membrane protein structure (Figs. 3 and 4). Interestingly, the loss of this membrane protein has no effect on silkworm survival or growth. This finding suggests that it may be possible to convert all silkworms to the BmBDV-resistant phenotype by simply disrupting the membrane proteins involved in the virus-susceptibility phenotype through a

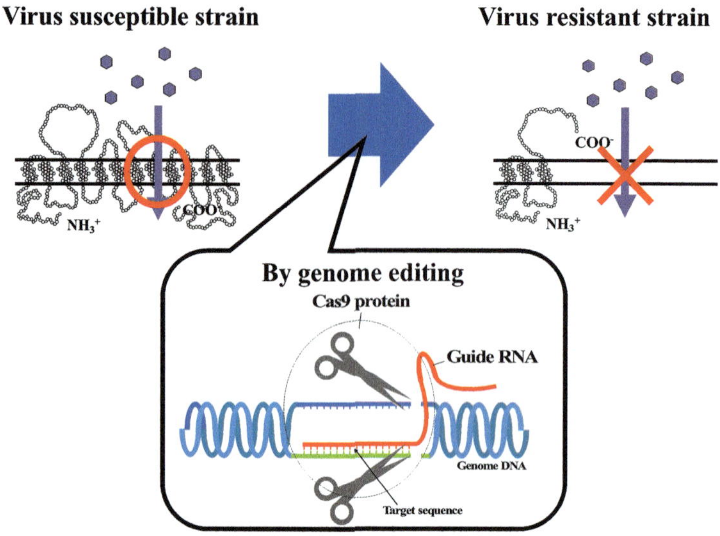

Fig. 8 New approach to generate novel BmBDV-resistant strains using genome editing. Genome editing methods can be used to artificially generate BmBDV-resistant silkworms

genome editing approach. We are currently conducting this experiment as part of an international collaboration between Japan and India (Fig. 8). Thus, we aim to make effective use of the viral resistance/susceptibility gene information we have already uncovered to produce more useful results in the sericultural farm and field. We strongly hope that our research will support the future development of the sericulture industry.

Acknowledgments We thank Dr. K. Kadono-Okuda (Dainippon Silk Foundation), Dr. K. Mita (Southwest University), Dr. T. Shimada (Gakushuin University), Dr. S. Katsuma (The University of Tokyo), Dr. J. Kobayashi (Yamaguchi University), Dr. V. Sivaprasad (Former Director Technical, Central Silk Board), Dr. C. M. KishorKumar (Former Director, Central Sericultural Research & Training Institute, Berhampore), Dr. K. M. Ponnuvel (Seribiotech Research Laboratory), and Dr. T. Gupta (Seribiotech Research Laboratory) for technical support and valuable discussion.

References

Adams MJ, Carstens EB (2012) Ratification vote on taxonomic proposals to the International Committee on Taxonomy of Viruses (2012). Arch Virol 157:1411–1422

Beumer KJ, Trautman JK, Bozas A, Liu JL, Rutter J, Gall JG, Carroll D (2008) Efficient gene targeting in *Drosophila* by direct embryo injection with zinc-finger nucleases. Proc Natl Acad Sci USA 105:19821–19826

Castagna M, Shayakul C, Trotti D, Sacchi VF, Harvey WR, Hediger MA (1998) Cloning and characterization of a potassium-coupled amino acid transporter. Proc Natl Acad Sci USA 95:5395–5400

Cotmore SF, Agbandje-McKenna M, Chiorini JA, Mukha DV, Pintel DJ, Qiu J, Soderlund-Venermo M, Tattersall P, Tijssen P, Gatherer D, Davison AJ (2014) The family *Parvoviridae*. Arch Virol 159:1239–1247

Daimon T, Kiuchi T, Takasu Y (2013) Recent progress in genome engineering techniques in the silkworm, *Bombyx mori*. Dev Growth Diff 56:12–25

Eguchi R, Nagayasu K, Ninagi O, Hara W (2007) Genetic analysis on the dominant nonsusceptibility to densonucleosis virus type 1 in the silkworm, *Bombyx mori*. Sanshi-Kontyu Bio TeC 76:159–163

Feldman DH, Harvey WR, Stevens BR (2000) A novel electrogenic amino acid transporter is activated by K^+ or Na^+, is alkaline pH-dependent, and is Cl^--independent. J Biol Chem 275:24518–24526

Fujii T, Daimon T, Uchino K, Banno Y, Katsuma S, Sezutsu H, Tamura T, Shimada T (2010) Transgenic analysis of the *BmBLOS2* gene that governs the translucency of the larval integument of the silkworm *Bombyx mori*. Insect Mol Biol 19:659–667

Furuta Y (1994) Susceptibility of Indian races of the silkworm, *Bombyx mori*, to the nuclear polyhedrosis virus and densonucleosis viruses. Acta Seric Entmol 8:1–10

Goldsmith MR, Shimada T, Abe H (2005) The genetics and genomics of the silkworm, *Bombyx mori*. Annu Rev Entomol 50:71–100

Gupta T, Kadono-Okuda K, Ito K, Ponnuvel KM (2015) Densovirus infection in silkworm *Bombyx mori* and genes associated with disease resistance. Invertebr Surviv J 12:118–128

Gupta T, Ito K, Kadono-Okuda K, Murthy GN, Gowri EV, Ponnuvel KM (2018) Characterization and genome comparison of an Indian isolate of Bidensovirus infecting the silkworm *Bombyx mori*. Arch Virol 163:125–134

Gupta T, Raghavendar G, Terenius O, Ito K, Mishra RK, Ponnuvel KM (2022) An investigation into the effects of infection and ORF expression patterns of the Indian bidensovirus isolate (*Bm*BDV) infecting the silkworm *Bombyx mori*. Virus Dis 33:76–83

Ito K, Kidokoro K, Sezutsu H, Nohata J, Yamamoto K, Kobayashi I, Uchino K, Kalyebi A, Eguchi R, Hara W, Tamura T, Katsuma S, Shimada T, Mita K, Kadono-Okuda K (2008) Deletion of a gene encoding an amino acid transporter in the midgut membrane causes resistance to a *Bombyx* parvo-like virus. Proc Natl Acad Sci USA 105:7523–7527

Ito K, Kidokoro K, Shimura S, Katsuma S, Kadono-Okuda K (2013) Detailed investigation of the sequential pathological changes in silkworm larvae infected with *Bombyx* densovirus type 1. J Invertebr Pathol 112:213–218

Ito K, Shimura S, Katsuma S, Tsuda Y, Kobayashi J, Tabunoki H, Yokoyama T, Shimada T, Kadono-Okuda K (2016) Gene expression and localization analysis of *Bombyx mori* Bidensovirus and its putative receptor in *B. mori* midgut. J Invertebr Pathol 136:50–56

Ito K, Kidokoro K, Katsuma S, Sezutsu H, Uchino K, Kobayashi I, Tamura T, Yamamoto K, Mita K, Shimada T, Kadono-Okuda K (2018) A single amino acid substitution in the *Bombyx*-specific mucin-like membrane protein causes resistance to *Bombyx mori* densovirus. Sci Rep 8:7430

Ito K, Ponnuvel KM, Kadono-Okuda K (2021) Host Response against Virus Infection in an Insect: Bidensovirus Infection Effect on Silkworm (*Bombyx mori*). Antioxidants 10:522

Jinek M, Chylinski K, Fonfara I, Hauer M, Doudna JA, Charpentier E (2012) A programmable dual-RNA-guided DNA endonuclease in adaptive bacterial immunity. Science 337:816–821

Kadono-Okuda K, Ito K, Murthy GN, Sivaprasad V, Ponnuvel KM (2014) Molecular mechanism of densovirus resistance in silkworm, *Bombyx mori*. Sericologia 54:1–10

Kawase S, Kurstak E (1991) In: Kurstak E (ed) Viruses of invertebrates. Marcel Dekker, New York, pp 315–343

Kobayashi M, Hashimoto Y, Mori H, Nagamine T (1986) Changes in DNA, RNA, protein and glycogen in the midgut of the silkworm, *Bombyx mori* (Lepidoptera: Bombydicae), during the infection with a densonucleosis virus. Appl Entomol Zool 21:486–489

Lü P, Pan Y, Yang Y, Zhu F, Li C, Guo Z, Yao Q, Chen K (2018) Discovery of anti-viral molecules and their vital functions in *Bombyx mori*. J Invertebr Pathol 154:12–18

Mang D, Toyama T, Yamagishi T, Sun J, Purba ER, Endo H, Matthews MM, Ito K, Nagata S, Sato R (2023) Dietary compounds activate an insect gustatory receptor on enteroendocrine cells to elicit myosuppressin secretion. Insect Biochem Mol Biol 155:103927

Mita K, Kasahara M, Sasaki S, Nagayasu Y, Yamada T, Kanamori H, Namiki N, Kitagawa M, Yamashita H, Yasukochi Y, Kadono-Okuda K, Yamamoto K, Ajimura M, Ravikumar G,

Shimomura M, Nagamura Y, Shin-I T, Abe H, Shimada T, Morishita S, Sasaki T (2004) The genome sequence of silkworm, *Bombyx mori*. DNA Res 11:27–35

Ogoyi DO, Kadono-Okuda K, Eguchi R, Furuta Y, Hara W, Nguu EK, Nagayasu K (2003) Linkage and mapping analysis of a non-susceptibility gene to densovirus (*nsd-2*) in the silkworm, *Bombyx mori*. Insect Mol Biol 12:117–124

Qin J, Yi WZ (1996) Genetic linkage analysis of *nsd-Z*, the nonsusceptibility gene of *Bombyx mori* to the Zhenjiang (China) strain densonucleosis virus. Sericologia 36:241–244

Seki H (1984) Mode of inheritance of the resistance to the infection with the densonucleosis virus (Yamanashi isolate) in the silkworm, *Bombyx mori*. J Seric Sci Jpn 53:472–475. (in Japanese with English summary)

Seki H, Iwashita Y (1983) Histopathological features and pathogenicity of a densonucleosis virus of the silkworm, *Bombyx mori*, isolated from sericultural farms in Yamanashi Prefecture. J Seric Sci Jpn 52:400–405. (in Japanese with English summary)

Shimizu T (1975) Pathogenicity of an infectious flacherie virus of the silkworm, *Bombyx mori*, obtained from sericultural farms in the suburbs of Ina city. J Seric Sci Jpn 44:45–48. (in Japanese)

Suzuki T, Tang S, Otuka H, Ito K, Sato R (2022) Nodule formation in *Bombyx mori* larvae is regulated by BmToll10-3. J Insect Physiol 142:104441

Tamura T, Thibert C, Royer C, Kanda T, Abraham E, Kamba M, Komoto N, Thomas JL, Mauchamp B, Chavancy G, Shirk P, Fraser M, Prudhomme JC, Couble P (2000) Germline transformation of the silkworm *Bombyx mori* L. using a *piggyBac* transposon-derived vector. Nat Biotechnol 18:81–84

Wang YJ, Yao Q, Chen KP, Wang Y, Lu J, Han X (2007) Characterization of the genome structure of *Bombyx mori*densovirus (China isolate). Virus Genes 35:103–108

Watanabe H, Kurihara Y (1988) Comparative histopathology of two densonucleoses in the silkworm, *Bombyx mori*. J Invertebr Pathol 51:287–290

Watanabe H, Maeda S (1978) Genetic resistance to peroral infection with a densonucleosis virus in the silkworm, *Bombyx mori*. J Seric Sci Jpn 47:209–214. (in Japanese with English summary)

Watanabe H, Maeda S (1981) Genetically determined nonsusceptibility of the silkworm, *Bombyx mori*, to infection with a densonucleosis virus (Densovirus).*J*. Invertebr Pathol 38:370–373

Watanabe H, Maeda S, Matsui M, Shimizu T (1976) Histopathology of the midgut epithelium of the silkworm, *Bombyx mori*, infected with a newly-isolated virus from the flacherie-diseased larvae. J Seric Sci Jpn 45:29–34. (in Japanese with English summary)

Watanabe H, Kawase S, Shimizu T, Seki H (1986) Difference in serological characteristics of densonucleosis viruses in the silkworm, *Bombyx mori*. J Seric Sci Jpn 55:75–76. (in Japanese)

Xia Q, Zhou Z, Lu C, Cheng D, Dai F, Li B, Zhao P, Zha X, Cheng T, Chai C, Pan G, Xu J, Liu C, Lin Y, Qian J, Hou Y, Wu Z, Li G, Pan M, Li C, Shen Y, Lan X, Yuan L, Li T, Xu H, Yang G, Wan Y, Zhu Y, Yu M, Shen W, Wu D, Xiang Z, Yu J, Wang J, Li R, Shi J, Li H, Li G, Su J, Wang X, Li G, Zhang Z, Wu Q, Li J, Zhang Q, Wei N, Xu J, Sun H, Dong L, Liu D, Zhao S, Zhao X, Meng Q, Lan F, Huang X, Li Y, Fang L, Li C, Li D, Sun Y, Zhang Z, Yang Z, Huang Y, Xi Y, Qi Q, He D, Huang H, Zhang X, Wang Z, Li W, Cao Y, Yu Y, Yu H, Li J, Ye J, Chen H, Zhou Y, Liu B, Wang J, Ye J, Ji H, Li S, Ni P, Zhang J, Zhang Y, Zheng H, Mao B, Wang W, Ye C, Li S, Wang J, Wong GK, Yang H, Biology Analysis Group (2004) A draft sequence for the genome of the domesticated silkworm (*Bombyx mori*). Science 306:1937–1940

Yuasa M, Kiuchi T, Banno Y, Katsuma S, Shimada T (2016) Identification of the silkworm quail gene reveals a crucial role of a receptor guanylyl cyclase in larval pigmentation. Insect Biochem Mol Biol 68:33–40

Biotechnological Approaches in Wild Silk Culture

Kaiho Kaisa, Jigyasha Tiwari, D. S. Mahesh, Suraj Shah, and Kallare P. Arunkumar

Abstract

Wild silk moths form a very important part of livelihood in rural areas of India and China. Silks produced by wild silk moths are distinct in nature with special characteristics of luster, texture, and color. The development of cost-effective DNA sequencing technologies and advanced gene editing tools even in non-model organisms in recent years has made it possible for the application of biotechnological tools in wild silk moths such as *Antheraea assamensis* (muga silkworm), *Samia ricini* (eri silkworm), *Antheraea mylitta* (tasar silkworm), *Antheraea pernyi* (Chinese oak tasar), and *Antheraea yamamai* (Japanese wild silkworm). A high-quality genome sequence is available for all these wild silk moths. Several tissue transcriptomes from different wild silk moths have been sequenced to obtain global gene expression profiles. The genomics and transcriptomics resources have been made easily accessible through databases, namely, *WildSilkbase* and *Vanya Silkbase*. The gene editing protocol has been tried and tested successfully in eri silkworm and is being tested in other silk moths as well. With these tools and reagents already available for studying the biology of wild silk moths, it is now possible to address some of the issues in economically important wild silk moths in Asia, especially in improving silk quantity and quality besides value addition through non-textile application.

Kaiho Kaisa and Jigyasha Tiwari contributed equally with all other contributors.

K. Kaisa · J. Tiwari · D. S. Mahesh · S. Shah · K. P. Arunkumar (✉)
Central Muga Eri Research and Training Institute (CMER&TI), Lahdoigarh, Jorhat, Assam, India
e-mail: arunkallare.csb@gov.in

R. V Suresh et al. (eds.), *Biotechnology for Silkworm Crop Enhancement*,
https://doi.org/10.1007/978-981-97-5061-0_10

Keywords

Wild silks · *Antheraea* · *Samia* · Silkbase and *Tasar*

Introduction

Wild silk moths include tasar silkworm, eri silkworm, oak-tasar silkworm, and muga silkworm. They are economically important as a large number of rural populaces depend on them for their livelihood in counted like India and China. Most of the research and development of technology on wild silk moths is confined to China, India, and Japan in Asia. With over 20 ecoraces, the Indian tasar silkworm *Antheraea mylitta* is a native species of tropical India. Tasar cocoons from China and India are used in enormous quantities to make a variety of fabrics. A wild silk moth known as *Antheraea assamensis* (muga silkworm) (n = 15) was first recorded in literature in 1662 BC. In Northeastern India, especially in the state of Assam, it is extensively practiced and spread. This semi-domesticated multivoltine species secretes muga silk, which is golden-yellow in color. Mejangkori silk is produced by larvae that are fed mejangkori leaves (*Litsea citrata*). This type of silk is prized for its longevity, luster, and creamy white hue. *Antheraea yamamai*, a native of Japan, is a 31-chromosome Japanese oak silkworm. It is unique in that it is used to produce pricey silk. The multivoltine silkworm *Samia cynthia ricini* (n = 13), sometimes known as the "eri silkworm," is distinguished by its white or brick-red eri silk. It is available in India's northeastern region. The remaining ecoraces (around sixteen) are dispersed over the Indo-Austral and Palaearctic biogeographic zones. Southern China is the origin of *Antheraea pernyi* (n = 49), which dates to the Han and Wei eras. *Antheraea roylei* (n = 303,132) is found in India's sub-Himalayan region. A successful hybrid of *Antheraea roylei* and its Chinese equivalent is *Antheraea proylei* (n = 49).

Silk fibers dating to around 2450–2000 BC have been discovered through a recent examination of artifacts discovered within copper-alloy ornaments from Harappa and steatite beads from Chanhu-daro, two significant Indus sites. Silk has been discovered by microscopic examination of ancient thread remains. The silk has sericin-coated twinned brins, or filaments, of fibroin, but it is not degummed. According to a micromorphological analysis, the silk is not obtained from the cultivated silkworm *Bombyx mori* but rather from wild silk moth species (Good et al. 2009). This research, which is roughly contemporary with the oldest Chinese evidence for silk, provides the earliest evidence for any silk found outside of China. These silks came from wild silk moths, which suggests that the history of wild silk is at least as long as that of mulberry silkworms.

The quick and affordable acquisition of many data kinds, made possible by new technology advancements, is a major driving force behind genome science. Even with non-model organisms, the application of modern technology has resulted in the collection of vast amounts of data. The development of genomic research has brought out more than just an abundance of resources and information for the study of molecular biology. It has been the catalyst for an astounding burst of interdisciplinary research. In the case of wild silk moths, also work on molecular biology and genomics was initiated more than two decades ago when the next generation sequencing (NGS) was still in infancy. In this review, the genetics and genomics resources generated in wild silk moths and genetic transformation studies that have been carried out in wild silks are reviewed.

Genomics and Transcriptomics Resources in Wild Silk Moths

The paucity of genetic and genomic data has led to a poor understanding of silk manufacturing. Omic data is crucial to comprehend the organism at the molecular level are few. An abundance of information on many plant and animal species has been produced by functional genomics, which has also produced innovative methods for separating useful features or genes from these creatures. If these methods are extended to economically valuable species, it could lead to significant improvements in the quality of their produce by providing guidance for future efforts in gene mining, genetic manipulation, or breeding. However, the majority of research on silk protein structure have only offered scant overviews, and as of yet, only low-coverage EST data are available for *A. assama*, *S. cynthia*, and *A. mylitta*. It is still unclear how complex the silk moth transcriptome is functionally for the *Saturniidae* family of silk moths.

Recent eukaryotic transcriptome applications of RNA-sequence technology have identified a growing number of unique transcripts and sequence variations. It has also been used to examine significant gene circuits connected to characteristics in animals, including those for which there isn't currently a reference genome. Dong et al. (2015) carried out comprehensive comparative transcriptomics of the silk glands of six *Saturniidae* silk moths in order to further identify the genetic basis behind the reported variances in the qualities of silk generated by these species: *A. yamamai*, *A. pernyi*, *A. assama*, *Ac. selene*, *S. cynthia*, and *R. newara*. They also conducted transcriptomics of the mulberry silkworm, *B. mori*. However, the limits of the traditional silk made by *B. mori* have caused a halt to the greater utilization of silk in current times. Since genomic data scarcity has been covered by de novo transcriptome sequencing of non-model organisms, the transcriptome ideally represents the complete set of RNAs transcribed in any organism.

For the first time, Chetia et al. (2017) reported a novel transcriptome of 5th instar larvae of *A. assamensis* grown on *Machilus bombycina*'s plant leaves and generated 121,433 transcripts; 71,397 simple sequence repeat (SSR) markers were also predicted. In addition, candidate antimicrobial peptides, namely, attacin, cecropin, defensin, moricin, gloverin, and gallerimycin that are known to provide immunity against bacteria, fungus, and parasite, were identified as well. Essential mechanistic insights into immune cell functions can also be obtained through transcriptomic studies by dissecting the immune system. Two main signaling pathways, *Toll* and *Imd*, are triggered when insects encounter microorganisms like bacteria or fungi. This leads to the transcription of genes that encode lectins, peptidoglycan recognition proteins (PGRPs), antimicrobial peptides (AMPs), lysozymes, and other immune-related molecules (De Gregorio et al. 2002). Thus, it is essential to comprehend how immune-related genes are activated.

A comparative analysis was conducted in order to identify immune-related genes from the fat body mRNA of *S. c. ricini* that are activated or repressed by microbial injection with *E. coli.* The investigation saw the transcriptional activation of substances such as enzymes, immune-inducible peptides, and pattern recognition agents. It's interesting to note that transcriptional suppression and acute stimulation of genes encoding AMPs were demonstrated by the EST frequencies of two important hemolymph storage protein genes, *ScSP1* and *ScSP2*(Meng et al. 2008). According to the results, *S. c. ricini* may possess a genetic program designed to suppress the transiently essential expression of genes in order to stimulate the acute and effective production of immune-related genes. Another essential conserved iron-binding protein, ferritins, which are mainly involved in the immune system, detoxification, and storage of iron, was taken into transcriptomic study. To comprehend the function of ferritin in the immune response, antioxidation, and iron homeostasis regulation, Yu et al. (2017) conducted analysis. After challenging with *Pseudomonas aeruginosa* and *Staphylococcus aureus*, the expression levels of the *S. c. ricini* ferritin heavy chain subunit (ScFerHCH) was shown to be upregulated. According to this transcriptome analysis, ScFerHCH is probably crucial for *S. c. ricini's* iron storage, antioxidation, and innate immunological protection against infections.

Eri silkworm is considered as a useful pharmacological model. It is a known fact that mulberry leaves are harmful to *S. c. ricini* because they contain the chemical 1-deoxynojirimycin (DNJ). Accordingly, Bhattacharya and Kaliwal (2005) conducted a study on DNJ's mechanism of action. In order to elucidate the molecular mechanism underlying the eri silkworm hemolymph reactions to DNJ, via transcriptome sequencing for the first time in the hemolymph of the insects to examine changes in the genes linked to lipid metabolism, energy metabolism, and glycometabolism after the insects were fed DNJ (Zhang et al. 2018). After 12 h, the study found that, in the 2% DNJ group, 288 DEGs were downregulated and 577 DEGs were upregulated in comparison to the control group (ddH2O), suggesting that DNJ causes notable changes to affect lipid metabolism, energy metabolism, and glycometabolism. This work establishes the groundwork for further investigation into the harmful impacts of DNJ on eri silkworm.

There have also been reports over the last 20 years of reductions in the wild populations of *A. assamensis*, which have been linked to ongoing deforestation, population fragmentation, and inbreeding, which suggests that this species has little genetic variety. Therefore, Arunkumar et al. (2009) developed several useful microsatellite markers that could be utilized in analyzing the genetic structure and phylogenetic status of this species in order to understand the phylogeography and the loss of genetic variety. The study concluded with the development of 10 polymorphic loci from expressed sequence tags and one from genomic library. Additionally, 97 individuals from the six populations were genotyped using 13 informative markers from the created collection of microsatellites, and their population structure and genetic variation were examined (Arunkumar et al. 2012). The study discovered a relatively high level of genetic diversity in the WWS-1 population, which originated from the West Garo Hills region of Meghalaya state.

The digital revolution in life science has been brought about by the combination of biology and information technology for data transmission and statistics for data analytics. The rapidly advancing techniques of imaging, analysis, sequencing, and other fields have resulted in a massive generation of data. In particular, NGS, ChIP-seq, and other faster and more effective sequencing methods have advanced to generate millions of sequence data at once. By deciphering the *A. assamensis* first ever complete mitogenome through NGS technology and contrasting it with other known lepidopteran mitogenomes, Singh et al. (2017) improved our comprehension of the evolutionary and comparative biology.

The production and control of steroid hormones have been modeled using the insect prothoracic gland (PG), an important endocrine organ (Ou et al. 2016). The insect PG first focused on characterization of specific genes, particularly those related to the circadian clock mechanism and steroid hormone production and regulation (Morioka et al. 2012). In order to identify certain common characteristics of the prothoracic gland (PGs) among a pool of about 30 silkworm larvae of the 5th instar of two lepidopteran species *B. mori* and *A. pernyi*, a comparative transcriptomic method was used (Bian et al. 2019). Using Illumina HiSeq 2500 technology, they produced more than 24 million high-quality sequencing reads that combined into over 50,000 transcripts. Through a search using the transcriptome data, nearly all of the genes involved in the manufacture of steroid hormones, the prothoracicotropic hormone receptor gene (torso), numerous genes associated to juvenile hormones, two ecdysone receptor genes, and a group of clock genes have been found in the PGs of two silkworms. The study found that the PG transcriptomes share an expressed-gene profile. This demonstrates that orthologs in non-model organisms may be identified using RNA-seq technology. Several advantageous genes were identified by combining transcriptomic analysis and population genomics with QTLs. These included 71 differentially expressed genes (DEGs) in the transcriptome of the 5th instar larval silk gland between *B. mori* and *Bombyx mandarina*, as well as 14 domestication-related genes (Fang et al. 2020). It also contributes to the understanding of the connections between the expression of significant candidate genes and offers fresh perspectives on the molecular mechanisms underlying intricate silkworm features. Therefore, this method could be effective in locating potential causative genes in silkworms.

The genetic resources of *S. ricini* are on par with or even greater than those of *B. mori:* at least 26 eco races of *S. ricini* have been identified (Lee et al. 2021a), and it is capable of hybridizing with wild *Samia* species, which are found throughout Asia, to produce fruitful offspring. Lee et al. (2021a) determined the whole genome sequence of *S. ricini* to enhance genetic research. With 155 scaffolds, the completed genome of *S. ricini* weighed 458 Mb, and its N50 length was roughly 21 Mb. In the assembly, 16,702 protein-coding genes were predicted.

An overview of genomic resources available for molecular biological studies in wild silk moths, with detailed information presented in Table 1 for easy reference.

Table 1 Genomic resources available for molecular biological studies in wild silk moths

Species	Common name	Genome size (Mb)	Reference
Antheraea assamensis	Muga silkworm	501.18 Mb	Dubey et al. (2023)
Antheraea mylitta	Tropical tasar	698.4 Mb	https://www.ncbi.nlm.nih.gov/datasets/genome/GCA_014332785.1/
Antheraea pernyi	Chinese oak tasar	720.67 Mb	Duan et al. (2020)
Samia ricini	Eri silkworm	458 Mb	Lee et al. (2021a, b)
Antheraea yamamai	Japanese wild silkworm	656 Mb	Kim et al. (2018)

Database Hosting Wild Silk Moth Sequences

Recent developments in the disciplines of sequencing and analysis have led to an enormous rise of data and the development of data science techniques. Numerous seri-related databases have been created because of this scientific progress, and they are of great assistance to the scientific community. Plenty of databases have been created for *B. mori*; nonetheless, the lack of genetic resources in wild silk moths is primarily responsible for their underrepresentation. Thus, only a meager amount of genetic data from wild lepidopterans has been started to be generated thus far.

Silk Database: An integrated representation of six draft genome sequences, cDNAs, EST clusters, transposable elements (TEs), and functional annotations of genes with assignments to InterPro domains and Gene Ontology (GO) terms is provided by the Silk database (SilkDB), commonly known as the "Silkworm Knowledgebase." For the ease of in-depth comparative research, this database aims to offer a thorough knowledgebase regarding the silkworm genome and related material in graphically systematic ways (Wang et al. 2005).

Arunkumar et al. (2008) describe WildSilkbase as a database of ESTs produced from many tissues at different developmental stages of three commercially significant saturniid silk moths: *Antheraea assama*, *A. mylitta*, and *Samia cynthia ricini*. It receives 60,000 ESTs in total at this time, 57,113 ESTs are sent to the database, and these are grouped and assembled into 10,019 singletons and 4019 contigs.

According to Kawamoto et al. (2022), *SilkBase* is an integrated database containing transcriptome and genomic resources of the silkworms *B. mori*, *B. mandarina*, *Trilocha varians*, *Ernolatiamoorei*, and *S. ricini*. An easy-to-use snapshot of the generation of *piRNAs*, histone modifications, alternative splicing, stage-, and tissue-specific expression of the genes can be obtained at the particular target locus with this integrated genome browser. There were installed 155 scaffolds of the *S. ricini* genome assembly (Lee et al. 2021a) and 171,159 assemblies de novo of RNA-seq data; also 78,839 ORFs were predicted from RNA-seq assemblies and 20,320 cDNA ends (Arunkumar et al. 2008).

SilkwormBase is a comprehensive genetic resource database that handles information about strains, races, varieties, etc. Transcriptomes and genes, in addition to genomes, are essential components of seri-databases. Listed in *SilkwormBase* were 608 genes and 481 wild silkworm strains' genetic resources (Priyadharshini and

Maria Joncy 2019). One significant character of this database is that, in order to close the genetic resource gap for wild silkworms, it also handles the collection, the preservation, and eventually the distribution of wild silkworms.

A Web-based resource called *Vanya Silkbase* houses transcriptome and genomic information derived from wild silk moths. As of right now, the database contains the transcriptome and genome sequences for six distinct silkworm species: *S. ricini, A. yamamai, A. assamensis, A. pernyi*, and *A. mylitta.* In addition, the database offers DEG mining, BLAST functionality, and other features. This resource also offers the SNPs from the genomes of wild silk moths. Additionally, users can search for genes of interest using the genome browse feature. This site aims to function as a one-stop shop for exploring the transcriptome and genome of wild silk moths (https://vanyasilkbase.cmerti.res.in/).

A silkworm genome database that includes a genetic map viewer and a genome viewer is called *KAIKObase*. The most recent version of *KAIKObase* (version 4) was built using the chromosomal-level genome assembly that was made public in 2019 (10.1016/j.ibmb.2019.02.002) in addition to the gene sets and genetic markers from the previous *KAIKObase* (Kawamoto et al. 2019).

In order to improve the functional and evolutionary study of wild silk moths in particular, this database is a valuable resource for learning about comparative genomics, functional genomics, and molecular evolution in general as well as gene discovery, transposable elements, gene organization, and genome variability of insect species in particular. Better research and the creation of more effective techniques for science in this area will be facilitated by this. Figure 1 showcases a

Fig. 1 Web-accessible databases available for genomics and transcriptomics of wild silk moths. *Silkworm database (SilkDb), WildSilkbase, SilkwormBase, Vanya Silkbase, SilkBase, KAIKObase*

comprehensive overview of Web-accessible databases for genomics and transcriptomics of wild silk moths, featuring key resources such as *SilkDb, WildSilkbase, SilkwormBase, Vanya Silkbase, SilkBase, and KAIKObase.*

Attempts to Make Transgenesis in Wild Silkworm

The origin of genetic engineering, specifically transgenic technology, has enabled us to alter the genetic makeup of an organism in numerous innovative ways. Transgenesis, also known as genetic modification, is the procedure of incorporating a gene of interest (also known as transgene) from one organism into the genome of another. The objective is for the ensuing transgenic organism to actively express the gene and manifest a novel property or characteristic. Extensive research suggests that transgenic technology holds promising potential across various domains, including the elucidation of gene function, enhancement of animal genetics, and the synthesis of valuable pharmaceutical proteins through the utilization of animal bioreactors. Furthermore, it extends to applications such as creating animal disease models and facilitating advancements in organ transplantation (Sharma et al. 2022). Hence, the scientists are harnessing the potential of recombinant DNA technology to advance the development of transgenics, as conventional breeding methods face limitations in effective selection of desired traits. The domesticated silkworm *B. mori* has emerged as a valuable model in *Lepidoptera*, playing an increasingly significant role in fundamental biological research. Tamura et al. (2000) achieved the initial creation of a transgenic silkworm, *B. mori*, through gene transfer using a vector derived from the piggyBac transposon. The establishment of the *B. mori* genomic map framework in 2004 marked the initiation of the era of *B. mori* genomics (functional) and transgenic technology, garnering widespread attention. Subsequently, transgenic silkworms have been engineered for the purpose of synthesizing recombinant human proteins, such as procollagen, cytokines, and monoclonal antibodies, within their cocoons and silk glands.

Hence, transgenic research has successfully incorporated favorable traits into silk and its related products. Spider silk serves as a notable example, significantly transforming the silk industries. With the standardization of the transgenesis process in the cultivated silkworm, over a thousand transgenic silkworm lines have been created using this proven method. Very fewer attempts have been made for the establishment of transgenic wild silkworms. Takeda et al. (2014) also suggested the incorporation of a diapause trait into the muga silkworm, a wild silkworm species, using transgenic methods. This approach aims to mitigate seed loss during unfavorable seasons and enhance muga silk production. Hosamani et al. (2015) conducted an inaugural systematic investigation showcasing the infection of eri silkworm (*S. ricini*) larvae experimentally with *Autographa californica multiple nucleopolyhedrovirus* (AcMNPV) highlighting its potential as a bioreactor for expressing heterologous recombinant proteins. As a result, they came to the conclusion that the eri silkworm larvae might be utilized as a viable substitute bioreactor for recombinant protein synthesis using the baculovirus system. Thus, genetic engineering technology in wild silkworms is still yet to be fully explored; hence, more research should be conducted on wild silkworms.

Gene Editing Tools

Genome editing tools are molecular technologies that enable precise modification of DNA within the genomes of living organisms. These tools have revolutionized genetic research and have broad applications in areas such as medicine, agriculture, and biotechnology. Any living thing's genomic DNA can have directed changes added to it using targeted genome editing methods, including deletions, insertions, and substitutions of genetic sequences (Moon et al. 2019). Newly developed genome editing tools, which rely on programmable nucleases including zinc finger nucleases (ZFNs) (Urnov et al. 2010), transcription activator-like effector nucleases (TALENs) (Joung and Sanders 2013), and the clustered regularly interspaced palindromic repeats (CRISPR)/CRISPR-associated (Cas) system (Cho et al. 2013; Hwang et al. 2013; Jiang et al. 2013), have been effectively utilized for genome editing in various organisms, such as bacteria, insects, plants, or human cells. Presently, through the application of engineered nucleases, significant progress is being achieved in the correction of genetic mutations, the regulation of gene expression, and the creation of therapeutic agents. These techniques are also used to improve our understanding of how genes work and the processes that lead to the emergence of particular genetic diseases.

ZFN: Zinc finger nucleases represent the first-generation tools employed in genome engineering (Urnov et al. 2010). These artificial enzymes comprise a zinc finger DNA binding domain coupled with the endonuclease domain derived from the FokI restriction enzyme The zinc finger domain is customizable to selectively bind to a particular genomic site, while the FokI domain induces a double-stranded break (DSB) at the designated target site (Daimon et al. 2014). This often leads to the occurrence of a brief insertion/deletion mutation, a consequence of the nonhomologous end joining (NHEJ) repair mechanism for double-stranded breaks (DSBs) (Daimon et al. 2014). The first achievement of gene targeting with zinc finger nucleases (ZFNs) in insects was documented in the fruit fly *D. melanogaster* (Bibikova et al., 2002). In 2008, a zinc finger consortium (www.zincfingers.org) introduced a selection-based platform known as oligomerized pool engineering (OPEN), which facilitated the tailored and accessible design of robust zinc finger proteins (Maeder et al. 2008).

Takasu et al. (2013) documented the pioneering instance of gene targeting in *Bombyx* through the utilization of zinc finger nucleases (ZFNs) mRNA injection, effectively inducing somatic and germline mutations in a target gene through the mechanism of nonhomologous end joining (NHEJ). They initially reported the effective manipulation of endogenous genomic loci, a long-standing objective in *B. mori* research. Following this, the *BmBLOS2* marker gene, which controls the development of uric acid granules in the larval epidermis, was introduced by Takasu et al. (2013). Zhu et al. (2017) also reported that this marker can be employed to confirm the efficacy of various genome editing tools in *B. mori* and other *Lepidoptera* insects.

TALENS: Transcription activator-like effector nucleases represent the second generation of tools for genome editing (Joung and Sander 2013). Similar to ZFNs, TALENs are composed of a DNA binding domain derived from bacterial TAL effectors and a FokI nuclease domain, which initiates double-stranded breaks (DSBs) at the target site (Christian et al. 2010; Miller et al. 2011). In comparison to ZFNs, TALENs, with their simple and programmable recognition code, possess numerous advantages. Consequently, they have supplanted ZFNs and are acknowledged as a more valuable and routine tool for genome editing.

Hence, TALENs have also been employed for targeted gene disruption in various organisms, encompassing animals, plants, and yeast (Joung and Sander 2013), and successful gene knockout experiments have been documented in *Bombyx* (Takasu et al. 2013; Ma et al. 2012; Sajwan et al. 2013). Ma et al. (2012) were the first to report the application of TALENs in *B. mori* by directly microinjecting mRNA from two synthetic TALEN pairs into embryos. The observed efficiency of both mosaic and heritable mutations was significantly higher and diverse.

CRISPR/Cas system: Clustered regularly interspaced palindromic repeats *(CRISPR)/Cas* is a recently developed genome editing tool for inducing targeted mutagenesis. In this technology, custom guide RNAs are generated in cultured cells or embryos to guide site-specific DNA cleavage performed by the *Cas9* endonuclease (Cho et al. 2013; Hwang et al. 2013; Jiang et al. 2013). This system relies on a bacterial immune system, where Cas9 is directed to its specific target DNA by CRISPR RNAs (crRNAs) and the transactivating CRISPR RNA (tracrRNA). Jinek et al. (2012) led the pioneering development of CRISPR by outlining all the principles and methods necessary for conducting in vivo studies. Numerous studies have explored the applications of the *CRISPR/Cas9* system across a range of organisms. In addition to classical model, organisms' genetic manipulation experiments have been conducted in important crops and animals. The successful application of the *CRISPR/Cas9* system in *B. mori* has been achieved through the utilization of various delivery strategies.

The introduction and implementation of genome editing have significantly deepened insights into *B. mori* research. Scientists have attempted to study wild silkworms in the past. Lee et al. (2018) successfully implemented a genome editing system in *Samia ricini*, a wild silkworm species. They utilized transcription activator-like effector nucleases (TALENs) and achieved the creation of several gene knockout lines. This accomplishment signifies that the functional analysis of genes of interest is now attainable. The use of genome editing tools still needs to be explored in another species of wild silkworm.

Non-Textile Application of Wild Silks

Humans have historically appreciated silk to be associated with luxury, and it has gradually become apparent that this protein-based fiber-forming material spun by living organisms is more than just a high-tech fiber. Of the desirable qualities that are highlighted for each application, the biomedical field has seen the greatest and most extensive use of these materials, particularly in sutures and protective clothing (Dong et al. 2015). Additional non-textile applications include clinical trials for wound healing, drug delivery, and tissue engineering, as well as new biomedical uses for silk in a variety of media, including films, scaffolds, hydrogels, electrospun materials, and solutions (Holland et al. 2019). Because of their hemostatic qualities, low inflammatory potential, permeability to oxygen and water, and capacity to act as a barrier against bacterial colonization, silk-based biomaterials hold special promise for the healing of skin wounds. With the advancement of research features, the usefulness of silk has been thoroughly investigated these days. Moreover, silk can regenerate into nanomembranes, nanofilms, and nanofibers when treated with specific agents, providing surfaces for creative functionalization. Functional silk has a distinct and bright future in optics and sensing, particularly in e-textiles and biosensing as well. In order to create blended fabrics, silk from *S. cynthia* is frequently combined with cotton, wool, hemp, or synthetic fiber. Specifically, in contrast with *B. mori*, the silk from *Antheraea* moths, like *A. yamamai*, *Ac. selene*, and *A. assamensis*, all exhibit noticeably greater tensile strength, tenacity, and overall toughness. The biocompatibility, remarkable mechanical strength, minimal immunogenicity, tunable degradability, and versatility of naturally derived silk protein make it a promising biomaterial for a varieties of tissue engineering applications (3D scaffolds, nanofibers, thin films, hydrogels, etc.) (Gupta et al. 2016; Kumar et al. 2018). Since these silk varieties have natural cell binding motifs (RGD), Ramachandran et al. (2020) showed that fibroin from the non-mulberry varieties of silk *A. assamensis*, *S. ricini*, can be utilized to culture corneal endothelial cells without requiring additional surface or structural modifications. Consequently, a significant clinical need for healthy donor tissues for transplantation is met.

Given that biomaterial molecules are complicated, standardizing process/manufacturing parameters and equipment is essential to the commercialization of silk functional goods. Naturally occurring biomaterials such as silk will undoubtedly find further use in higher-tech applications as more bio-based economies and industries spring up in response to the increasing demand for more environmentally friendly products and materials.

Future Prospects and Conclusion

The utilization of novel biotechnologies to tackle enduring problems in wild silk farming is crucial, as it contributes to a general rise in silk yield. The development of new diagnostic instruments that can more accurately identify illnesses in wild silk moths while lowering costs and labor should make use of advances in electronics, molecular biology, microscopy, and optics. These technologies can also aid in high throughput testing. The development of novel biomaterials utilizing wild silks need to be prioritized. It is necessary to discover fresh uses for wild silk and its waste using modern biological methods in the fields of cosmetics, medicine, and other related fields. There is a lot of promise for the use of this method in other wild silkworms as well, thanks to the discovery of a genome editing process in eri silkworms (*S. cynthia and S. ricini*).

Three major fields of biotechnology genome editing, transgenesis, and omics technologies have seen extraordinary advancements in the last 20 years. The farmed silkworm *B. mori* has been used to test all three, and attempts have previously been done with wild silkworms. The way they are used to solve the persistent issues in this industry and expand the uses of silk outside of textiles will ultimately decide the eventual outcome of wild silk culture.

Acknowledgment This work was supported by the Department of Biotechnology, Government of India, through the project BT/NER/143/SP44417 to AKP.

References

Arunkumar KP, Tomar A, Daimon T, Shimada T, Nagaraju J (2008) WildSilkbase: an EST database of wild silkmoths. BMC Genom 9:1–9

Arunkumar KP, Kifayathullah L, Nagaraju J (2009) Microsatellite markers for the Indian golden silkmoth, *Antheraea assama* (Saturniidae: Lepidoptera). Mol Ecol Resour 9:268–270

Arunkumar KP, Sahu AK, Mohanty AR, Awasthi AK, Pradeep AR, Urs SR (2012) Genetic diversity and population structure of Indian golden silkmoth (Antheraea assama). PLoS ONE 7(8):e43716. https://doi.org/10.1371/journal.pone.0043716

Bhattacharya A, Kaliwal BB (2005) Biochemical content of the fat body and haemolymph of silkworm (*Bombyx mori L.*) larvae fed with mulberry leaves fortified with mineral magnesium chloride (MgCl2). Biol Agric Food Sci 88(3):359–362

Bian HX, Chen DB, Zheng XX, Ma HF, Li YP, Li Q (2019) Transcriptomic analysis of the prothoracic gland from two lepidopteran insects, domesticated silkmoth *Bombyx mori* and wild silkmoth *Antheraea pernyi*. Sci Rep 29;9(1):531. https://doi.org/10.1038/s41598-019-41864-0

Bibikova M, Golic M, Golic KG, Carroll D (2002) Targeted chromosomal cleavage and mutagenesis in Drosophila using zinc-finger nucleases. Genetics 161(3):1169–1175

Chetia H, Kabiraj D, Singh D, Mosahari PV, Das S, Sharma P (2017) De novo transcriptome of the muga silkworm, *Antheraea assamensis* (Helfer). Gene https://doi.org/10.1016/j.gene.2017.02.021

Cho SW, Kim S, Kim JM, Kim JS (2013) Targeted genome engineering in human cells with the *Cas9* RNA-guided endonuclease. Nat Biotechnol 31(3):230–232

Christian M, Cermak T, Doyle EL, Schmidt C, Zhang F, Hummel A (2010) Targeting DNA double-strand breaks with TAL effector nucleases. Genetics 186(2):757–761

Daimon T, Kiuchi T, Takasu Y (2014) Recent progress in genome engineering techniques in the silkworm, *Bombyx mori*. Dev Growth Differ 56(1):14–25

De Gregorio E, Spellman PT, Tzou P, Rubin GM, Lemaitre B (2002) The *Toll* and *Imd* pathways are the major regulators of the immune response in Drosophila. EMBO J 3; 21(11):2568–2579. https://doi.org/10.1093/emboj/21.11.2568. PMID: 12032070

Dong Y, Dai F, Ren Y, Liu H, Chen L, Yang P (2015) Comparative transcriptome analyses on silk glands of six silkmoths imply the genetic basis of silk structure and coloration. BMC Genom 17;16(1):203. https://doi.org/10.1186/s12864-015-1420-9.

Duan J, Li Y, Du J, Duan E, Lei Y, Liang S (2020) A chromosome-scale genome assembly of *Antheraea pernyi* (Saturniidae, Lepidoptera). Mol Ecol Resour 20(5):1372–1383

Dubey H, Pradeep AR, Neog K, Debnath R, Aneesha PJ, Shah SK (2023) Genome Sequencing and Assembly of Indian Golden Silkmoth, *Antheraea* Assamensis Helfer (Saturniidae, Lepidoptera). *Antheraea Assamensis Helfer (*Saturniidae, Lepidoptera*)*. Preprint

Fang SM, Zhou QZ, Yu QY, Zhang Z (2020) Genetic and genomic analysis for cocoon yield traits in silkworm. Sci Rep 10(1):5682. https://doi.org/10.1038/s41598-020-62507-9

Good IL, Kenoyer JM, Meadow RH (2009) New evidence for early silk in the indus civilization. Archaeometry 51(3):457–466

Gupta P, Adhikary M, Joseph C, M., Kumar, M., Bhardwaj, N., & Mandal, B.B. (2016) Biomimetic, osteoconductive non-mulberry silk fibre reinforced tricomposite scaffolds for bone tissue engineering. ACS Appl Mater Interf 8:30797–30810

Holland C, Numata K, Rnjak-Kovacina J, Seib FP (2019) The biomedical use of silk: past, present, future. Adv Healthcare Mater 8(1):e1800465. https://doi.org/10.1002/adhm.201800465. Epub 2018 Sep 20. PMID: 30238637

Hosamani M, Basagoudanavar SH, Sreenivasa BP, Inumaru S, Ballal CR, Venkataramanan R (2015) Eri silkworm (*Samia ricini*), a non-mulberry host system for AcMNPV mediated expression of recombinant proteins. J Biotechnol 216:76–81

https://vanyasilkbase.cmerti.res.in/

https://www.ncbi.nlm.nih.gov/datasets/genome/GCA_014332785.1/

Hwang WY, Fu Y, Reyon D, Maeder ML, Tsai SQ, Sander JD (2013) Efficient genome editing in zebrafish using a CRISPR-Cas system. Nat Biotechnol 31(3):227–229

Jiang W, Bikard D, Cox D, Zhang F, Marraffini LA (2013) RNA-guided editing of bacterial genomes using *CRISPR-Cas* systems. Nat Biotechnol 31(3):233–239

Jinek M, Chylinski K, Fonfara I, Hauer M, Doudna JA, Charpentier E (2012) A programmable dual-RNA–guided DNA endonuclease in adaptive bacterial immunity. Science 337(6096):816–821

Joung JK, Sander JD (2013) TALENs: a widely applicable technology for targeted genome editing. Nat Rev Mol Cell Biol 14(1):49–55

Kawamoto M, Jouraku A, Toyoda A, Yokoi K, Minakuchi Y, Katsuma S (2019) High-quality genome assembly of the silkworm, *Bombyx mori*. Insect Biochem Mol Biol 107:53–62., ISSN 0965-1748. https://doi.org/10.1016/j.ibmb.2019.02.002

Kawamoto M, Kiuchi T, Katsuma S (2022) *Silk Base*: an integrated transcriptomic and genomic database for *Bombyx mori* and related species. Database 2022:baac040. https://doi.org/10.1093/database/baac040

Kim SR, Kwak W, Kim H, Caetano-Anolles K, Kim KY, Kim SB (2018) Genome sequence of the Japanese oak silk moth, *Antheraea yamamai*: the first draft genome in the family Saturniidae. Gigascience 7(1):gix113

Kumar M, Gupta P, Bhattacharjee S, Nandi SK, Mandal BB (2018) Immunomodulatory injectable silk hydrogels maintaining functional islets and promoting anti-inflammatory M2 macrophage polarization. Biomaterials 187:1–17

Lee J, Kiuchi T, Kawamoto M, Shimada T, Katsuma S (2018) Accumulation of uric acid in the epidermis forms the white integument of Samia ricini larvae. PLoS One 13(10):e0205758

Lee J, Nishiyama T, Shigenobu S, Yamaguchi K, Suzuki Y, Shimada T (2021a) The genome sequence of *Samia ricini*, a new model species of lepidopteran insect. Mol Ecol Resour 21(1):327–339

Lee J, Nishiyama T, Shigenobu S, Yamaguchi K, Suzuki Y, Shimada T et al (2021b) The genome sequence of Samia ricini, a new model species of lepidopteran insect. Mol Ecol Resour 21(1):327–339

Ma S, Zhang S, Wang F, Liu Y, Liu Y, Xu H (2012) Highly efficient and specific genome editing in silkworm using custom TALENs. PLoS ONE 7(9):e45035. https://doi.org/10.1371/journal.pone.0045035

Maeder ML, Thibodeau-Beganny S, Osiak A, Wright DA, Anthony RM, Eichtinger M (2008) Rapid "open-source" engineering of customized zinc-finger nucleases for highly efficient gene modification. Mol Cell 31(2):294–301

Meng Y, Omuro N, Funaguma S, Daimon T, Kawaoka S, Katsuma S (2008) Prominent down-regulation of storage protein genes after bacterial challenge in eri-silkworm, *Samia cynthia ricini*. Arch Insect Biochem Physiol 67(1):9–19. https://doi.org/10.1002/arch.20214

Miller JC, Tan S, Qiao G, Barlow KA, Wang J, Xia DF (2011) A TALE nuclease architecture for efficient genome editing. Nat Biotechnol 29(2):143–148

Moon SB, Kim DY, Ko JH, Kim YS (2019) Recent advances in the CRISPR genome editing tool set. Exp Mol Med 51(11):1–11

Morioka E, Matsumoto A, Ikeda M (2012) Neuronal influence on peripheral circadian oscillators in pupal Drosophila prothoracic glands. Nat Commun 3:909

Ou Q, Zeng J, Yamanaka N, Brakken-Thal C, O'Connor MB, King-Jones K (2016) The insect prothoracic gland as a model for steroid hormone biosynthesis and regulation. Cell Rep 16(1):247–262

Priyadharshini P, Maria Joncy A (2019) Silkworm databases and its applications. J Int Acad Res for Multidiscipl 7(11) ISSN: 2320–5083

Ramachandran C, Gupta P, Hazra S, Mandal BB (2020) In vitro culture of human corneal endothelium on non-mulberry silk fibroin films for tissue regeneration. Translat Vision Sci Technol 9(4):12. https://doi.org/10.1167/tvst.9.4.12

Sajwan S, Takasu Y, Tamura T, Uchino K, Sezutsu H, Zurovec M (2013) Efficient disruption of endogenous Bombyx gene by TAL effector nucleases. Insect Biochem Mol Biol 43(1):17–23

Sharma A, Gupta RK, Attri K, Afreen S, Bandral RS, Bali K (2022) Transgenesis in silkworm: An overview. Int J Entomol Res 7(1):73–80

Singh D, Kabiraj D, Sharma P, Chetia H, Mosahari PV, Neog K (2017) The mitochondrial genome of Muga silkworm (*Antheraea assamensis*) and its comparative analysis with other lepidopteran insects. PLoS One 15;12(11):e0188077. https://doi.org/10.1371/journal.pone.0188077. PMID: 29141006; PMCID: PMC5687760.

Takasu Y, Sajwan S, Daimon T, Osanai-Futahashi M, Uchino K, Sezutsu H (2013) Efficient TALEN construction for Bombyx mori gene targeting. PLoS One 8(9):e73458

Takeda M, Kobayashi J, Kamimura M, Prakash PJ (2014) Introduction of Diapause Trait into the Muga Silkmoth, *Antheraea assamensis*: a scientific collaboration over the Himalayas. Int J Wild Silkmoth Silk 18:39–54

Tamura T, Thibert C, Royer C, Kanda T, Eappen A, Kamba M (2000) Germline transformation of the silkworm Bombyx mori L. using a *piggyBac* transposon-derived vector. Nat Biotechnol 18(1):81–84

Urnov FD, Rebar EJ, Holmes MC, Zhang HS, Gregory PD (2010) Genome editing with engineered zinc finger nucleases. Nat Rev Genet 11(9):636–646

Wang J, Xia Q, He X (2005) *SilkDB:* a knowledgebase for silkworm biology and genomics. Nucl Acids Res 33:399–402

Yu HZ, Zhang SZ, Ma Y, Fei DQ, Li B, Yang LA (2017) Molecular characterization and functional analysis of a ferritin heavy chain subunit from the Eri-silkworm, *Samia cynthia ricini.* Int J Mol Sci 18(10):2126

Zhang SZ, Yu HZ, Deng MJ, Ma Y, Fei DQ, Wang J (2018) Comparative transcriptome analysis reveals significant metabolic alterations in eri-silkworm (*Samia cynthia ricini*) haemolymph in response to 1- deoxynojirimycin. PLoS ONE 13(1):e0191080. https://doi.org/10.1371/journal.pone.0191080

Zhu GH, Peng YC, Zheng MY, Zhang XQ, Sun JB, Huang Y (2017) CRISPR/Cas9 mediated BLOS2 knockout resulting in disappearance of yellow strips and white spots on the larval integument in *Spodoptera litura*. J Insect Physiol 103:29–35

An Overview of Climatic and Genetic Influences on the Emergence of *Antheraea* spp.

Shuddhasattwa Maitra Mazumdar, Nabanita Banerjee, Khasru Alam, Rupa Harsha, and T. Selvakumar

Abstract

Antheraea mylitta Drury is integral to vanya sericulture contributing about 14.73% of total vanya raw silk production (2021–2022) in India reared across ten tropical and subtropical Indian states. Understanding the impact of genetic and nongenetic factors on the emergence of *A. mylitta* has been imperative as the bottleneck in successful rearing is observed to be nonseasonal erratic emergence of *A. mylitta* during the preservation period (diapausing phase of the cocoons). The seasonal months are June to November for the trivoltine and July to October for the bivoltine generations. Nonseasonal unsynchronized emergence leads to failure of successful mating as the male-female ratio becomes close to 1 or often the proportion of females is more leading to crisis of male moths. Evaluation has led to possibilities that the factors regulating emergence were both nongenetic (abiotic) and genetic factors. The abiotic factors predominantly consist of temperature, humidity, and rainfall. The extent to which each factor contributes has been summarized by various workers. Nonseasonal months were further subdivided into zones of minimal emergence (December to February) and zones of

S. M. Mazumdar (✉)
Basic Seed Multiplication and Training Centre, Central Silk Board, Kathikund, India

N. Banerjee
Department of Zoology, The University of Burdwan, Burdwan, India

K. Alam
Central Sericulture Research and Training Institute, Central Silk Board, Berhampore, India

R. Harsha
Balurghat College, Balurghat, India

T. Selvakumar
Basic Tasar Silkworm Seed Organization, Central Silk Board, Bilaspur, India

R. V Suresh et al. (eds.), *Biotechnology for Silkworm Crop Enhancement*, https://doi.org/10.1007/978-981-97-5061-0_11

unsynchronized or erratic emergence (March to May). During the seasonal months, continuous emergence with 60–70% of males and 30–40% of females occurs resulting in successful mating. However, nonseasonal months witness sudden emergence interrupted by prolonged periods of no-emergence. Of the climatological factors, humidity and precipitation were more deterministic in regulating the emergence. During nonseasonal months, relative humidity is more crucial to emergence; contrary to it, rainfall contributes equally with humidity to bring about emergence in the seasonal months. A stenothermic ambience ranging from 21 to 35 °C throughout the year across the tropical and subtropical India contributes synergistically with either relative humidity or rainfall to cause emergence in *A. mylitta* population. Temperature alone has non-significant effect to shift the paradigm of *A. mylitta* emergence.

The abiotic factors influence the internal genetic factors in regulating breaking of diapause and subsequent eclosion of adult *Antheraea* spp. Although species specific information on the genetics of *A. mylitta* emergence remains unavailable, duality of *timeless* and *period* genes in maintaining the biological rhythmicity of *Antheraea* spp. has been key to emergence. Association of PER protein to *Antheraea pernyi* circadian clock specific neurons has been deciphered. Diapause and voltinism are the key parameters characterizing the biological or circadian clock of *Antheraea* spp. The stimulations for break of diapause and emergence of adult moths have been directly linked to PER proteins. PER-positive neurons were connected to eclosion hormones and prothoracicotropic hormones. Expressions of these hormones are seasonal which is extrapolated in *Antheraea pernyi.* The entire pathway is in a feedback loop. Seasonal phenomenon like change in photoperiodicity directly controls the internal circadian rhythm of the moth. The plausibility of erratic emergence is inversely related to synchronized release of PER and timeless proteins which depends on the synchronization in expression of concerned genes. The expression of genes in turn depends on external climatic conditions which initiates ethological modulations in silkworm from diapause to pre-eclosion to eclosion specific behavior of tasar silkworm.

Keywords

emergence · climatological factors · behavioral modulations · genetic factors · *Antheraea* spp.

Introduction

Vanya sericulture is a subject of predominance in India. The term *vanya or wild silk* defines sericulture practices confined to the following species of lepidoptera (Saturniidae): tropical tasar: *Antheraea mylitta*, oak tasar (temperate): *Antheraea pryolei*, Muga: *Antheraea assama*, Eri: *Attacus ricini*. The association of man and silk dates back to the era of the Vedic period. In Indian subcontinent, the earliest evidence of sericulture was documented from Chanhudaro and Harappa along the banks of the Indus River from 2450 to 2000 BC. Scanning electron microscopic examination of threads retrieved from Harappa indicated that they belonged to *A. mylitta* and *A. assamensis*, whereas those from Chanhudaro were from *Attacus ricini*. Greece and Persia had been using wild silk from ancient times which was termed as "Amorgina," and the garments made were termed as "Amorgian garments" (Singh et al. 2014). Of all the vanya silkworms cultured in India, *A. mylitta* makes a difference in terms of raw silk production and revenue generation. As per Annual Report 2021–2022, India produced 1315 MT of tasar silk out of the 36,582 MT total silk production, which is almost 3.60% of total silk production including mulberry and 14.36% of vanya silk production. *Antheraea mylitta* (Drury 1773) has been reared across ten tropical Indian states. The states are Chhattisgarh, Jharkhand, West Bengal, Orissa, Andhra Pradesh, Telangana, Maharashtra, Madhya Pradesh, Bihar, and Uttar Pradesh. The states are located in eastern and central India. Tropical tasar *A. mylitta* ranges from 12–31°N to 72–96°E (Jolly et al. 1974). More or less 3.5 lakh tribal farmers from the most remote of the difficult forest regions are associated with tasar sericulture. A total of 44.50 million Ha of forest area supports tasar sericulture (Mazumdar et al. 2023). The Bengal district Gazetteers, during 1910, described tasar cocoon collection as an inferior form of occupation of the forest dwellers also known as indigenous people. Growing demand for tasar silk, coupled with the popularization of tasar sericulture by the Government of India during the 1970s, had prompted several farmers to take up tasar sericulture as an alternative source of income. Various projects like the United Nations Development Programme (UNDP), Catalytic Development Programme (CDP), SGSY (Swarnjayanti Gram SwarozgarYojana), Mahila Kisan Sashaktikaran Pariyojana (MKSP), and Tribal Development Fund (Tribal Sub Plan—TSP) have also demonstrated the successful seed production models involving local communities and NGOs (Amarnath et al. 2007; Sathyanarayana et al. 2016). The District Mineral Foundation (DMF) Amendment Act 2015 has also promoted tasar sericulture in the mining-affected district, which has a poverty headcount ratio of 47% (Shalya and Agarwal 2018). Recently, many progressive rearers in Bihar and Jharkhand have shifted tasar sericulture from forest to their agricultural land through community-based organization as a successful small-scale producer organization by involving women and youths (Sathyanarayana et al. 2016). The polyphagous nature of silkworms provides a multifaceted ecological benefit to humankind through the food web, nutrient recycling, and carbon sequestration (Hussain et al. 2021; Jolly et al. 1968; Mevada et al. 2021). The non-diapause cocoons are generally called seed cocoons and are used for the preparation of dfls (disease-free laying) after thorough monitoring for pebrine disease. The diapause cocoons are used for both reeling and dfls production (only disease-free cocoons considered). Therefore, the silkworm-rearing (outdoor), grainage (dfls preparation house), and reeling (silk threads separation) activities are simultaneously operating in most of the districts. Also, these activities are interconnected livelihood practices of many tribal populations (Rathore et al. 2019). Besides tropical tasar, India also practices temperate tasar sericulture, commonly known as oak tasar, along the terrains of Uttarakhand, Jammu and Kashmir, and Manipur. The silkworm responsible for oak tasar is *Antheraea proylei* Jolly. The silkworm has its predecessors from China. A hybrid of *Antheraea pernyi* and *Antheraea roylei* was made to obtain *A. proylei* which has been adopted as oak-reared silkworms by Indian farmers across the foothills. Equivocality in nomenclature exists vide ICZN opinion of 2027 as it is a hybrid (Morrone 2000; Peigler 2012). In 1920, W. Tutt has already crossed and named female *pernyi* x male *roylei* as *moorie* and male *pernyi* x female *roylei* as *kirbyi* (Peigler 2012). The concept of priori stands irrelevant as all the lineages are hybrids following ICZN (Morrone 2000). *Antheraea pernyi*, *Antheraea roylei*, *Antheraea yamamai*, *Antheraea mylitta*, and *Antheraea assamensis* are the commercially exploited Lepidoptera belonging to the genus *Antheraea* across Indo-China for obtaining non-mulberry silks. Understanding the lineages of the aforementioned species belonging to *Antheraea* across the Indo-China region is of significant importance before we discuss the influence of genetic, physiological, and environmental factors on the emergence of silkworms. *Antheraea pernyi* is the tussah silk moth of China which has been hypothesized to have evolved from *A. roylei*. Peigler (2012) remarked that cytological and morphotaxonomic evidence opined little possibility of establishing *pernyi* and *roylei* as separate species. Rather it was an artificial selection of *A. roylei* that resulted in *A. pernyi* (d'Abrera 2012). *A. pernyi* has been introduced to several places across the world since the mid-1800s (Peigler 2012). Hence, relatedness of *roylei*, *pernyi*, *proylei*, and possibly *mylitta* does exist. Maximum research on the evaluation of genetic, physiological, and behavioral studies in the make and break of diapause, eclosion, and genetic basis of biological rhythms that ultimately leads to emergence has been done on *A. pernyi*. Extrapolation of the information generated on *A. pernyi* and its relatedness to *A. mylitta* (Jolly et al. 1971) would enable us to understand the emergence dynamics in *A. mylitta*. Hymenopterans, hemipterans, and noninsect predators like birds, monkeys, and squirrels were the major predators recorded on the silkworm.

Hemipteran predators like *E. furcellata* and *S. collaris* were recorded more frequently within the rearing field. In addition, *Rhynocoris marginatus* (Fabricius) was also recorded for attacking silkworms. Paper wasp, *Polistes olivaceus* (Deg.) (Hymenoptera: Vespidae), were the visiting predators of silkworms. *Polister sigma*, *Delta conoideum*, *Sceliphron madraspatanum*, and *Chalybion bengalense* were also recorded in the silkworm-rearing field. Common birds like crow, Indian koel, black drongo, red-vented bulbul, great coucal, myna, shikra, pigeon, common cuckoo, sparrow, crane, hoopoes, and jungle babler were also recorded in the tasar silkworm-rearing fields. However, rufous treepie, black drongo, and crow were the common bird predators of silkworms. Other predators like spiders (grass spiders, yellow sac spiders, jumping spiders, and wolf spiders), snakes, rats, lizards, etc. were also recorded by different sericulture scientists. About 41.49% ± 14.10% mortality of silkworms was recorded due to the disease-causing organisms during rearing (50.38% and 39.09% during first and second generations, respectively). Out of which, nearly 24.01% ± 10.72% and 16.91% ± 5.69% of silkworm mortality were associated with viral and bacterial pathogens, respectively. The impacts of both the pathogens were statistically on par with each other ($P > 0.05$). However, fungal infection (0.56% ± 0.73%) was substantially low compared to virus and bacterial infection ($F = 17.63$; $df = 2,15$; $P \leq 0.01$). Further, viral ($F = 13.70$; $df = 4,15$; $P \leq 0.01$) and bacterial ($F = 18.76$; $df = 4,15$; $P \leq 0.01$) infections also varied significantly among the different instars of the silkworms compared to fungal infection ($P = 0.465$). Mortality of silkworms was considerably high during the fifth and fourth instar stages. Usually, mechanical protection of silkworms from the first to early third instar was done by erecting a nylon net on the hosts' plants during the rearing. Enhanced cocoon yield was observed in such cases. The average cocoon yield per dfl was 54.13 ± 7.19 (±SE) in an open field rearing (farmers) practice. Meanwhile, cocoon yield per dfls was 104.44 ± 5.41 when the early instar silkworms were protected using nylon net, which was nearly 92.94% higher yield than the farmer practice.

Species belonging to the silkworm-producing genus of *Antheraea* across the Indian subcontinent and surroundings are holometabolous with egg, larva, pupa, and adult stage (Singh et al. 2014). The adults are semelparous. The emergence of *Antheraea* is usually univoltine, bivoltine, or multivoltine (Peigler 2012). However, the problem lies in the synchronization of the emergence pattern, which adversely affects the rearing of silkworms when exploited commercially (Jolly et al. 1971). Globally significant research on the linkage between genetical, physiological, and other environmental factors regulating the emergence of silkworms has been done. Voltinism and diapause are the two functional events regulating insect emergence. The genes and release of hormones affecting these phenomena are very much quasi-dependent (Hall 1995). Seasonal emergence of *Antheraea* is a phenomenon dependent on biological and circadian rhythm. *Per* and *timeless* are the two of the most important genes regulating the events. In the light of abiotic influence on moth emergence, Mazumdar et al. (2023) clustered events into synchronized and unsynchronized pattern. The abiotic factors stimulate the internal physiological factors into further subsequent events like eclosion, molting, and emergence.

Influence of Abiotic Factors on the Emergence of *Antheraea* spp.

The emergence of *Antheraea* is a phenomenon of importance when it comes to commercial mass rearing as observed in the case of *A. mylitta* in the Indian context. Jolly et al. (1979) remarked that proper coordination between emergence, oviposition, rearing, and mating is key aspect for success in sericulture of *A. mylitta*. In tropical tasar silkworms, emergence is observed across the year although few emerge during the winter seasons; however, the percentage of moths coupling and their fecundity, oviposition, and hatching vary (Shapiro 1993). Usually in the Indian tropical template, the population of *A. mylitta* is bivoltine and trivoltine in nature. The states practicing bivoltine populations are Bihar, Jharkhand, and Keonjhar region in Orissa. Contrary to *Bombyx mori*, the major difficulty in rearing *A. mylitta* is the erratic emergence observed across the season. Since *A. mylitta* is semidomesticated in nature, abiotic factors mainly comprising photoperiod, temperature, humidity, and rainfall contribute significantly to the voltinism process undergone by the silkworm *A. mylitta*. It is termed semidomesticated as half of its life cycle undergoes in confinement and rest in open under natural field condition. The grainage activities of the *A. mylitta* (i.e., preservation, oviposition) occur in a closed place, whereas the rearing activity of silkworms is undergone in the open. In the trivoltine population, three generations are observed, the first from July to August, second from September to October, and third from November to December. Emergence occurs after mating of adults emerging from diapausing pupa, and rearing commences after oviposition. The diapause in the first generation lasts for 15 days, in second generation lasts for 20 days, and in third generation is close to 150 days (Shapiro 1993). In bivoltine, two generations are observed: first generation in July to August and second generation in September to October. The diapause period of the first generation is 9–10 days, and that of the second generation is 180 days approximately. The rearing and oviposition parameters of the three generations vary greatly which is attributed to the climatic conditions. The importance of climatological factors in the mating and reproduction of tasar silk moth is evident (Jolly et al. 1974; Nayak and Dash 1989). Information on effect of the abiotic factors on the life history trait of the tasar silkworm (*A. mylitta*) is important for developing a reservoir place for the storage of cocoons with proper simulations. In India, such study on *A. mylitta* has been undertaken, and observation has been reported from the State Sericulture farm in Durgapur, Orissa (Shapiro 1993), and Central Silk Board seed farm in Kathikund and Kharsawan, Jharkhand. Siddaiah et al. (2014) established a strong correlation between the bioclimatic factors and emergence of insects. Messenger (1958) in the narrative mentions the importance of temperature, humidity, and rainfall in the emergence of insects including silk moths. In India, since the tasar is practiced over a wide geographical region, subtle to prominent difference in generation time, brushing schedule, and cocoon harvested is observed. Each region practicing tasar is unique as exhibited by the normalized difference vegetation index (NDVI) (Fig. 1). The observation has been based on the clustering of NDVI values through Agglomerative Hierarchical Clustering (AHC) with dissimilarity index based on Euclidean distance and Ward's method (Fig. 2), the sites were classified into the following subsets; subset 1: Kharsawan (Jharkahnd) and Bhandara (Maharastra); subset 2: Sundergarh (Orissa) and Balaghat (Madhya Pradesh); subset 3: Patelnagar (West Bengal) and Boirdadar (Chattisgarh). Kota (Chhattisgarh) has been more close to subset 1, whereas Kathikund (Jharkhand) has been located more close to subset 2 and 3. The places were also clustered in terms of the production of silkworm cocoons based on AHC (Fig. 3). The clustering was exhibited: subset 1, Balaghat, Boirdadar, Sundargarh, and Patelnagar;

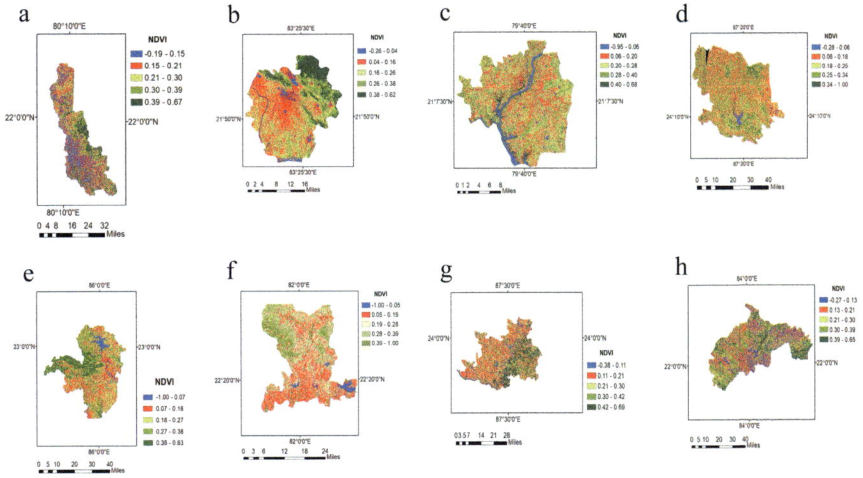

Fig. 1 Normalizeds difference vegetation index mapping of the regions practicing sericulture of *Antheraea mylitta* Drury in India: (**a**) Balaghat (Madhya Pradesh), (**b**) Boirdadar (Chhattisgarh), (**c**) Bhandara (Maharashtra), (**d**) Kathikund (Jharkhand), (**e**) Kharsawan (Jharkhand), (**f**) Kota (Chhattisgarh), (**g**) Patel Nagar (West Bengal), (**h**) Sundargarh (Orissa)

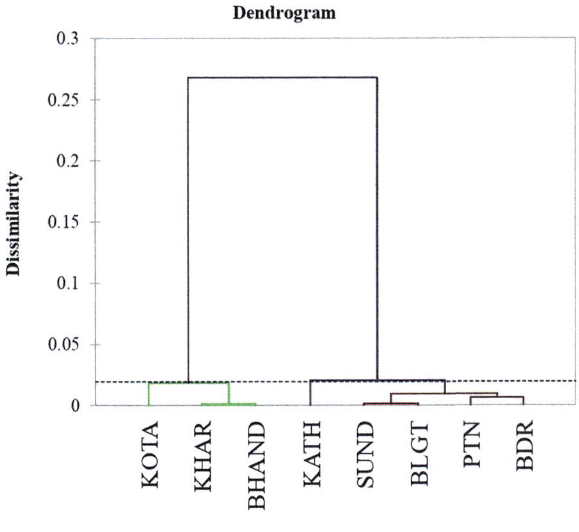

Fig. 2 Agglomerative hierarchical clustering of the rearing sites estimated from the normalized difference vegetation index values based on Euclidean distance with Wards' method. *KHAR* Kharsawan, *BHAND* Bhandara, *KATH* Kathikund, *SUND* Sundargarh, *BLGT* Balaghat, *PTN* Patel Nagar, *BDR* Boirdadar

subset 2: Kharsawan, Kathikund, and Bhandara. The performance of Kota was different from the rest. Hence, the widely diversified geographical uniqueness of each region calls for a meticulous analysis of the effect of temperature, humidity, and rainfall on the life history traits of *A. mylitta* based on which emergence of silk moths may be early or delayed (Bhatia and Yousuf 2014).

Mazumdar et al. (2023) in a study have done an exhaustive analysis of the effect of temperature, humidity, and rainfall on the emergence of *A. mylitta* in the seasonal and nonseasonal months from Kathikund and Kharsawan located in the Indian state

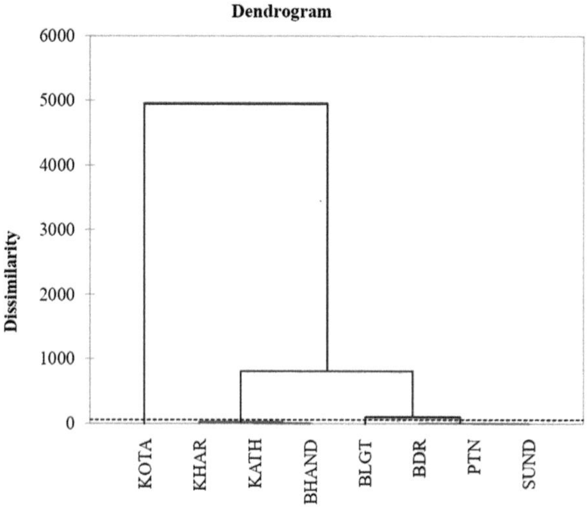

Fig. 3 Agglomerative hierarchical clustering of the rearing sites estimated from the cocoon yield per DFLs based on dissimilarity index. *KHAR* Kharsawan, *BHAND* Bhandara, *KATH* Kathikund, *SUND* Sundargarh, *BLGT* Balaghat, *PTN* Patel Nagar, *BDR* Boirdadar

of Jharkhand. The degree of the effect of all the environmental variabilities on emergence was quantified. Based on the time of emergence, the population appears to be segregated sub-locally. Gradually these small changes accumulate with time which ultimately leads to intraspecific differences sufficient to bring about speciation as a result of interruption in gene flow. Mazumdar et al. (2023) clustered the months into three groups based on the emergence of the bivoltine eco-race of *A. mylitta*. The seasonal months were June, July, September, and October, and the nonseasonal months with unsynchronized emergence were March, April, and May. Nonseasonal months with negligible emergence were November, December, January, and February (Fig. 4). The emergence during nonseasonal months was termed erratic or nonseasonal emergence. Significant differences in the mean value of rainfall, temperature, and humidity were observed between seasonal and nonseasonal months. For rainfall, in seasonal months, the mean value was 221.43 ± 26.47 mm with a range of 140–328 mm of rainfall. Relative humidity observed during seasonal months (i.e., June to October): Mean value was 79.91 ± 2.92% with a range of 66.7–86.7%. The average temperature was 30.79 °C with a range of 29.5–32.5 °C. During the nonseasonal months (i.e., November to May), the average rainfall was 16.14 ± 5.31 mm with a range of 1–78 mm of rainfall. Relative humidity was 55.39 ± 2.7% with a range of 36.7–68%. Temperature ranged from 21 to 33.5 °C with a mean value of 26.89 °C (Fig. 5). The proportional difference in *A. mylitta* emergence between seasonal and nonseasonal months was that *A. mylitta* emergence was 50 times more during the seasonal month compared to the nonseasonal months (Fig. 6). Correlation analysis exhibited significant relation when annual emergence irrespective of the seasonal bifurcations was plotted against

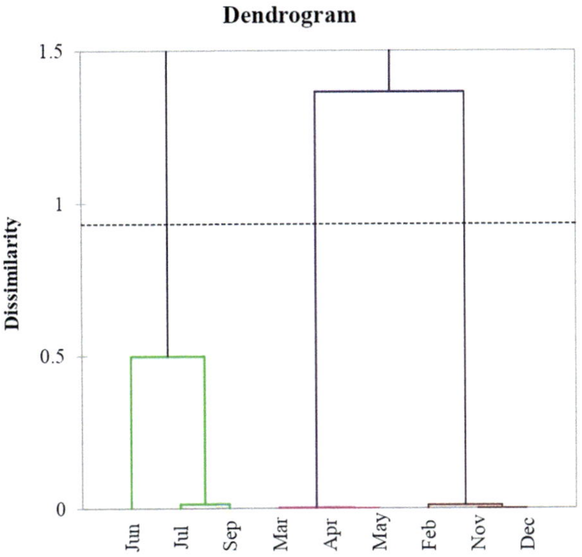

Fig. 4 Dendrogram depicts the dissimilarities (Euclidean distance based on Ward's method) between different monthly emergences of *A. mylitta*. The solution resulted in the three classes: class (i) June, July, September; (ii) March, April, May; (iii) November to February (Mazumdar et al. 2023)

humidity (0.75; $p < 0.001$) and rainfall (0.81; $p < 0.01$). However, the effect of temperature (0.21; $p > 0.05$) on emergence was not much of significant in the tropical and subtropical templates. When the analysis was further bifurcated into seasonal and nonseasonal months, the correlation matrix exhibited rainfall humidity was more crucial to the emergence of *A. mylitta* compared to temperature. During seasonal months, the effect of relative humidity (0.75; $p < 0.0001$) on emergence was more crucial relative to the effect of rainfall (0.32; $p < 0.001$). However, during the seasonal months, rainfall (0.41; $p < 0.05$) and humidity (0.64; $p < 0.0001$) were equally impactful on the emergence of tasar silkworms. The observations of the correlation study were further subjected to cluster analysis for further validation. Factor loading values suggested humidity and rainfall as the two significant parameters regulating the emergence of *A. mylitta* population (Mazumdar et al. 2023). The difference in the impact of humidity and rainfall on the emergence of silk moths between seasonal and nonseasonal months was because in seasonal months, humidity was positively correlated to rainfall whereas during nonseasonal months, rainfall was less dependent on the increase in humidity (nonpositive correlation) (Mazumdar et al. 2023). Further, Mazumdar et al. (2023) summarized that temperature was constant as a variable with little role in increasing or decreasing the emergence of *A. mylitta* across the seasonal or nonseasonal months. The temperature was pretty constant since the tasar sericulture zones in India fall under tropical and subtropical template. Average temperature in nonseasonal months was 26.9 °C compared to 30.8 °C in seasonal months. According to Rogers and West-brook (1985), a

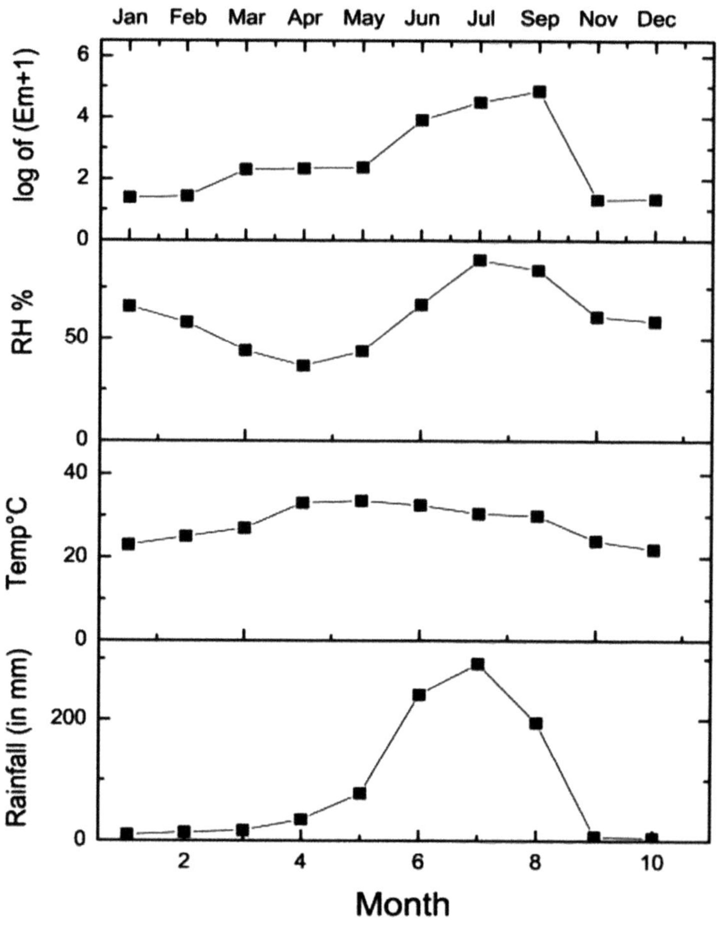

Fig. 5 Monthwise variations in humidity (RH), temperature (Temp), and rainfall with the emergence (Em) of *A. mylitta* observed at Kathikund and Kharsawan, Jharkhand, India, during 2019 and 2020. (Mazumdar et al. 2023)

minimum difference of 10 °C is necessary to delay or fasten the process of emergence as observed in *Homoeosoma electellum*. In the present study, there was a difference of 4 °C between seasonal and nonseasonal months. It may be understood that constant temperature works in synergy with humidity and rainfall to bring about the emergence of *A. mylitta* (Mazumdar et al. 2023). In September, increased rainfall during the monsoon brings about the eclosion of pupa to adult. The rainfall and humidity during this month is maximum which gradually decreases thereafter. Moreover, the duration of the photoperiod available during this month also has an effect on stimulation and release of eclosion hormones which causes the final emergence (Truman 1971a). The emergence of insects is an imprinted behavior regulated

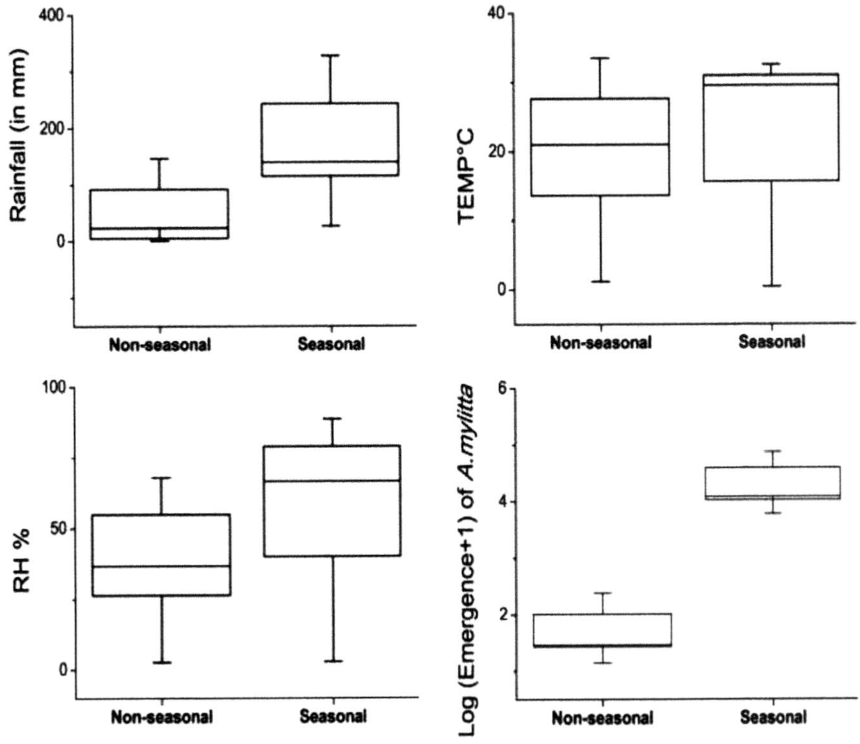

Fig. 6 A comparative box plot representation of humidity, rainfall, temperature, and emergence of *A. mylitta* between seasonal and nonseasonal months at Kathikund and Kharsawan during 2019–2020 (Mazumdar et al. 2023)

by climatological conditions (Morohoshi 2001). Thus, abiotic and biotic factors act in coordination to bring about the emergence of *A. mylitta*.

Similarly, Shapiro (1993) made an exhaustive account of trivoltine species of *A. mylitta* bionomics. Emergence and coupling were highest from July to October. Temperature and humidity ranged between 27 and 30 °C and 90 and 98%, respectively. The emergence of *A. mylitta* peaked in September with a temperature of 27.6 °C and relative humidity of 98.7%. The emergence and coupling gradually decreased from November to June. Although a significant difference between the emergence pattern of male and female tasar silk moths was observed, climatological factors exhibited a positive correlation irrespective of sex across the seasons. Further season-wise differences in fecundity, percent hatching, and percent coupling of silkworms with weather parameters were noted. Fecundity was highest during January which decreased gradually to May. Similarly hatching percentage was highest in November and gradually decreased in May. Shapiro (1993) analyzed the findings attributing that temperature and humidity being maximum was the reason for bringing about the highest emergence in September. The emergence peaked due to a conducive temperature and humidity simulating change in reproductive physiology. The gradual decrease in temperature and humidity from October onward resulted in the lowest emergence in November. In summer months, low emergence was attributed to high temperature and lower relative humidity (Shapiro 1993). The climatological extremes during winter result in the population of *A. mylitta* entering diapause.

Mansingh (1972) made a detailed account of the emergence and development of *A. pernyi* to seasonal changes from central New York and southern Ontario. It was reported that the developmental physiology of the larval instars and pupa and finally eclosion to adult *A. pernyi* depended on seasonal changes associated with the host plant. The nutritional quantity of oak leaves (the host plant of *A. pernyi*) affected the voltinism process undergone by the silkworms. In Ontario, the larval developmental phase feeding on young oak leaves was delayed during early summer. Each larval instar was delayed to twice that of the New York population. On an average, the larval developmental time at New York of three instars was first generation, 22 ± 2; second generation, 23 ± 3; and third generation, 23 ± 2 days, whereas the larval developmental time of the Ontario population was first generation, 57 ± 3; second generation, 68 ± 3; and third generation, 83 ± 5 days (Mansingh 1972). Despite the differences in larval developmental time between the locations, significant differences in the pupal weight were not observed. The deleterious effect of aging leaves on larval development was evident. The damage was extreme as larva feeding on aging leaves has a retarded growth leading to high mortality. Further, it was observed that mortality of larvae due to quality and age of leaves was higher during the third generation population in a year. The nutritional inadequacy of leaves was due to the absence of biochemical, alkaloid, and secondary metabolite derivatives present in leaves. The significant being was the low amount of tannin during early summer resulting in swift growth and development of larvae (Southwood 1961); however, in late summer (July onward in North America), increased quantity of tannin in leaves slowed the process of larval development (Feeny 1970) which adversely affected the subsequent biology like emergence and survivability of the insects. The quality of leaves taken as food by the larva during the development is crucial in the later diapausing phase and subsequent emergence. The quantity of glucose, trehalose, glycogen, sorbitol, and glycerol in the diapausing pupa determines the appropriateness of subsequent physiological events. The feasibility of the locations supporting rearing depends on the microclimatic variability such as temperature, humidity, soil nutrient, and day length which favors the completion of various stages in the life cycle of insects (Mansingh 1972). The role of day length is crucial as the photo spectral quality brings about the photochemical reactions in leaves ultimately leading to the phenomenon of aging (Humphries and Wheeler 1963).

Jolly et al. (1971) made a comprehensive study on the initiation and termination of pupal diapause regulated by photoperiod in the bivoltine *A. mylitta*. Although it was mentioned in the manuscript regarding role of temperature, his findings suggested more of day length and night (i.e., photophase and scotophase) during rearing period responsible for break and make of diapause. Tanaka (1950, 1951) has already mentioned on the influence of short photoperiod initiating diapause of

A. pernyi. Jolly et al. (1971) remarked that first generation pupa is prevented from diapausing as the silkworm larvae are exposed to long photophase during rearing in the month of July and August. Thus, prevention of diapause means continuous emergence cycle will persist with the first generation pupa. The second generation pupa enters diapause, thereby pausing the emergence cycle. It may be attributed to the fact that second generation silkworm larva is exposed to short day and long night conditions during rearing in the month of September, October, and November. Jolly et al. (1971) also reported that continuous photoperiod of 18 h is the maximum which terminated diapause beyond which an increase in day length fails to initiate any provocation. It was reported that larval stages were sensitive to photoperiod with maximum during the final two instars (Tanaka 1950, 1951). Williams and Adkisson (1964) reported that long-day conditions can be simulated by putting pupa of *A. pernyi* in 2–3 °C chilling. However, such experimental work needs to be replicated in *A. mylitta*.

Physiology of *Antheraea* spp. Rhythmicity in the Context of Emergence

Although no information is available on *A. mylitta* or other *Antheraea* species related to vanya sericulture in India, Truman (1972) elaborately assessed the circadian rhythmicity associated with the emergence of *A. pernyi*. The connection of the photoperiodic cycle with the photoreceptors of the circadian rhythm was determined. The role of optic lobes, compound eyes, subesophageal ganglion, corpora allata, and corpora cardiac in maintaining rhythmicity on gating eclosion was made. Transplantation and implantation experiments were carried out to localize the center for circadian clock rhythmicity controlling eclosion of the *A. pernyi*. It has been a known fact that behavioral and physiological changes are under the influence of the intrinsic circadian clock (Truman 1971a). The location of the clock and understanding the pathways have been an area of interest for the researchers across the globe. Researchers on photoperiodism pioneered the research on photosensitivity of insect brains. Lees (1964) exhibited the facilitation of the brain in bringing about the photoperiodic response. The photosensitive nature of *A. pernyi* brain was proven by Williams and Adkisson (1964). Eidmann (1956) made an impactful observation on the lesser involvement of the insect eye in the circadian clock. Similarly, eclosion and the subsequent emergence of *A. pernyi* have been enforced by the circadian clock in the brain of silk moth (Truman and Riddiford 1970). The circadian clock directs the release of eclosion hormone by the neurosecretory cells which then stimulates the central nervous system to switch on the eclosion behavior from pre-eclosion behavior (Truman 1971b; Truman and Sokolove 1972). The significance of the brain in controlling gated eclosion was studied through ganglionic extirpations of nervous sections and debrained pupae with transplanted brain. It was noted that in the brain, an extra-optic photoreceptor stimulates the eclosion rhythm in *A. pernyi*. The receptor was highly specific to cerebral lobes in its location. Although various assumptive and deterministic studies suggested minimal involvement of the insect eye, however, their absolute abstain remains undeniable. It was evident that the insect eye modifies the output of rhythmic expression. The onset of daylight or "lights-on" phase during experiments was perceived through compound eyes which is communicated through neural connections to the brain area. Thus, the synchronization of photoreceptivity depended on the clock mechanism located in the brain and the immediate perception of daylight stimulation communicated through compound eyes. Truman (1972) made an account of observations based on experiments on the ability of the brain when transplanted as circadian clock regulator. Truman and Riddiford (1970) remarked that transplantation of brains between different silk moths resulted in different eclosion gates which signifies that rhythmicity is intrinsic and unique to the brains. The extirpation of the optic lobe as a whole resulted in different expressions compared to when extirpation was done after the cerebral lobe was subdivided and removed. The implantation of brain transects into lateral neurosecretory cells failed to evoke response capable of gating eclosion. The findings opine against the presence of a circadian clock in the lateral portion of the cerebral lobe (Truman 1972). The photosensitive portion of the brain-centered circadian clock regulates the diapause and emergence in *A. pernyi* (Williams and Adkisson 1964). Surgical maneuverings with the brain of *A. pernyi* resulted in a differential response. It was observed that the exclusion of the optic lobe from the brain of the diapausing pupa partially affected its photoperiod-dependent emergence dynamics (Williams and Adkisson 1964). Similarly, when the brain was regionated into lateral and median, the failure of the median portion to discriminate the day lengths was evident (Truman 1972).

Behavioral Modulations in the Circadian Clock Related to the Emergence of *Antheraea* spp.

Although no data is available on *A. mylitta* or other *Antheraea* silkworm species (Saturniidae) exploited in Indian sericulture, the information generated in *A. pernyi* was discussed in this segment. It was already discussed that the emergence of silk moths from pupa depended on the circadian clock located in the brain. In this segment, behavioral synchronization of insect emergence in response to different photoperiod was discussed. The emergence of insects comprises of eclosion and ecdysis. The development of adult insects does not follow any fixed time in a day; however, eclosion process is scheduled for a fixed period; hence, the term "gated" is used for eclosion (Pittendrigh 1966). Truman (1971a) made a detailed study on the impact of photoperiod in gating eclosion of *A. pernyi*. The circadian clock of pupa was subjected to certain conditions: The clock was put into a free run of continuous darkness to test the effect on "gate" of eclosion, a complete run against different regimens of photoperiod varying from 23 L(light):1D(dark) to 4 L:20D. In the experiments by Truman (1971a), diapausing pupa was subjected to 17 L:7D to break the diapause. Adult eclosion in *A. pernyi* is stimulated by the release of neurosecretory hormones from the brain 1.5 h before emergence. It is the hormonal release gated by the circadian clock which determines the synchronization of later events leading to emergence (Truman 1971a). *A. pernyi* usually emerged throughout for 3.5 h at a certain time of the day (Truman 1971a). The males emerged around 1 h earlier than the females. In the continuous darkness, eclosion hormones were stimulated for release at the end of the free-running cycle. The termination of the running time of eclosion clock was determined by the onset of photophase (i.e., lights-on) even when the photoperiod is as less as 4 h. The circadian clock is again restarted with the onset of darkness (i.e., lights-off). Truman (1971a) observed that the width of the eclosion gate varied with the duration of the photoperiod. Regimens with very short scotophase (23 L:1D) resulted in a wide gate. With the increase in duration of exposure to scotophase, the gate width decreased rapidly up to 3.5-h window as the lowest constant. Further increase in the period of darkness had little effect on the gate width. The accuracy of the clock at different photoperiods was estimated as the number of moths that emerged during the gate divided by the total number of moths. During the extremes, i.e., 23D:1 L, the clock has a maximum free-running time with an accuracy of 0.96 (maximum up to 1), whereas in 23 L:1D, the clock has a minimum free-running time of 1 h with an accuracy of 0.38 (Truman 1971a). Hence, it was observed that a minimum free run time of the circadian clock little over 6 h initiated into maximum accuracy of around 0.96. Increase in free run time duration beyond that has little effect on increase in accuracy (Truman 1971a). The entire duration of the free-running time of the circadian clock (period of darkness) is subdivided into the early synchronization phase and the later dark decay phase (Truman 1971a). In the 22D:2 L phase, the first 2 h is the synchronization phase whereas the remaining 20 h is the decay phase. Almost 50% of the events responsible for the emergence of the silk moth *A. pernyi* are attained in this phase. The events in this phase contribute maximally to the rhythmic synchronization of emergence in the silk moth. The free run of the circadian clock is terminated with the onset of daylight or "lights-on." The clock again restarts with the next dark phase. The eclosion hormone is released after the onset of the lights-on phase. The exception was observed in a regimen with 23D:1 L, where the eclosion hormone was released before the light phase (Truman 1971a).

Influence of Genetic Factors on the Emergence of *Antheraea* spp.

The environmental factors acted as a clue to which the biological receptors within the insect body responded to initiate the process of eclosion. Further introspection has led to the relatedness of the environmental factors and their implication on the insect genetics. The molecular events occurring within *A. pernyi* unfolded the tale of two genes: "*period*" and "*timeless*." The hypothesis on their interdependency has been proven (Reppert and Sauman 1995). Hardin et al. (1992) remarked that the mechanism of action of PER protein in maintaining the circadian clock is through negative regulation of its transcription. The entry of *per* mRNA into the nucleus is an environmental temperature-dependent phenomenon (Reppert and Sauman 1995). Besides another gene *tim* also facilitates entry of *per* mRNA into the nucleus. Similar to *per*, *tim* is temperature-dependent autoregulatory in its mode of action as a transcription regulator (Sauman and Reppert 1996). Oscillation of *per* mRNA in the brain and eyes of *A. pernyi* has been documented. Photosensitivity of *per* receptors in the nucleus was tested over 24-h cycle based on purified anti-PER antibody. Thereafter, the photoreceptor nuclei obtained at different hours of stages after lights-on were stained. Minimum immunoreactivity of PER was observed between 12 and 18 h after lights were on. A circadian rhythm of photosensitive immunoreactivity was exhibited by the PER. RNA from the eyes of silk moth *A. pernyi* exhibited maximum reactivity between 16 and 18 h after lights-on and minimum between 6 and 10 h of lights-on (Sauman and Reppert 1996). The central portion of the *A. pernyi* brain houses the circadian clock that is responsible of eclosion (Truman and Riddiford 1970). Eight PER-positive cells were noted consistently during the adult as well as pupal stages. The cytoplasm of the cells was heavily stained whereas no stain was observed in the nucleus. The cytoplasmic staining was intense in cells at 16–22 h and low at 4–8 h after lights-on. Sauman and Reppert (1996) concluded that PER-containing and PER-producing cells were the same. The PER is heavily localized in the cytoplasm of the cell. In the eye of *A. pernyi*, per mRNA and protein oscillation manifested rhythmicity with a temporal difference of 4–6 h between the two rhythms. The oscillation of PER was due to its occurrence in nuclei of photoreceptor cells. The circadian rhythm controller is located in the dorsolateral portion of the cerebrum not in the eye of *A. pernyi*. Each cerebral hemisphere consists of four PER-containing cells. PER is present in dendritic as well as axonal regions of these cells (Sauman and Reppert 1996). Each cell is strategically so aligned that the axons extend close to one side of corpora allata and influence the release of prothoracicotrophic hormone (PTTH) and eclosion hormone (EH). The mechanism however remains unknown whether PER is released and translocated by vehicles to corpora allata to initiate further stimulus or it stimulates the release of neuropeptides which are translocated to corpora allata for further sequential events. The latter explanation appears more likely owing to the larger size of PER proteins. In *A. pernyi*, PER present in axon of these cells is more responsible for regulating the circadian rhythm (Sauman and Reppert 1996). Although the temporal difference in per mRNA expression and PER protein level is observed in the eye, no difference is observed in the brain of *A. pernyi*. Hence, it might be concluded that the posttranslational and co-translational events are highly tissue-specific. In the brain, the mRNA cycle directly influences the protein cycle without much delay. Evidence supporting autoregulation of expression of *per* mRNA is lacking, thereby suggesting the alternative antisense *per* mRNA in limiting rhythmic activity of *per* mRNA. Silk moth antisense RNA rhythm counters the circadian cycle of sense RNA expression. Moreover, the *per*-expressing cells also produce TIM protein which co-localizes with PER as the second important molecule in PER-driven rhythmicity of the clock. Mutation in TIM also disrupts circadian rhythm as it affects the posttranscriptional activity of *per* gene. The entry of PER into the nucleus is hindered (Vosshall et al. 1994) in *tim* mutant organisms. However, there is no homology between *per* and *tim*. PER and TIM form a codependent feedback loop. The rhythmicity of PER and TIM are in synchronization with each other as observed in the head of the organism to the extent that the timing of their peak activity was indistinguishable (Sehgal et al. 1995). Thus, it is the dual tango of tim and per gene that guides the circadian rhythm controlling the eclosion and emergence of *A. pernyi.*

Although information on various life history traits and genetics of commercially exploited Indian silkworms belonging to the genus *Antheraea* is scanty, the findings in *A. pernyi* could be very much extrapolated to these Indian species. Sauman and Reppert (1996) have stated that cellular localization and mechanism of action of PER expression are not unique in silkworm moths belonging to *A. pernyi* as other saturniid moths exhibit the sameness where similar results were inferred. The localization of PER is restricted to eight neurosecretory cells in dorsolateral cerebrum of all saturniid moths. Similarly, histology and histochemistry exhibited PER-positive staining in the cytoplasm and axonal regions of the cells. Jolly et al. (1971) remarked that *A. pernyi* is close to *A. mylitta* and mentioned subtle difference in physiology, emergence pattern, circadian clock-cycle might exist. Thus, we conclude that possibly a similar mechanism in controlling of circadian clock will be exhibited in *A. mylitta*, which has been noted as close relative of *A. pernyi* (Jolly et al. 1971).

Conflict of Interest The authors declare no conflict of interest.

References

Amarnath S, Sathyanarayana K, Khanna RP (2007) Tasar silkworm seed production in private sector under special SGSY projects in Bihar and Jharkhand - a breakthrough. In: Abstracts of international conference "sericulture challenges in the 21' century" (Serichal 2007) & the 3'1 BA GSA meeting, 18–21 September 2007, Vratza, Bulgaria, pp 23–24

Annual Report (2021–22) Central Silk Board. https://csb.gov.in/wp-content/uploads/2023/04/CSB-Annual-Report-2021-22.pdf. Accessed 2 Nov 2023

Bhatia NK, Yousuf M (2014) Effect of rearing season, host plants and their interaction on economical traits of tropical tasar silkworm, *Antheraea mylitta* Drury-an overview. Int J Ind Entomol 29:93–119

d'Abrera B (2012) Saturniidae mundi: saturniid moths of the world, part II. Hill house, Melbourne, p xlix + 178

Drury D (1773) Illustrations of natural history, vol II. B. White, London, 90 pp

Eidmann H (1956) Uber rhythmische Erscheinungen bei der Stabheuschrecke *Carausius morosus.* Br Z vergl Physiol 118:370–390

Feeny P (1970) Seasonal changes in oak leaf tannins and nutrients as a cause of spring feeding by winter moth caterpillars. Ecology 51:565–582

Hall JC (1995) Tripping along the trail to the molecular mechanism of biological clock. Trends Neurosci 18:230–240

Hardin PE, Hall JC, Rosbash M (1992) Circadian oscillations in period gene mRNA levels are transcriptionally regulated. Proc Natl Acad Sci 89(24):11711–11715

Humphries EC, Wheeler AW (1963) Physiology of leaf growth. Annu Rev Plant Physiol 14:385–409

Hussain S, Hussain S, Guo R, Sarwar M, Ren X, Krstic D, Aslam Z, Zulifqar U, Rauf A, Hano C, El-Esawi MA (2021) Carbon sequestration to avoid soil degradation: a review on the role of conservation tillage. Plan Theory 10(10):2001

Jolly MS, Chaturvedi SM, Prasad SA (1968) Survey of Tasar crops in India. Indian J Seric 1:50–58

Jolly MS, Sinha SS, Razdan JL (1971) Influence of temperature and photoperiod on termination of pupal diapause in the Tasar silkworm, *Antheraea mylitta*. J Insect Physiol 17(4):753–760
Jolly MS, Sen SK, Ahsan MM (1974) Tasar culture. Ambika Publishers, Bombay, pp 105–124
Jolly MS, Sen SK, Sonwalker TN, Prasad GK (1979) Non-mulberry silks. Food and Agriculture Organization of the United Nations, Rome, pp 1–179
Lees AD (1964) The location of the photoperiodic receptors in the aphid *Megoura viciae* Buckton. J Exp Biol 41(1):119–133
Mansingh A (1972) Developmental response of Antheraea pernyi to seasonal changes in oak leaves from two localities. J Insect Physiol 18(7):1395–1401
Mazumdar SM, Reddy BT, Chandrashekharaiah M, Chowdary NB, Chattopadhyay S, Rathore MS, Sathyanarayana K (2023) Influence of abiotic factors on seasonal and nonseasonal emergence of Tasar silkworm, *Antheraea mylitta* Drury. J Environ Biol 44:445–451
Messenger PS (1958) Bioclimatic studies with insects. Annu Rev Entomol 4(1):183–206. https://doi.org/10.1146/annurev.en.04.010159.001151
Mevada RJ, Nayak D, Patel DP, Tandel MB (2021) Potential of tasar silkworm (*Antheraea mylitta*) excreta as fertilizer on growth, yield and quality of rice. J Environ Biol 42:1070–1077
Morohoshi S (2001) Development physiology of silkworms. Science Publishers, p 287
Morrone JJ (2000) International code of zoological nomenclature. Acta Zool Mex:253–255
Nayak BK, Dash AK (1989) Effect of refrigeration on egg incubation period of the tasar silk insect a. *mylitta* (Saturniidae). J Lepid Soc 43:152–153
Peigler RS (2012) Diverse evidence that *Antheraea pernyi* (Lepidoptera: Saturniidae) is entirely of sericultural origin. Trop Lepid Res:93–99
Pittendrigh CS (1966) Light pulse PRCS and entrainment in Drosophila. Z Pflanzenphysiol 54:275–307
Rathore MS, Chandrashekharaiah M, Sinha RB (2019) Linkage of rearers for production of commercial DFLs in tropical tasar silkworm, *Antheraea mylitta* D. J Entomol Zool Stud 7(6):1048–1051
Reppert SM, Sauman I (1995) Period and timeless tango: a dance of two clock genes. Neuron 15(5):983–986
Rogers CE, West-brook JK (1985) Sunflower moth (Lepidoptera, Pyralidae) overwintering and dynamics of spring emergences in the Southern Great Plains. Environ Entomol 14:607–611. https://doi.org/10.1093/ee/14.5.607
Sathyanarayana K, Nandi S, Shamshad Alam M (2016) Cost benefit analysis of commercial seed production of tasar silkworm (*Antheraea mylitta* D.) in the central Indian tribal belt. Sericologia 56(1):18–25
Sauman I, Reppert SM (1996) Circadian clock neurons in the silkmoth *Antheraea pernyi*: novel mechanisms of period protein regulation. Neuron 17(5):889–900
Sehgal A, Rothenfluh-Hilfiker A, Hunter-Ensor M, Chen Y, Myers MP, Young MW (1995) Rhythmic expression of timeless: a basis for promoting circadian cycles in period gene autoregulation. Science 270(5237):808–810
Shalya C, Agarwal S (2018) Tasar rearing may bail out 600 mining affected families in Keonjhar. https://www.downtoearth.org.in/news/mining/tasar-rearing-may-bail-out-600-mining-affected-families-in-keonjhar-62260. Accessed 2 Dec 2023
Shapiro AM (1993) Effect of temperature and relative humidity on certain life history traits in *Antheraea Mylitta* (SATURNIIDAE). J Lepid Soc 47(3):244–247
Siddaiah AA, Prasad R, Rai S, Dubey O, Satpaty S, Sinha R, Prasad S, Sahay A (2014) Influence of abiotic factors on seasonal incidence of pests of tasar silkworm *Antheraea mylitta* D. J Ind Entomol 29:1355–1144
Singh RN, Sinha MK, Sinha AK, Tikader A (2014) Tasar culture. APH Publishing Corporation, pp 8–9
Southwood TRE (1961) The number of species of insects associated with various trees. J Anim Ecol 30:1–8
Tanaka Y (1950) Studies on hibernation with special reference to photoperiodicity and breeding of the Chinese Tussar-silkworm-I. J Seric Sci Jpn 13:1147–1162

Tanaka Y (1951) Studies on hibernation with special reference to photoperiodicity and breeding of the Chinese Tussar-silkworm-I. J Seric Sci Jpn 20:132–138
Truman JW (1971a) Hour glass behavior of the circadian clock controlling eclosion of the silkmoth Antheraea pernyi. Proc Natl Acad Sci 68:595–599
Truman JW (1971b) Physiology of insect ecdysis. I. The eclosion behavior of silkmoth and its hormonal control. J Exp Biol 54:805–814
Truman JW (1972) Physiology of insect rhythm. J Comp Physiol 81:99–114
Truman JW, Riddiford LM (1970) Neuroendocrine control of ecdysis in silkmoths. Science 167(3925):1624–1626
Truman JW, Sokolove PG (1972) Silk moth eclosion: hormonal triggering of a centrally programmed pattern of behavior. Science 175(4029):1491–1493
Vosshall LB, Price JL, Sehgal A, Saez L, Young MW (1994) Block in nuclear localization of period protein by a second clock mutation, timeless. Science 263(5153):1606–1609
Williams CM, Adkisson PL (1964) Physiology of insect diapause. XIV. An endocrine mechanism for the photoperiodic control of pupal diapause in the oak silkworm, *Antheraea pernyi*. Biol Bull 127(3):511–525

The Journey of Biotechnology in Tasar Sericulture: Past Experiences, Current Strategies, and Future Horizons

Immanual Gilwax Prabhu, Vikas Kumar, and Narisetty Balaji Chowdary

Abstract

This chapter delves into the molecular research that has illuminated pathways and genetic markers critical for enhancing silk production in silkworm, notably tropical tasar silkworm, *A. mylitta*. The chapter highlights the significance of genomic studies and molecular breeding, emphasizing how these approaches have shifted from traditional breeding methods to more precise, gene-based selections. Through a detailed analysis of breeding techniques, including conventional and molecular breeding, and transgenic research, it demonstrates the advancements in developing tropical tasar silkworm with superior silk yield and quality. Transitioning to breeding strategies, the chapter contrasts conventional breeding with molecular breeding, underscoring the limitations of traditional methods in achieving specific character improvements. It elaborates on the advantages of molecular breeding, particularly marker-assisted selection, in enhancing selection efficiency, reducing breeding cycles, and precisely establishing desirable traits. The discussion extends to the breeding of host plants for silkworms, highlighting the challenges and proposed strategies to overcome these obstacles, such as developing varieties with enhanced leaf yield and resistance to stress factors. Further, the chapter explores the realm of transgenic research in tasar silkworms, focusing on the integration of foreign genes to impart desired characteristics like enhanced silk quality or disease resistance. It discusses the advancements in germline transformation techniques and the potential of transgenic silkworms in producing value-added proteins alongside tasar silk. Additionally, it emphasizes the potential of CRISPR-Cas9 technology in precise genome editing for trait improvement in silkworms. The narrative then shifts to genetic diversity analysis using molecular markers, detailing various

I. G. Prabhu (✉) · V. Kumar · N. B. Chowdary
Central Tasar Research and Training Institute, Central Silk Board, Ranchi, Jharkhand, India

R. V Suresh et al. (eds.), *Biotechnology for Silkworm Crop Enhancement*,
https://doi.org/10.1007/978-981-97-5061-0_12

markers developed for tasar silkworms and their applications in understanding genetic variation and assisting in selective breeding programs. The chapter underscores the importance of molecular markers in improving host plant quality and the potential of next-generation sequencing technologies in enhancing our understanding of the genetic basis of important traits in *A. mylitta*. In summary, this chapter presents a comprehensive overview of the current state and future directions in the molecular breeding of *A. mylitta* and host plants. It emphasizes the integration of molecular biology techniques, genetic resources, and advanced breeding strategies as pivotal to unlocking the genetic potential of *A. mylitta* for increased tasar silk production, thereby contributing to the sustainability and growth of the sericulture industry.

Keywords

Tasar silkworm · Genomics · Molecular breeding · Transgenics · Genetic diversity · Molecular markers · CRISPR-Cas9

Introduction

The tasar silkworm, scientifically known as *Antheraea mylitta*, holds a significant position in the realm of sericulture, contributing extensively to silk production. This species, indigenous to India and various regions of Southeast Asia, has been cultivated for thousands of years, playing a pivotal role in the industrial, cultural, and economic landscapes of these areas. Distinguished from the commonly known *B. mori* silkworms, tasar silkworms belong to the Saturniidae family. The Central Tasar Research and Training Institute (CTRTI) in Ranchi has embarked on exploratory surveys across 16 Indian states and the Union Territory of Dadra & Nagar Haveli, aiming to catalog the various ecoraces of *A. mylitta* Drury. To date, these efforts have unveiled 45 distinct ecoraces. The diet of these silkworms, primarily the leaves of certain host plants, plays a crucial role in determining the quality of silk produced, showcasing the ecological interdependence within sericulture (Thangavelu 1992).

Tasar silkworms are celebrated for their adaptability to diverse climatic conditions, thriving in both tropical and temperate zones. This versatility allows them to be cultivated across various geographical landscapes, contributing to the sericulture sector's robustness. The tradition of weaving tasar silk, passed down through generations, enriches the tapestry of traditional crafts. Moreover, the tasar silkworm industry plays a vital role in biodiversity conservation, promoting the symbiosis of different plant and insect species, thus supporting ecosystem sustainability and the preservation of native flora.

The historical narrative of tasar silk production illuminates the intricate web of traditions, trade networks, and cultural significance embedded in this unique sector. However, shifts in land use, climate change, and evolving socioeconomic dynamics pose challenges to traditional sericulture practices. Addressing these issues, innovations in biotechnology, sustainable agricultural practices, and community-driven initiatives are pivotal in preserving the essence of tasar silk production while navigating contemporary challenges.

Understanding the molecular intricacies of silk production is essential for enhancing silk yield, particularly in silkworm strains with varying silk output. Detailed investigations into the genetic underpinnings of silk yield variations are crucial to bridging the knowledge gaps in the molecular mechanisms at play. Molecular biology research holds the key to unlocking valuable insights that could empower local silkworm breeders to develop high-yielding breeds. The potential of marker-assisted selection and breeding programs is immense, with techniques like RNA sequencing (RNA-seq) paving the way for identifying markers associated with desirable traits such as silk yield. Access to genetic resources from exotic strains of *B. mori* and databases like SilkDB, KAIKObase, SilkTransDB, SilkSatDb, and BmTEdb is invaluable, providing a wealth of genomic and transcriptomic data essential for research and breeding efforts. RNA-seq, in particular, offers a powerful tool for understanding gene function, biological pathways, and gene expression dynamics, facilitating the discovery of new transcripts and variants. Recent applications of RNA-seq in studying the impact of environmental factors on gene expression further underscore its utility in advancing our comprehension of silk production mechanisms and enhancing silk yield through targeted breeding strategies.

Genomic Studies and Molecular Breeding

Conventional Breeding and Molecular Breeding

In the field of sericulture, breeding stands as a pivotal strategy for the preservation and enhancement of silkworm races, focusing on their distinct traits and phenotypes. Conventional breeding methods have laid the foundational groundwork for the development of new silkworm breeds, tailored to specific environmental or climatic conditions. However, these traditional approaches do not necessarily guarantee improvements in specific characteristics within organisms. As Sinha (1994) elucidate, the success of breeding, and consequently the quality and yield of silk, is contingent upon the health of the silkworms, which in turn is influenced by the quality of feed and the absence of disease within silkworm stocks. Over time, breeding techniques have been instrumental in strain enhancement, leading to the creation of novel, high-yielding silkworm breeds. Yet, the yield of cocoons per diseased female line (dfl) showcases variability, indicating that conventional methods may be nearing the biological limits of the silkworm in terms of fertility, silk yield, and growth rate. Given these constraints, there is an escalating imperative to employ advanced methodologies such as genetic engineering and biotechnological research to expedite the development of superior silkworm strains and innovative technologies for a variety of bioactive products, aiming to bolster the tasar silk industry efficiently.

Molecular breeding presents a stark contrast to these traditional techniques by leveraging molecular markers associated with desirable phenotypes. This approach facilitates the establishment of specific traits within a species, ultimately enhancing the production of those phenotypic characteristics within the sericulture sector. Marker-assisted selection (MAS), a cornerstone of molecular breeding, offers numerous advantages over conventional breeding methods. It allows breeders to identify and select for desirable traits early in the development process, independent of environmental influences. By providing precise genetic information, MAS significantly reduces the time required for trait development and accelerates the breeding cycle. This method enhances selection efficiency by correlating genotype directly with phenotype, enabling focused selection through molecular markers and proving particularly beneficial in reducing the number of generations and population size needed for trait fixation.

Breeding in Silkworms

The cultivation of host plants for sericulture is primarily driven by the need for leaves, which serve as the principal food source for silkworms and play a vital role in silk production. The development of host plant varieties capable of high leaf yield is therefore of paramount importance. However, host plant breeding faces several challenges, including limited understanding of genetics and inheritance patterns, absence of pure lines, unclear parentage, dioecious traits, inbreeding depression, perennial characteristics with extended juvenile phases, and a scarcity of genetic markers and efficient screening methods (Gargi et al. 2015). Selective breeding of host plants for tasar silkworms (*A. mylitta*) seeks to enhance the qualities of the plants that constitute the silkworms' primary diet. Typically, these host plants are species within the *Terminalia* genus.

A study assessing the rearing performance of 79 accessions of *Terminalia tomentosa* and *Terminalia arjuna* revealed significant variations in cocoon weight, with *T. arjuna* showing a shorter larval period and a higher single cocoon weight compared to *T. tomentosa*. This investigation into the genetic diversity of these plant species utilized random amplified polymorphic DNA (RAPD) markers, uncovering high levels of genetic diversity (91.89% for *T. arjuna* and 96.38% for *T. tomentosa*) (Gargi et al. 2010; Kumar et al. 2009). The researchers employed the unweighted pair group method with arithmetic mean (UPGMA) to visually represent the genetic relationships between these accessions based on their molecular profiles. This hierarchical clustering method produced a dendrogram illustrating the genetic relatedness between the *T. arjuna* and *T. tomentosa* accessions, offering valuable insights into the genetic diversity and relationships among different plant groups, thereby highlighting the critical role of host plant selection in tasar silk production.

Breeding of Host Plants for Tasar Silkworm

Breeding strategies for *Terminalia* species, characterized by their perennial and heterozygous nature, encounter specific challenges, including the complication of small flower sizes that hinder effective hybridization processes. To surmount these obstacles, a multifaceted approach to breeding is advocated, encompassing a range of methodologies aimed at enhancing the genetic potential and adaptability of *Terminalia* species.

- Rapid Growth and Early Maturity: Initiatives to develop varieties with a reduced gestation period are pivotal, involving the screening of germplasm for attributes such as high seedling vigor. Additionally, the induction of variability through mutation breeding or distant hybridization represents a significant strategy for generating diversity and accelerating the breeding cycle.
- Abiotic Stress Tolerance: Addressing the resilience of *Terminalia* species to environmental stressors, efforts focus on identifying and utilizing germplasm with inherent variability in drought-responsive traits. This strategy aims to breed varieties capable of thriving under water-limited conditions, thereby enhancing sustainability and adaptability to changing climates.
- Low-Input Adaptability: The pursuit of low-input responsive varieties emphasizes the evaluation of germplasm under marginal land conditions, focusing on performance and stability. This approach seeks to optimize resource use efficiency and ensure consistent yields in less fertile or degraded soils.
- Pest and Disease Resistance: The systematic screening for and exploration of resistant genes within related species are crucial for developing varieties resistant to gall insects and diseases. This proactive measure aims to safeguard *Terminalia* crops from biotic stressors, thus ensuring healthier growth and higher productivity.
- Enhanced Leaf Yield and Quality: The development of hybrids, leveraging heterozygous parents, stands as a strategy to enhance leaf yield and quality. Ongoing efforts to create viable hybrids within *Terminalia* species are indicative of the dynamic nature of breeding programs aimed at maximizing genetic gains.
- Polyploid Advantages: The exploration of polyploid development highlights potential benefits such as heterosis (hybrid vigor), gene redundancy for enhanced trait expression, and opportunities for asexual reproduction, contributing to genetic stability and diversity.
- Superior Germplasm for Plantation Success: Selecting superior germplasm with excellent rooting ability is critical to meet the increasing demand for seedlings in block plantation setups, ensuring robust establishment and growth.

These sophisticated breeding strategies underscore the necessity for advanced biotechnological interventions, including tissue culture, anther culture, genetic editing, and protoplast fusion. Such cutting-edge techniques are instrumental in navigating the complexities of perennial, heterozygous species, thereby unlocking their genetic potential for improved traits, stress tolerance, and overall productivity. The integration of these approaches into *Terminalia* breeding programs promises significant strides in the development of varieties optimized for enhanced characteristics, environmental resilience, and increased yield, marking a progressive era in the cultivation and utilization of *Terminalia* species.

Transgenic Research in Tasar Silkworm

The field of contemporary biotechnology has seen significant advancements, enabling scientists to precisely identify and manipulate specific genes. This intricate process involves isolating a particular gene, cutting it from its original location, and then inserting or incorporating it into a different organism's genetic makeup, where it can be altered and expressed in new and innovative ways. This sophisticated form of genetic manipulation is broadly referred to as genetic engineering, a cornerstone of modern biotechnological applications. Thanks to the leaps made in biotechnology, our comprehension of the biological mechanisms that govern life has dramatically improved. This deeper understanding has opened up new possibilities for applying biotechnological innovations across a wide spectrum of human activities, including but not limited to enhancing food production, advancing forestry and fisheries, improving livestock and animal husbandry practices, and refining horticultural methods. Through the application of genetic engineering, we are now capable of creating organisms that carry specific genes from entirely unrelated species, marking a significant milestone in biotechnological applications.

This capability is particularly beneficial in contexts where traditional breeding methods fall short. For instance, in the case of tasar silkworms, conventional breeding techniques have been unable to introduce certain desirable traits. As a result, scientists are increasingly relying on recombinant DNA technology to produce transgenic organisms, a revolutionary approach that involves integrating foreign genes into an organism to imbue it with specific characteristics or traits. This transgenic technology has been explored extensively in the context of silkworms for several purposes, including but not limited to the production of value-added proteins, the enhancement of disease resistance, and the improvement of silk quality in tropical tasar silkworm, *A. mylitta*. To enhance the tensile strength, quality, and various other attributes of silk, researchers have delved into modifying the genetic codes responsible for silk protein production in tasar silkworms. By either introducing new genes or altering existing ones, scientists can engineer silk fibers with tailored properties, meeting specific needs or creating novel functionalities. This has been achieved through the application of transgenic techniques, which have allowed for the incorporation of genes encoding for recombinant proteins directly into tasar silkworms. Consequently, the silk glands of these genetically modified silkworms are capable of producing not just silk but also these recombinant proteins, facilitating the extraction of high-value bioproducts alongside traditional silk.

Furthermore, genetic modification techniques are being explored as a strategy to bolster disease resistance in tasar silkworms. By introducing genes that encode for immune responses or antimicrobial peptides, scientists aim to enhance the innate defense mechanisms of these silkworms, potentially providing them with a greater resistance to prevalent diseases. This innovative approach not only promises to improve the health and viability of tasar silkworm populations but also to increase the overall quality and yield of silk production, showcasing the profound impact of genetic engineering on both agricultural practices and biotechnological research.

Transgenic Silkworm

Recent advancements in the field of sericulture research have spotlighted the potential of germline transformation techniques, not just in the domesticated silkworm *B. mori* L. but also in the tasar silkworm, *A. mylitta*, a species of significant commercial and ecological value. The methodology, originally detailed by Tamura et al. (2000) using microinjection of PiggyBac (PB) derived vectors for *B. mori*, has been explored for its applicability to *A. mylitta*. This approach allows for the precise introduction of target genes into the chromosomes of tasar silkworms, mirroring the successes seen in *B. mori* for the enhancement of silk production and quality. Utilizing the silk gland as a biofactory, this innovative technique has the potential to revolutionize the tasar silk industry by enabling the production of transgenic tasar silkworms. Such transgenics could produce various recombinant proteins and enhanced silk types, thereby broadening the scope of sericulture to include not only textile production but also biomedical applications such as the production of animal pharmaceuticals, treatments for rare diseases, and monoclonal antibodies for cancer therapy. The development of recombinant human serum albumin (HSA) in transgenic cocoons and glycoproteins with reduced antigenicity, achievements previously noted in *B. mori* by Minagawa et al. (2018) and Qian et al. (2018), illustrate the vast potential of this research direction in enhancing the utility of silk produced by *A. mylitta*.

The exploration of transgenic tasar silkworms opens new pathways for improving the characteristics of host plants and the quality of tasar silk, aligning with the innovative strides seen in *B. mori* research. However, the application of transgenic research in *A. mylitta* also introduces unique challenges and ethical considerations. Concerns about the potential escape of transgenes into the wild, with the consequent risk of gene flow into natural populations of *A. mylitta* or related species, highlight the need for careful management of transgenic technologies. Such gene flow could inadvertently introduce unwanted traits into these populations, leading to ecological and biological ramifications that could disrupt local ecosystems and biodiversity. Accordingly, the responsible development and deployment of transgenic technologies in the context of *A. mylitta* require a comprehensive evaluation of environmental impacts and a consideration of the broader implications for biodiversity conservation. The goal is to harness the benefits of transgenic research to enhance sericulture and silk production while ensuring the ecological integrity and sustainability of tasar silkworm populations and their natural habitats. This balanced approach underscores the importance of advancing scientific innovation in sericulture within an ethical framework that prioritizes environmental stewardship and the preservation of biodiversity.

Deciphering Genetic Diversity Through Molecular Markers

The *A. mylitta* D. displays notable phenotypic variations such as fertility, voltinism, cocoon weight, and silk ratio, which are critical for its cultivation and silk production efficiency. These variations, along with the selection of host plants, contribute to the complexity of tasar silkworm breeding and conservation efforts. A pressing challenge in the conservation and breeding of this non-mulberry silkworm is the identification and preservation of its diverse ecoraces, whose numbers have been steadily declining. This underscores the necessity for advanced tools for identification and genetic analysis. In this context, the development of molecular markers has emerged as a pivotal advancement for identifying and understanding the genetic relationships among the wild silkworm ecoraces. According to Saha and Kundu (2006), these markers play a crucial role in the conservation and study of genetic diversity within *A. mylitta* populations. Various types of molecular markers have been developed for tasar silkworms, facilitating a deeper understanding of their genetic makeup. These include restriction fragment length polymorphism (RFLP), fluorescent inter simple sequence repeat PCR (FISSR-PCR) as highlighted by Nagaraju et al. (2002), random amplified polymorphic DNA (RAPD) as detailed by Chatterjee et al. (2004a), and single nucleotide polymorphisms (SNPs) which are particularly useful for high throughput genotyping of diverse silkworm strains.

The application of these molecular markers has revolutionized the way researchers and breeders approach the study and conservation of tasar silkworms. Unlike traditional morphological and biochemical methods, molecular markers enable the rapid and precise identification of breeding lines, hybrids, cultivars, and species, thereby enhancing the efficiency of genetic diversity analysis. This precision is further exemplified in Fig. 1, which illustrates the utilization of a comprehensive database and molecular markers for various applications in tasar silkworm biology. This flowchart outlines a systematic approach for employing genetic information in the selection of ecoraces, understanding genetic relationships, and enhancing breeding strategies for the improvement of silk production. By integrating such molecular tools with traditional breeding practices, researchers and breeders can achieve a more nuanced understanding of genetic diversity and its implications for tasar silk production. This integration not only aids in the preservation of existing genetic resources but also paves the way for the development of new, more resilient, and productive tasar silkworm strains. Through the application of molecular markers and the strategic use of genetic databases as depicted in Fig. 1, the field of tasar silkworm biology is poised to overcome the challenges of ecorace identification and decline, heralding a new era of innovation and sustainability in tasar silk production.

DNA Markers Used in Sericulture

DNA markers have become indispensable tools for exploring gene-related studies, facilitating genotype screening, gene identification, and germplasm collection characterization, among other applications. Seminal works by Lichun et al. (1996) and Lou et al. (1998) have highlighted the utility of these markers in comprehensive genetic investigations. The array of DNA markers includes single nucleotide polymorphisms (SNPs), inter simple sequence repeats (ISSRs), simple sequence repeats (SSRs), amplified fragment length polymorphisms (AFLPs), and restriction

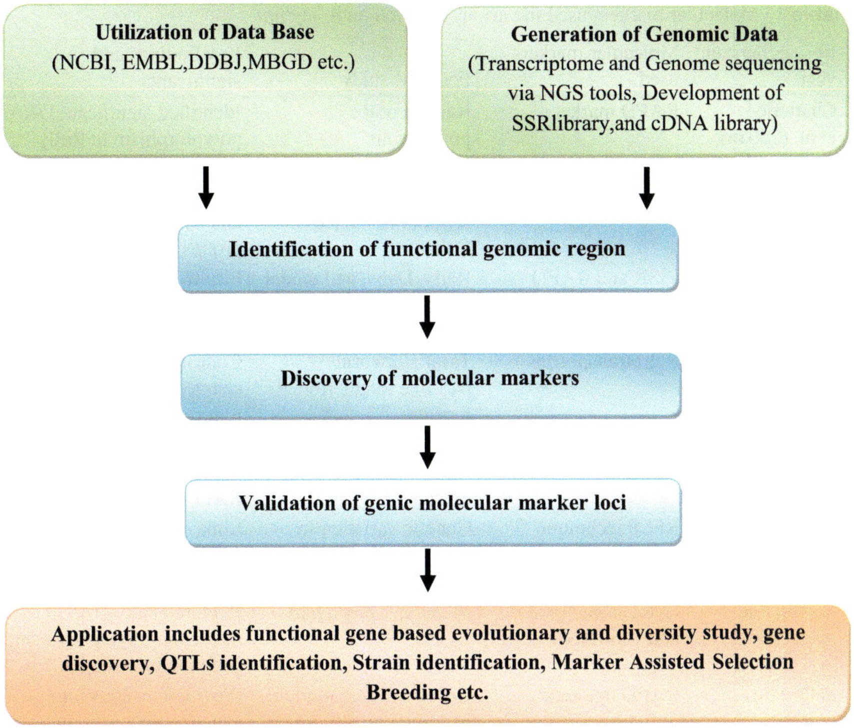

Fig. 1 Depicting the use of database and molecular marker for various applications in tasar silkworm biology

fragment length polymorphisms (RFLPs), with each offering unique advantages for genetic analysis (Lusser et al. 2012) (Table 1). Compared to conventional breeding methods, marker-assisted selection (MAS) presents several benefits that significantly enhance the breeding process. It enables breeders to select desirable hybrids at an early seedling stage, unaffected by environmental factors, with a high degree of accuracy and efficiency, thus saving valuable time. MAS facilitates early selection, necessitating fewer generations and smaller population sizes, thereby streamlining the breeding cycle and enhancing selection efficiency (Staub et al. 1996).

In the context of host plant improvement, molecular markers have proven to be extremely valuable across various critical domains, including assessing genetic diversity, molecular characterization of germplasm and varieties, development of linkage and quantitative trait locus (QTL) maps, association mapping, devising parental selection schemes, and implementing marker-assisted selection (MAS). These areas benefit from the strategic application of molecular markers to improve selection accuracy and efficiency (Khurana and Checker 2011). The significance of molecular markers extends to the enhancement of tasar silkworms, with several studies employing markers such as internal transcribed spacer (ITS) (Xuan et al. 2019), start codon targeted polymorphism (SCoT) (Rohela et al. 2020), cleaved

Table 1 Molecular markers used in various research on *A. mylitta*

Researchers and year	Molecular markers used	Focus of study	Key findings and significance
Chatterjee et al. (2004b)	RAPD marker	Raily ecorace populations	Identified significant DNA polymorphism in Raily ecoraces
Kar et al. (2005)	ISSR markers	Genetic diversity in semidomesticated Daba	Notable variations within and between Daba ecorace groups
Vijayan et al. (2005)	ISSR and RAPD markers	Raily, Daba, and modal ecoraces	Limited genetic variation among ecoraces, establishing stability
Ghosh et al. (2005)	Retrotransposons	Tasar silkworm pao-like long terminal repeats	Contributed to molecular characterization of retrotransposons
Saha and Kundu (2006)	SCAR and RAPD markers	Molecular identification of ecoraces	Discriminated eight out of ten ecoraces using RAPD-selected bands
Mahendran et al. (2006a)	RFLP technique	Genetic variation in nine ecoraces	Linked genetic variations with phenotypic and geographical differences
Mahendran et al. (2006b)	Cytochrome oxidase and 16S rRNA	Molecular phylogeny	Explored silkworm molecular phylogeny using genomic information
Saha et al. (2008)	RAPD markers	Genetic differentiation in commercial ecoraces	Provided insights into genetic variability for breeding strategies
Chakraborty et al. (2015)	Microsatellite markers	Indian tasar silk moth population	Enhanced resolution in identifying genetic variation and population structure
Renuka and Shamitha (2016)	SSR and ISSR markers	Phylogenetic analysis and genetic variation	Contributed to understanding genetic structure and evolutionary links
Niranjan et al. (2018)	RAPD-SCAR markers	Daba BV high shell weight line	Marker development for silk yield enhancement
Prabhu et al. (2023)	RAPD-SCAR markers	Thermotolerant tasar silkworms	Molecular markers for selecting thermotolerant lines

amplified polymorphic sequences (CAPS) (Arora et al. 2017), and genomic-SSR sequence-related amplified polymorphism (SRAP) (Krishnan et al. 2014). These markers have been instrumental not only in uncovering the molecular diversity of both wild and cultivated mulberry but also hold the promise for elucidating the genetic diversity and improving the breeding strategies for tasar silkworms.

This holistic approach to understanding and enhancing the genetic foundation of both host plants and tasar silkworms underscores the transformative potential of molecular markers in sericulture. By applying these advanced genetic tools, researchers and breeders can achieve unprecedented insights into the genetic composition of these organisms, paving the way for the development of improved varieties and strains that exhibit enhanced qualities, resilience, and productivity. This scientific endeavor not only contributes to the advancement of sericulture but also ensures the sustainability and growth of this vital agricultural sector.

Genetic Diversity Analysis in Tasar Silkworm, *A. mylitta*, Through Molecular Marker

The exploration of molecular markers, specifically SSRs and SNPs, across various *Antheraea* species, has significantly advanced our understanding of genetic diversity and molecular profiling among different germplasm collections of these species. Such studies have been pivotal in characterizing the ecoraces of *A. mylitta* across diverse tropical forest zones, shedding light on the genetic diversity within tasar silkworm populations. SSR markers, in particular, have been extensively employed in genetic diversity assessments due to their high level of polymorphism and informativeness, as highlighted by Vijayan et al. (2005). Traditional methods, including morphological, physio-genetic, behavioral, and biochemical techniques, have been used to study *A. mylitta*, although distinguishing between genetic and environmental influences on observed variability poses challenges. The discovery of shared bands between these ecoraces suggested a close genetic relationship, offering new insights into the genetic connections among various tasar silkworm populations. Building on this, Chatterjee et al. (2004b) delved into DNA polymorphism within the Raily ecorace populations, uncovering substantial genetic variation within these groups.

Further extending the scope of molecular exploration, Kar et al. (2005) examined the genetic diversity within semidomesticated groups of the Daba ecorace. Utilizing ISSR markers, they identified significant genetic variations both within and between groups, indicating that semidomestic populations may exhibit distinct genetic differences. This finding underscores the importance of understanding genetic variability for conservation and systematic breeding programs. A more comprehensive approach was later adopted by Vijayan et al. (2005), who assessed genetic diversity within and among the Raily, Daba, and Modal ecoraces collected from various Indian states. Their findings suggested that the genetic variation among ecoraces was not sufficient to prompt significant genetic drift in the near future, establishing a foundation for understanding the genetic structural stability across tasar silkworm populations.

Saha and Kundu (2006) advanced the molecular identification efforts by utilizing RAPD-sequence characterized amplified region (SCAR) markers. They successfully discriminated eight out of ten ecoraces using seven RAPD-selected bands, demonstrating the efficacy of molecular techniques for precise identification and differentiation of tasar silkworm ecoraces (Fig. 2). The subsequent sequencing of these identified RAPD segments and the creation of SCAR markers validated the potential of molecular methods for accurate characterization and conservation of tasar silkworm genetic diversity. This body of work collectively highlights the transformative impact of molecular markers in enhancing our understanding of the genetic landscape of tasar silkworm populations, paving the way for targeted conservation strategies and the development of breeding programs informed by genetic insights.

Mahendran et al. (2006a, 2006b) undertook a pivotal study on the genetic variation within nine ecoraces of the tasar silkworm, employing the RFLP technique to elucidate the phylogenetic relationships underpinned by both phenotypic traits and geographical isolations. This investigation successfully correlated genetic variations with the observable phenotypic and geographical disparities among the ecoraces, enhancing our understanding of the evolutionary dynamics within this species. Concurrently, Ghosh et al. (2005) delved into the molecular characterization of tasar silkworm pao-like long terminal repeat retrotransposons, contributing to the molecular genetic landscape of this silkworm. Additionally, Mahendran et al. (2006a) expanded the scope of molecular phylogeny studies by analyzing cytochrome oxidase and 16S ribosomal RNA, providing a deeper insight into the genetic underpinnings of tasar silkworm diversity.

Building on this foundation, Saha et al. (2008) utilized RAPD markers to explore the genetic differentiation among ten commercially significant ecoraces of the tasar silkworm. Their findings shed light on the genetic variability within these groups, offering invaluable data for the formulation of targeted breeding and conservation strategies. RAPD markers proved to be instrumental in identifying distinct genetic profiles among ecoraces, facilitating their potential use in selective breeding programs aimed at enhancing silk production (Chatterjee and Pradeep 2003). In a similar vein, Niranjan et al. (2018) and Prabhu et al. (2023) explored the utility of RAPD markers to develop RAPD-SCAR markers tailored for specific traits. Niranjan et al. aimed to identify a Daba BV line with high shell weight, an essential trait for increasing silk yield. Prabhu et al. (2023) pursued a novel approach by generating molecular markers for the identification of thermotolerant tasar silkworm lines, a trait of increasing importance in the face of global warming and changing climate conditions. Their work exemplifies the effectiveness of RAPD-SCAR markers in precise selection for desirable traits, offering a promising avenue for genetic improvement in tasar silkworms.

The application of SSR and ISSR markers for phylogenetic analysis and understanding genetic variation among tasar silkworm ecoraces has been significantly advanced by the work of Renuka and Shamitha (2016). Their research clarified the genetic structure and evolutionary connections between various populations, establishing a solid framework for the identification and preservation of genetically distinct ecoraces. Adopting an alternative approach, Chakraborty et al. (2015) conducted a comprehensive genetic study of the Indian tasar silk moth population using microsatellite markers. Owing to their high degree of codominance and polymorphism, microsatellites offered enhanced resolution for delineating genetic variation and population structure. Their contributions have been integral to informing conservation and management strategies, thereby enriching our understanding of genetic diversity within tasar silkworm populations. This body of research collectively underscores the critical role of molecular markers in advancing our

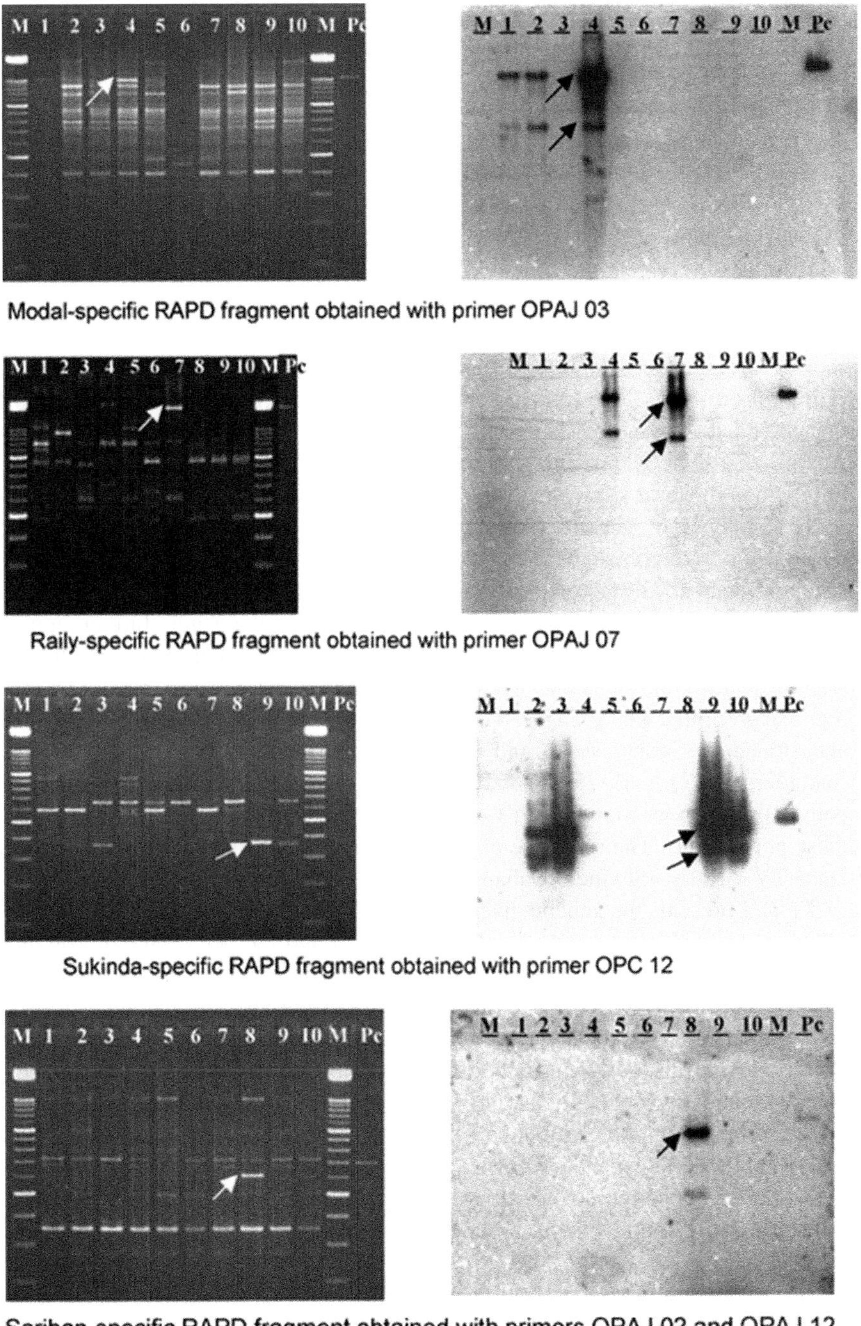

Fig. 2 RAPD fragment of various ecoraces of *A. mylitta*. Source: Saha and Kundu 2006

knowledge of tasar silkworm genetics, paving the way for informed conservation efforts and the development of genetically optimized strains for improved silk production.

In their seminal work, Renuka and Shamitha (2016) embarked on a comprehensive genetic characterization of seven genotypes of the tasar silkworm (*A. mylitta*), employing microsatellites, or SSRs, as their primary tool. Microsatellites, characterized by their short DNA sequence motifs, are known for their high allelic diversity and suitability for analysis via PCR. The study utilized ten SSR primers to delve into the genetic variation within and among 16 populations of 7 geographically distinct ecoraces of *A. mylitta*. Their findings highlighted substantial genetic polymorphism, with 420 out of 887 bands (47.3%) being polymorphic, thereby underscoring the genetic diversity present within these populations. Among the SSR primers used, Amysat023 emerged as the most polymorphic, followed by a sequence that included Amysat025, Amysat041, Amysat038, Amysat001, Amysat040, and Amysat013, thereby proving these primers' utility in phylogenetic and genetic research on the tasar silkworm.

The analysis of genetic diversity within the ecoraces revealed variations in degrees of polymorphism, with the Bhandara ecorace displaying the highest level of polymorphism at 72.41%, while the Sukinda and Andhra local ecoraces showed the lowest at 55.17%. Utilizing Nei's genetic distance, the study further applied UPGMA (unweighted pair group method with arithmetic mean) analysis to construct a dendrogram. This phenogram delineated two distinct clusters without clear regional grouping, where Cluster 4 contained samples from Karimnagar (Telangana) and Bhandara (Maharashtra) and Cluster 3 grouped samples from Karimnagar, Sukindergarh (Orissa), Keonjhar (Orissa), and Bastar (Chhattisgarh). Cluster 2, comprising samples from Daba TV and Sukinda, indicated genetic proximity among these populations. The sequence of genetic closeness was summarized as follows: Daba TV < Raily <Sukinda<Daba BV < Modal < Andhra local.

Table 2 presents the genetic diversity analysis in the ecoraces of *A. mylitta*, providing a detailed breakdown of polymorphic loci and the percentage of polymorphism across different geographical locations. This study's insights into tasar silkworm ecoraces' genetic diversity were further enriched by the application of SSR markers derived from expressed sequence tags (ESTs) for genetic characterization. EST analysis, a potent method for transcription profiling, homologous gene comparison, and gene identification, was exemplified in the characterization of the cDNA library from the Chinese oak silkworm, *Antheraea pernyi* (Li YuPing et al. 2009). While EST sequences furnish a comprehensive genomic data source for phylogenetic and genomic studies, the tradition of molecular systematics has often relied on PCR amplification of universal genes for phylogenetic information (Rudd 2003).

Focusing on five tasar silkworm ecoraces, including Andhra local, Daba TV, Daba BV, Modal, and Sukinda, the study developed EST-derived SSR markers using specific primers. Electropherogram analysis detected nucleotide variations, with polymorphisms at 60 locations identified through alignment with reference primers. Notably, the Daba BV ecotype exhibited the highest nucleotide variation,

Table 2 Genetic diversity analysis in the ecoraces of *A. mylitta*. Source: Renuka and Shamitha (2016)

Ecorace	Geographical location	Number of polymorphic loci	Percentage of polymorphic loci
Andhra local	Karimnagar	16	55.17
Daba TV	Karimnagar	20	68.97
Daba BV	Warangal	20	68.97
Modal	Keonjhar	20	68.97
Sukinda	Sukinadergarh	16	55.17
Raily	Bastar	19	65.52
Bhandara	Bhandara	21	72.41

aligning with its bivoltine nature as opposed to the trivoltine nature of the other ecoraces. Unique nucleotide signatures were observed in Andhra local and Daba TV, emphasizing their distinctiveness. The transition over transversion ratio, indicative of the type of nucleotide changes, was found to be 1.8, hinting at a bias toward transitions. This comprehensive research elucidates the genetic diversity among tasar silkworm ecoraces and provides invaluable data for future breeding and conservation efforts, showcasing the advanced capabilities of SSR markers derived from ESTs in analyzing genomic variation and underlining the critical role of sericulture research.

In a related study, Velu et al. (2008) utilized inter simple sequence repeat (ISSR) primers to investigate the genetic relationships among mutant silkworm stocks. Through the application of UPGMA and cluster analysis, the study effectively discriminated 20 mutant strains, delineating 6 subclusters and 1 major cluster, thus revealing considerable interstrain variability. This research demonstrated the efficacy of ISSR markers in probing heterozygosity and evolutionary relationships among silkworm populations, marking a significant advancement in the understanding of silkworm genetic diversity. Furthermore, Xia et al. (2009) conducted a comprehensive genomic analysis identifying 16 million single nucleotide polymorphism (SNP) markers across 40 farmed and wild silkworm strains, thereby elucidating the episodes of domestication and subsequent genetic differentiation in *B. mori*. This genomic exploration provided a clearer picture of the genetic shifts associated with silkworm domestication processes.

In pursuit of identifying SNP DNA markers related to cocoon characteristics in mulberry silkworms, Sreekumar et al. (2011) analyzed 240 primer pairs, of which 48 exhibited parental polymorphism as confirmed by their codominant expression in the F1 generation. Among these, a single base pair variation (04124) was distinctly observed in the amplified products, showcasing the potential of SNPs in enhancing silkworm breeding practices. Akkir et al. (2010) distinguished Turkish silkworm breeds using RAPD-PCR, differentiating between diapausing and nondiapausing local breeds. The investigation revealed a lower number of polymorphism loci in diapausing breeds compared to a higher occurrence in nondiapausing breeds, insights that could significantly influence future breeding efforts.

Turning the focus to *A. mylitta*, Prasad and Nagaraju (2003) explored the conservation and evolutionary evolution of mariner-like elements (MLEs) in this Indian saturniid silk moth. Their study posited *A. mylitta* MLEs within the *Cecropia* subfamily, suggesting vertical transmission from a common ancestor. This investigation not only sheds light on the genetic heritage and conservation of MLEs in *A. mylitta* but also highlights the potential of mariner-based vectors for the genetic transformation of silk moths, aiming to introduce economically valuable foreign genes into these species.

In an analogous vein to the work conducted by Yamamoto et al. (2008) on the silkworm *B. mori*, similar methodologies and objectives have been pursued within the context of *A. mylitta*, aiming to enhance our understanding of its genetic architecture. Yamamoto and colleagues successfully mapped 1755 SNP markers derived from the terminal sequences of bacterial artificial chromosomes (BAC) across 28 silkworm linkage groups. This mapping, achieved through the utilization of a recombining male backcross population, revealed an average inter-SNP distance of 0.81 cM, equivalent to approximately 270 kilobases.

Their application of a synteny test unveiled that among the silkworm linkage groups, six had eight or more orthologs with *Apis* (the honeybee), whereas 20 linkage groups showed similarities with *Tribolium* (the red flour beetle). This comprehensive integrated map, covering about 10% of the anticipated genes for silkworms, has been instrumental in facilitating genome scaffolding, annotation, and microsynteny testing within Lepidoptera, as well as promoting comparative genomic studies across insects.

Translating this approach to *A. mylitta*, a similar exploration would not only augment the assembly of whole-genome shotgun data but also streamline gene annotation processes. Such a resource would be invaluable for gene discovery within Lepidoptera, thereby significantly advancing our comprehension of *A. mylitta*'s genetic framework. The methodology and findings from Yamamoto et al.'s study provide a blueprint for conducting comprehensive genetic mapping and annotation in *A. mylitta*, thereby offering insights into its genome organization, evolutionary biology, and potential for genetic improvement. Through leveraging SNP markers and understanding the synteny relationships with other species, researchers can uncover critical genes and pathways relevant to *A. mylitta*'s unique biological traits, silk production capabilities, and adaptability to diverse environmental conditions. This not only enriches the genetic knowledge base of *A. mylitta* but also contributes to the broader field of lepidopteran genomics and the application of this knowledge in sericulture, conservation, and biotechnological advancements.

For instance, echoing the methodology of Li et al. (2007), who employed ISSR markers to investigate genetic variability among various silkworm ecoraces, a similar approach can be instrumental for *A. mylitta*. Li and colleagues discovered origin-based grouping in their subjects, underscoring the efficacy of ISSR markers in discerning genetic variability within silkworm populations. This technique offers a promising pathway for delineating the genetic landscape of *A. mylitta*, facilitating the identification of distinct genetic clusters and enhancing our understanding of its genetic diversity.

Further contributing to the discourse on genetic variability, Arunkumar et al. (2012) conducted an assessment of genetic variability within Indian golden saturniid silk moth populations, identifying pronounced diversity within specific populations. Their findings underscore the critical need for conservation efforts aimed at preserving this genetic diversity. Translating this focus to *A. mylitta*, such studies underscore the importance of implementing conservation strategies to safeguard the genetic heritage and variability of these populations, which are vital for the resilience and adaptability of the species.

Sharma et al. (1999) provided insights into the genetic relatedness within and among silkworm varieties, identifying clusters based on diapausing and nondiapausing varieties. This revelation not only highlights the genetic structuring within silkworm populations but also points to the potential of crossbreeding programs in producing heterotic hybrids. For *A. mylitta*, understanding the genetic basis of diapause can significantly impact breeding strategies, potentially leading to the development of strains with desirable traits such as enhanced silk production or improved adaptability to environmental stresses.

Renuka et al. (2018) screened for polymorphism by using EST-derived SSR markers for the assessment of genetic structure of *A. mylitta* population. The DNA profiles based on these markers suggest that they could be effectively utilized for identifying the genetic variability among tasar ecoraces. The alignment of sequences obtained from genomic PCR products has identified potential EST-SSR marker to recognize single nucleotide polymorphism by comparing various tasar ecoraces (Fig. 3).

Like agriculture, important traits in sericulture, including those relevant to *A. mylitta*, are often polygenic or quantitative, influenced by multiple genes in conjunction with environmental factors. This complexity necessitates a more sophisticated approach from geneticists and breeders working with silkworms to understand and manipulate these traits. Recent advances in silkworm genome studies have unveiled new methodologies and tools that can be leveraged to dissect and manage these complex traits. By integrating genomic insights with traditional breeding techniques, geneticists and breeders can enhance their ability to unravel and manipulate the intricate genetic underpinnings of desirable traits in *A. mylitta*. This integrative approach promises not only to refine the selection and breeding of *A. mylitta* for optimal silk production but also to contribute to the conservation and sustainable management of this economically and culturally significant species.

Isozymes

In the context of *A. mylitta*, the application of isozymes significantly enhances the precision of estimating genetic diversity beyond what is possible through mere observation of physical features. Electrophoresis is used to detect variations (alleles) at the loci encoding enzymes known as isozymes or allozymes, playing a crucial role in uncovering the allelic diversity within populations. One of the primary advantages of allozyme loci is their codominant nature, which allows for the direct

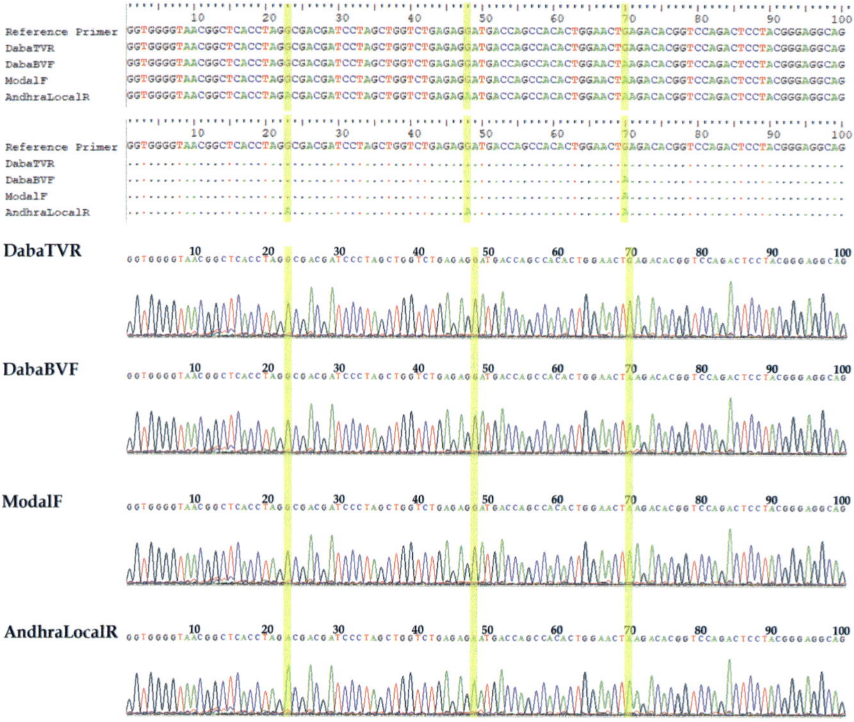

Fig. 3 Genetic relatedness within and among tasar silkworm varieties

scoring of heterozygotes, offering a clear advantage in genetic studies by showcasing allelic differences and their various catalytic capacities. Isozyme studies have become integral to understanding the genetic makeup of individuals within populations. This approach has been employed in silkworm diversity research, where the examination of protein profiles, enzymes, and isozymes—including carbonic anhydrase, glucose 6-phosphate dehydrogenase, amylase, phosphoglucomutase, aspartate-aminotransferase, malate dehydrogenase, and acid phosphatase—has provided insights into the genetic variability present within silkworm genotypes (Suetsugu et al. 2013). These isozyme markers serve as valuable tools for exploring the genetic diversity found in silkworm populations.

Ashok Kumar et al. (2011) leveraged metabolic enzyme profiles to assess the genetic diversity and thermotolerance among 21 different silkworm races, including those of *A. mylitta*. Their analysis enabled the classification of these heterogeneous genotypes into eight distinct clusters, with a strong correlation observed between the clusters' formation, the geographical origins, and the genetic makeup of the genotypes. Notably, the study highlighted the CSR18 race's high protein stability, attributing its enhanced thermotolerance to this feature and identifying it as a thermotolerant race with origins in the Kashmir region. The utilization of isozyme analysis in this study not only revealed a clear linkage between genetic diversity, morphological traits, and geographical origins but also emphasized the utility of metabolic profiles in accurately gauging genetic diversity. This approach proves particularly beneficial in the selection of parent stocks for the creation of superior hybrid silkworms, underscoring the importance of isozymes and electrophoresis in the genetic study and improvement of *A. mylitta*. Such methodologies afford researchers and breeders the opportunity to delve deeper into the genetic foundations of desirable traits, facilitating informed decisions in conservation and breeding strategies and ultimately enhancing the silk production capabilities of *A. mylitta*.

Production of Disease-Resistant Silkworm Through RNA Interference

The production of disease-resistant silkworms through the application of RNA interference (RNAi) represents a cutting-edge approach within the field of biotechnology, particularly emphasizing its implementation in *A. mylitta*. RNAi is a sophisticated biological process that mediates the downregulation of specific genes associated with disease susceptibility by facilitating the degradation of messenger RNA (mRNA) molecules. This degradation process effectively prevents the translation of proteins that could otherwise contribute to the pathogenic vulnerability of the organism. In the context of *A. mylitta*, RNAi can be extensively explored as a method to bolster the organism's defenses against a spectrum of bacterial, viral, and parasitic infections.

Research has identified certain genes within sericigenous moths that correlate strongly with increased susceptibility to prevalent diseases affecting silkworms, particularly those of bacterial and viral origin. These genes have become the focal point for RNAi interventions, employing constructs such as hairpin RNA sequences or short interfering RNA (siRNA) molecules, as documented in studies by Xu et al. (2016). The methodological approach to introducing these RNAi constructs into the silkworms has varied, with techniques ranging from dietary incorporation to direct microinjection.

Subsequent research has demonstrated the efficacy of RNAi in silencing the targeted genes within lepidopteran insects, a process verified through molecular assays like quantitative PCR, as noted by Guan et al. (2018). The silkworms subjected to these RNAi-based modifications have been tested for disease resistance by exposing them to specific infections, thereby assessing their health, symptomatology, and survival outcomes in response to pathogenic exposure.

The iterative optimization of RNAi constructs and their delivery methods is a focal point of research, aiming to enhance the precision of gene silencing and, by extension, the disease resistance of *A. mylitta*. The potential impact of RNAi-mediated enhancements in sericulture, particularly concerning resistance to silkworm diseases, underscores the significant implications for the silk industry. This research not only seeks to improve the health and survival rates of silkworm populations but also to contribute to the sustainability and economic viability of silk production on a global scale.

Disease Diagnosis Through Antibody, PCR, and LAMP Based Techniques in Tasar Silkworm Industry

An array of molecular and immunological diagnostic techniques have been employed to ascertain and manage disease prevalence. Among these, polymerase chain reaction (PCR), loop-mediated isothermal amplification (LAMP), and antibody-based methodologies stand out due to their efficacy in the detection of pathogenic agents. Antibody-based strategies, such as the immunofluorescence assay (IFA) and enzyme-linked immunosorbent assay (ELISA), are pivotal in the specific identification of pathogen-associated antigens. These methods leverage antibodies to bind to unique antigens, enabling the detection of specific diseases with high specificity.

PCR technology has revolutionized pathogen detection by facilitating the amplification and subsequent identification of specific DNA sequences, thereby allowing not only for the detection of the presence of pathogens but also for differentiation among closely related strains. This method's sensitivity and specificity are particularly valuable in the tasar silkworm industry, where the accurate diagnosis of pathogenic strains can significantly influence disease management strategies.

LAMP offers a complementary approach by enabling the rapid and cost-effective amplification of target DNA under isothermal conditions. This method's efficiency and simplicity make it an attractive option for on-site diagnostics, providing a quick turnaround for disease identification without the need for complex thermal cycling equipment.

Collectively, these diagnostic tools—PCR, LAMP, and antibody-based assays—constitute a comprehensive approach to disease management in *A. mylitta*, balancing sensitivity, specificity, and expedience. This multifaceted diagnostic strategy is crucial for the early detection of pathogens, thereby facilitating timely intervention and management of diseases within the tasar silkworm populations. The integration of these techniques, as evidenced in the literature through studies in other sericigenous insects by Liu et al. (2018) for PCR-based pathogen detection and Arunkumar et al. (2014) for rapid LAMP diagnostics, underscores the importance of regular monitoring and the application of advanced diagnostics for effective disease management in the tasar silkworm industry. Such practices not only contribute to the health and viability of *A. mylitta* but also to the sustainability and economic efficiency of the tasar sericulture sector.

Addressing Virulent *Nosema mylitta* Infection in Tasar Silkworms

The viability and productivity of *A. mylitta* populations, a cornerstone of the tasar silkworm industry, are gravely compromised by the virulence of *Nosema* species. These pathogens not only disrupt silk production but also pose a significant threat to the broader sericulture ecosystem. Mitigating the impact of these virulent *Nosema* strains necessitates a multifaceted and integrated strategy. This approach encompasses genetic improvement, advanced molecular diagnostics, vigilant disease surveillance, stringent biosecurity protocols, adaptive cultural practices, targeted antimicrobial interventions, comprehensive educational outreach, and a commitment to continual research and development. The differentiation between virulent and avirulent strains of *Nosema* species in *A. mylitta* hinges on distinct molecular signatures, including variations in genetic sequences, gene expression patterns, and potential functional disparities. Techniques such as DNA sequencing, PCR amplification, and comparative genomics are pivotal in molecular characterization efforts, aiming to pinpoint the subtle genetic determinants that influence the pathogenicity of these microsporidian parasites. Such molecular insights are instrumental in elucidating the complex dynamics of the parasite-host interaction, paving the way for targeted disease management strategies and the enhancement of sericulture practices.

Selective breeding emerges as a crucial strategy for cultivating tasar silkworm strains with inherent resistance to *Nosema* pathogens. This involves identifying individuals exhibiting natural resistance traits and implementing controlled breeding and selection protocols. The efficacy of these breeding efforts is significantly enhanced by molecular tools like marker-assisted selection (MAS), which facilitate the identification and propagation of specific genetic markers associated with resistance. Consequently, breeding programs can more effectively target and expedite the selection of desirable traits. The early detection and management of *Nosema* infections are critically supported by ongoing surveillance and monitoring initiatives. Molecular diagnostic techniques, such as LAMP and PCR, offer precise and rapid detection capabilities, enabling timely interventions that curtail the spread of infectious strains among silkworm populations.

In parallel, the establishment of rigorous biosecurity measures is essential to prevent the introduction and proliferation of *Nosema* species within sericulture operations. Such measures should encompass the maintenance of sanitary rearing conditions and the enforcement of quarantine protocols for new silkworm stocks. Optimizing cultural practices for tasar silkworm rearing can significantly mitigate stress and reduce vulnerability to *Nosema* infections. Proper environmental controls, nutritional support, and hygienic rearing conditions are vital in bolstering the overall health and disease resistance of silkworms.

Genetic Characterization of Iflavirus Infection in *A. mylitta*

A. mylitta is periodically afflicted by a pathogenic condition colloquially referred to as "virosis" by practitioners in the field of sericulture. Despite the nomenclature, definitive evidence pinpointing a viral etiology has not been established. Nonetheless, the symptomatic manifestations and epidemiological pattern of this disease strongly implicate a viral agent as the culprit. The disease trajectory is marked by a constellation of distinctive and progressively worsening symptoms, invariably leading to larval mortality. This ailment exhibits a pronounced prevalence during the autumnal months of September and October, significantly impacting the tasar silk production by its widespread occurrence and resultant damage. Ponnuvel et al. (2022)

identified a transcript closely aligned with the positive strand of an Iflavirus, an RNA virus class known for its pathogenicity in insect populations, within an expressed sequence tag (EST) database from a preceding investigation. Comprehensive analysis revealed the virus's genome to be comprised of 9728 nucleotides, encoding a singular, extensive open reading frame (ORF) that is book-ended by non-translated regions (NTRs) at both the 5′ and 3′ termini, alongside a naturally occurring polyadenylate (poly A) tail at the 3′ end.

The ORF is architecturally organized to encode for a polyprotein that is 2967 amino acids in length, with a bifurcation in its coding strategy: The N-terminal region is dedicated to encoding structural proteins (VP1–VP4), while the C-terminal region encodes nonstructural proteins including a helicase, an RNA-dependent RNA polymerase, and a 3C protease. This viral entity demonstrates a pervasive tissue tropism, being detectable within various anatomical sites of the host including the midgut, fat body, trachea, Malpighian tubules, and silk gland. Additionally, a vertical transmission pathway has been elucidated, whereby the virus can be passed from infected gravid females directly to their progeny. Given the accumulated genetic and biological evidence, this virus represents a novel addition to the Iflaviridae family. Consequently, it is proposed that this virus be designated as *A. mylitta* Iflavirus (AmIV), marking it as a distinct viral entity within this taxonomic classification and underscoring its significance in the pathology of *A. mylitta* and its consequential impact on the tasar silk industry.

Antimicrobial Protein of *A. mylitta*

Antimicrobial proteins (AMPs), characterized by their small size and cationic nature, play a pivotal role in the innate immune response of hosts by exerting a broad-spectrum antimicrobial activity against a diverse array of pathogens including bacteria, viruses, parasites, and fungi. These proteins are synthesized predominantly in the fat body and hemocytes of insects and are secreted into the hemolymph in response to microbial invasion. A detailed study performed by Chowdhury et al. (2021) involving the tasar silkworm, *A. mylitta*, particularly its bacterially immunized fifth instar larvae, hemolymph was extracted and subjected to a purification process to isolate AMPs. This process entailed organic solvent extraction, followed by meticulous purification steps involving size exclusion and reverse-phase high-pressure liquid chromatography (HPLC). Analytical techniques such as thin-layer chromatography (TLC) and sodium dodecyl sulfate-polyacrylamide gel electrophoresis (SDS-PAGE) were employed to verify the purity of the isolated AMP. The molecular mass of the AMP was precisely determined to be 14 kDa using matrix-assisted laser desorption ionization-time of flight (MALDI-TOF) mass spectrometry, leading to its designation as AmAMP14. Further molecular characterization through peptide mass fingerprinting of trypsin-digested AmAMP14 and subsequent de novo sequencing of a peptide fragment via tandem mass spectrometry unveiled its amino acid sequence as CTSPKQCLPPCK. A comprehensive search in existing databases revealed no homologous sequences, suggesting AmAMP14 represents a novel antimicrobial protein.

The antimicrobial efficacy of AmAMP14 was quantitatively assessed by determining its minimum inhibitory concentrations (MICs) against a panel of pathogenic organisms, including *Escherichia coli*, *Staphylococcus aureus*, and *Candida albicans*, with MIC values recorded at 30, 60, and 30 μg/ml, respectively. The application of electron microscopy to examine the structural integrity of cells treated with AmAMP14 disclosed significant membrane damage and the extrusion of cytoplasmic contents, indicative of its potent antimicrobial action. Collectively, these findings underscore the discovery of AmAMP14, a novel 14 kDa antimicrobial protein, synthesized within the hemolymph of *A. mylitta*, which contributes to the organism's defensive arsenal against microbial infections. This protein not only enriches the repertoire of known antimicrobial agents but also offers potential for therapeutic exploitation given its efficacy against a spectrum of pathogenic microbes.

Characterization of Cocoonase of *A. mylitta*

The enzyme designated as cocoonase, with the enzymatic classification number 3.4.21.4, operates analogously to trypsin, embodying characteristics of a serine-trypsin protease. This proteolytic enzyme is integral to the biochemical processes that facilitate the degradation of peptide bonds, situating itself within the broader category of proteases. Its critical function is evidenced in the context of the tasar silkworm, *A. mylitta*, particularly during the transition from pupal to adult stages, wherein cocoonase plays a pivotal role in softening the cocoon shell. Approximately 600–900 μl of cocoonase, with a concentration of 221 μg/ml, targets the sericin protein, responsible for binding silk fibroin filaments, without affecting the fibroin itself. This is achieved through its targeted proteolytic cleavage of sericin, a protein that envelops the silk fiber, while concurrently preserving the integrity of fibroin, the core protein constituent of silk. The specificity of cocoonase for sericin over fibroin not only preserves the fibrous core but also enhances the silk's luster and texture, attributes that are particularly valued in the silk industry. This enzymatic process, known as degumming, underlines the potential of cocoonase as a valuable yet underexploited by-product in silk production.

Despite its utility, the comprehensive characterization of cocoonase across various commercially bred ecoraces of *A. mylitta*, such as Modal, Sukinda, Laria, and Raily, remains an area ripe for further scientific inquiry. Traditional cocoon softening methods, which typically employ alkaline solutions containing detergents or alkalis, have been shown to detrimentally affect the silk's color, texture, and organic qualities. This recognition of the limitations inherent in conventional softening techniques has spurred interest in the development of enzymatic alternatives that could offer a more sustainable and less deleterious approach to enhancing silk quality.

Given the dual-protein composition of *A. mylitta* silk, which consists of a fibroin core encased in a sericin coating, the targeted action of cocoonase on sericin positions it as a promising candidate for applications not only within the silk industry for softening processes but also potentially in biomedical fields, courtesy of its proteolytic capabilities. The shift toward enzymatic methods for cocoon softening is advocated as a superior and environmentally benign alternative to traditional practices, highlighting the need for a nuanced understanding of cocoonase's activity across different silk-producing ecoraces.

Research elucidated the molecular dynamics of cocoonase, including its secretion process, volume, concentration changes, purification via sephadexG100 column, and characterization, including a molecular weight determination of 25–26 kDa through SDS-PAGE. The study also explored cocoonase's application in cocoon softening, which resulted in significant surface variations in tasar silk fibers as observed through SEM studies. Comparative analysis with cocoonase from other sericigenous insects, using MaldiTof-Tof MS, revealed a close harmony with nature, suggesting cocoonase's utility in developing eco-friendly cocoon processing technologies. In-depth characterization included RNA isolation, cDNA preparation, PCR amplification, and in silico analysis, revealing a 26% similarity to *A. pernyi* cocoonase and confirming its classification within the chymotrypsin-like serine protease superfamily. The study successfully generated a polyclonal antibody against cocoonase, demonstrating its specificity through Western blot analysis across various *A. mylitta* ecoraces. Findings from CD spectroscopy and DSC underscored cocoonase's structural stability, while kinetic studies highlighted its optimal activity conditions and potential for pharmaceutical applications due to its antioxidant and fibrinolytic properties. This comprehensive investigation not only advances our understanding of cocoonase's biological role and molecular structure but also paves the way for its application in sustainable silk production and beyond.

Characterization of Arylphorin of *A. mylitta*

The arylphorin protein specific to *A. mylitta* was meticulously isolated from the silk glands of fifth instar larvae, undergoing a sequence of purification steps including ammonium sulfate precipitation, ion-exchange chromatography, and gel filtration chromatography by Dutta et al. 2020. Initial analysis via N-terminal sequencing of a decamer peptide (NH2-SVVHPPHHEV-COOH) revealed a notable sequence homology with the arylphorin protein from *Antheraea pernyi*. Leveraging the sequence information from both the N-terminal and C-terminal regions of *A. pernyi* arylphorin, specific primers were synthesized, facilitating the cloning of *A. mylitta* arylphorin cDNA through reverse transcription polymerase chain reaction (RT-PCR) from mRNA extracted from the silk gland. The comprehensive sequencing of the resultant cDNA, inclusive of 25 nucleotides spanning the 5′ untranslated region (UTR) acquired via 5′ rapid amplification of cDNA ends (RACE), unveiled an open reading frame (ORF) comprising 2115 nucleotides. This ORF was predicted to encode a protein consisting of 704 amino acids, characterized by a preponderance of aromatic residues and exhibiting a molecular weight of approximately 83 kDa.

Subsequent homology modeling, utilizing the *A. pernyi* arylphorin structure as a template, was conducted to infer the tertiary structure of the *A. mylitta* arylphorin.

The cloned arylphorin cDNA was expressed in *Escherichia coli*, leading to the production of a recombinant His-tagged protein, which was subsequently purified through Ni-NTA affinity chromatography. A detailed examination of the arylphorin's tissue-specific expression, conducted via quantitative real-time PCR, identified a predominant expression in the fat body, followed by the silk gland and integument. The 5′ flanking region, encompassing 759 base pairs (bp) of the arylphorin gene, was amplified using inverse PCR techniques, and the full-length gene, spanning 5359 nucleotides and inclusive of 5 exons and 4 introns, was cloned from *A. mylitta* genomic DNA and sequenced. Further, a polyclonal antibody was generated against the purified arylphorin, enabling the purification of a more native form of the arylphorin protein (approximately 500 kDa) from the fat body via antibody affinity chromatography. Investigative studies into the mitogenic effects of both the native arylphorin and its chymotrypsin hydrolysate on various insect cell lines posited the protein's potential utility as a serum substitute in the in vitro cultivation of insect cells, underscoring its biological significance and versatility in biotechnological applications.

Key Gene Identification

The identification of key genes related to silk production, disease resistance, and adaptability in *A. mylitta* is crucial for advancing sericulture practices and developing improved strains (Table 3).

Further tasar silkworm genomic research is expected to identify more genes linked to silk production, disease resistance, and adaptability. The discovery of these genes opens the door to focused interventions and improvements in sericulture techniques by offering a thorough understanding of the molecular mechanisms behind crucial biological processes and characteristics in tasar silkworms.

Functional Genomics and Transcriptomics in Understanding Tasar Silkworm Biology

The utilization of functional genomics and transcriptomics has emerged as a pivotal methodology for delineating the molecular architecture driving a plethora of biological processes inherent to this species. These sophisticated techniques provide an exhaustive delineation of gene expression dynamics, delineate the functionalities of discrete genes, and elucidate the complex regulatory matrices orchestrating a myriad of biological phenomena in tasar silkworms. Through the lens of transcriptomic analyses focused on the silk glands of *A. mylitta*, an in-depth comprehension of the genetic constituents instrumental in silk biogenesis is attainable. Such investigations shed light on the gene expression trajectories throughout various

Table 3 Genes of *A. mylitta* and their respective function in silkworm

Gene	Function
Serine protease genes	Serine proteases play a critical role in silk protein synthesis. They are involved in the cleavage of fibroin and sericin proteins during silk gland processing. Genomic studies have identified serine protease genes expressed in the silk glands of tasar silkworms, contributing to our understanding of the silk production pathway
Fibroin and sericin genes	Fibroin and sericin are the major proteins composing silk fibers. Fibroin provides tensile strength, while sericin acts as a glue to hold fibers together. Genomic analyses have uncovered the genes responsible for the synthesis of fibroin and sericin, revealing insights into the regulation of silk protein expression
Transcription factors	Transcription factors control the expression of genes involved in silk production. Understanding these regulators is crucial for manipulating silk quality and quantity. Genomic studies have identified transcription factors that modulate the expression of silk protein genes, providing potential targets for genetic engineering
Chitinase genes	Chitinases are enzymes that degrade chitin, a component of the peritrophic membrane in the midgut. They are involved in silk gland processing. Genomic studies may identify chitinase genes that contribute to silk production and pupation
P25 and P50 protein genes	P25 and P50 proteins are structural proteins associated with the silk fibers, contributing to silk stability. Genomic analyses may reveal the genes responsible for encoding P25 and P50 proteins
Immune-related genes	Genes associated with the immune response play a key role in defending tasar silkworms against pathogens. Genomic studies have identified immune-related genes, including those encoding antimicrobial peptides and proteins involved in the immune signaling pathway
Pattern recognition receptors (PRRs)	PRRs recognize pathogen-associated molecular patterns (PAMPs) and activate immune responses. Genomic analyses have revealed the presence of PRR genes in tasar silkworms, contributing to our understanding of their innate immune system
Esterase and cytochrome P450 genes	Detoxification enzymes, such as esterases and cytochrome P450, are involved in defense against insecticides and toxins produced by pathogens. Genomic studies have identified genes encoding esterases and cytochrome P450, providing insights into the tasar silkworm's detoxification mechanisms
C-type lectin genes	C-type lectins recognize and bind to microbial surfaces, initiating immune responses. Genomic analyses may reveal C-type lectin genes involved in the recognition of pathogens in tasar silkworms
Serine protease inhibitor genes	Serine protease inhibitors regulate the activity of serine proteases involved in immune responses. Genomic studies may uncover serine protease inhibitor genes that modulate immune reactions
Heat shock proteins (HSPs)	HSPs play a role in cellular protection during stress, including heat stress. Genomic analyses have identified HSP genes in tasar silkworms, contributing to our understanding of their thermal adaptability

(continued)

Table 3 (continued)

Gene	Function
Olfactory receptor genes	Olfactory receptors are essential for detecting environmental cues, including host plants and pheromones. Genomic studies have revealed olfactory receptor genes, shedding light on the tasar silkworm's adaptability to diverse host plants
Cuticular protein genes	Cuticular proteins are involved in the formation of the insect cuticle, providing protection against environmental stressors. Genomic research has identified cuticular protein genes, contributing to our understanding of the tasar silkworm's adaptation to different climates
Antioxidant enzyme genes (e.g., superoxide dismutase)	Antioxidant enzymes protect cells from oxidative stress and contribute to environmental adaptability. Genomic analyses may identify genes encoding antioxidant enzymes that play a role in the tasar silkworm's adaptability
Desaturase genes	Desaturases are involved in lipid metabolism and can play a role in cold adaptation by influencing membrane fluidity. Genomic studies may reveal desaturase genes that contribute to the tasar silkworm's adaptability to different temperatures
Heat shock transcription factor genes	Heat shock transcription factors regulate the expression of heat shock proteins in response to stress. Genomic analyses may uncover genes encoding heat shock transcription factors, contributing to the tasar silkworm's heat stress response

developmental stages of the silk glands, paving the way for a granular understanding of the molecular cascades culminating in silk fiber production.

Moreover, the exploration of functional genomics furnishes a nuanced understanding of the genetic foundations underlying immune response mechanisms in tasar silkworms, thereby augmenting our grasp of their defense stratagems against microbial assailants. The elucidation of immune-centric genes and pathways heralds the advent of strategic interventions aimed at amplifying innate immunity. Transcriptomic explorations into the metabolic schema of *A. mylitta* vis-à-vis environmental adaptability unveil the genetic underpinnings of their resilience, spotlighting genes implicated in metabolic circuitries, heat shock proteins, and stress amelioration responses.

The deployment of functional genomics facilitates the reconnaissance of genetic markers associated with desirable phenotypes such as enhanced silk quality, disease resistance, and superior environmental adaptability. This, in turn, enhances the efficacy of marker-assisted selection (MAS) strategies by pinpointing genes correlating with traits of economic significance. Additionally, transcriptomic studies illuminate the complex interplay between tasar silkworms and their host plants, unearthing genes pivotal for olfactory perception, detoxification processes, and host plant identification, thereby enriching our understanding of their ecological dynamics.

Genome Study in *Antheraea* sp.

Genome of *Antheraea assamensis*, a semidomesticated lepidopteran insect crucial for the silk industry in Northeast India, was studied by Dubey et al. 2023. The study utilized two sequencing platforms and three library strategies, resulting in a 501.8 Mb genome with notable features. The analysis identified genes related to silk fiber production, revealing unique characteristics compared to other silk moths. Comparative genomics highlighted both shared and unique genes in *A. assamensis*, providing insights into its evolutionary history. The study also explored duplicated gene families, emphasizing the role of negative selection. The evolution of key gene families associated with feeding behavior in insects, such as GST, ABC transporter, and CYP450, was examined. The authors underscore the significance of the generated genomic resource for future molecular interventions in improving *A. assamensis* and advocate for future comparative genomic studies on wild silk moths, considering their distinct behaviors.

Next-Generation Sequencing of *A. mylitta* Population

The tasar silk genome's intricate details can be fully understood with the help of next-generation sequencing (NGS), which provides high-throughput, affordable, and quick sequencing capabilities. NGS technologies, like Illumina and PacBio, make it possible to analyze genomic DNA in great detail, which makes it easier to understand genetic variations, identify genes involved in the manufacture of silk, and explore the entire genomic landscape. The speed and efficiency of NGS empower researchers to conduct large-scale studies, including transcriptomics and epigenomics, providing valuable insights into the molecular mechanisms governing tasar silk production. The application of NGS in tasar silk genomics accelerates advancements in sericulture, guiding targeted research initiatives and contributing to the sustainable and improved production of this commercially significant silk.

A study conducted by (Gattu and Vodithala 2024) aims to characterize tasar silkworm, *A. mylitta*, and ecoraces across tropical forest zones, providing a foundation for the identification and assessing genetic diversity. The focus is on the Andhra local ecorace, aiming to guide a comprehensive breeding program for conserving the declining tasar silkworm population. Leveraging next-generation sequencing (NGS) technology, which revolutionizes plant genotyping and breeding with its ultra-throughput sequencing capabilities, this research aligns with the continuous demand for NGS analysis. The evolving landscape of alignment and variant calling tools in NGS remains an active area of exploration, ensuring advancements in genomic analysis for effective conservation strategies (Vestergaard et al. 2021). The study employed Illumina sequencing for eight samples, ensuring the suitability of the sequencing library with fragment sizes ranging from 200 to 700 bp. After adapter adjustments, the effective user-defined insert size was determined as 80–580 bp.The samples were then sequenced, and populations were identified along with SNPs at specific locations. Cluster analysis revealed that all eight samples were of the same length and clustered together, exhibiting unique GC-rich regions. The GC content varied among samples. Gattu and Vodithala (2024) also highlighted the utility of next-generation sequencing (NGS) protocols for SNP discovery and genotyping in crop improvement. The cost-effectiveness of genotyping-by-sequencing (GBS) makes it an attractive option for high-density SNP marker saturation in mapping and breeding populations. Findings contribute to the broader field of genomics and molecular characterization, emphasizing the potential of NGS technologies in silkworm genetics and breeding. Despite the advantages, the study acknowledged potential limitations of NGS platforms, such as underrepresentation of regions and nucleotide substitution errors. The application of nine trimming algorithms on different datasets demonstrated variable quality parameters, emphasizing the need for careful data processing.

Whole-Genome Sequence of *A. mylitta*

There was an absence of genomic sequence data for *A. mylitta*. Addressing this gap, initial efforts involved shallow sequencing, yielding 65 GB of DNA sequence data. The analysis revealed that the total guanine-cytosine (GC) content stood at 35.9%. Comparative genomic analysis indicated a substantial similarity, with 77.4% of the *A. mylitta* genome aligning with that of *Antheraea yamamai* and only 11.3% with *B. mori*, suggesting a closer genetic relationship with *A. yamamai*.

Building upon the shallow sequencing, comprehensive de novo whole-genome sequencing was undertaken employing both PacBio and Illumina technologies, marking the first successful attempt at sequencing the entire genome of *A. mylitta*. The process culminated in the submission of the sequencing data, including the Sequence Read Archive (SRA), to the National Center for Biotechnology Information (NCBI), thus facilitating accessibility for the research community. The PacBio platform contributed 10.22 GB of data. Concurrently, sequencing of 10X chromium libraries via the Illumina HiSeq X 10 system, utilizing 2x150 pair-end chemistry, generated 106,443.28 MB with a GC content of 37.68% and a high-quality score (Q30) of 93.305%.

The assembly of this genomic data was executed which resulted in an assembled genome size of approximately 707.76 MB pairs. The assembly achieved a remarkable genome coverage of 184.6x, incorporating 20,891 scaffolds with an N50 scaffold length of 5,182,261 base pairs and 33,698 contigs with an N50 contig length of 136,984 base pairs. The finished genome completeness was estimated at 97.50%. Gene annotation was conducted utilizing a suite of bioinformatics tools including Ganopht, *A. yamamai* protein databases, eggNOG (v5.0.0), and Gene Ontology (GO), leading to the annotation of 25,544 genes. *A. mylitta* whole-genome sequence related information uploaded on NCBI with accession no. JACEII000000000.

A detailed BLAST search comparison between the proteins of *A. mylitta* and *A. yamamai* revealed a significant degree of similarity, with 86.27% of *A. mylitta* proteins showing hits with *A. yamamai* proteins. Orthology analysis, conducted with OrthoFinder (v2.3.11), identified 14,317 ortholog groups among *B. mori*,

A. yamamai, and *A. mylitta*, further elucidating the evolutionary relationships and functional genomics across these species. This comprehensive genomic characterization of *A. mylitta* not only enhances our understanding of its genetic architecture but also contributes valuable insights into the evolutionary biology of sericigenous insects, functional genomics, and gene discovery for productive traits.

CRISPR-Cas9

Three noteworthy technologies have gained prominence in the field of genome editing in recent years: Clustered regularly interspaced short palindromic repeat-associated nucleases (CRISPR-Cas9), transcription activator-like effector nucleases (TALENs), and zinc finger nucleases (ZFNs) are examples of programmable nucleases (Guha et al. 2017; Li et al. 2020). A novel gene-editing method known as CRISPR-Cas9 allows precise modification of an organism's DNA. It is composed of two main components: the Cas9 protein, which acts as a molecular scissors, and a guide RNA (gRNA), which points the Cas9 protein toward the target DNA sequence. The Cas9 protein then breaks DNA, and when the cell repairs these breaks, it may lead to alterations in genes (Haque et al. 2018) (Fig. 4). Reversing genetic mutations, controlling gene expression, and creating therapeutic medicines are all being greatly aided by engineered nucleases. According to Schuijff et al. (2021), these methods also advance our knowledge of gene functions and the mechanisms behind a variety of illnesses and hereditary disorders. In the field of genome editing, it is essential to minimize off-target consequences. Due to its great fidelity, the CRISPR-Cas system exhibits exceptional reliability in its overall outcomes (Schuijff et al. 2021; Zhang et al. 2020). The CRISPR-Cas system, which was first discovered in bacteria, has steadily developed into a useful tool for scientists, enabling the editing of genomes in a wide range of animals, including bacteria, insects, plants, and human cells (Li et al. 2021a). Fibrin and sericin are the two main proteins that make up the silk proteome. The anterior (ASG), middle (MSG), and posterior (PSG) silk glands are the three anatomical regions that make up the silk gland. The MSG assembles sericin, whereas the PSG synthesizes fibroin. According to (Faure et al. 2019), this mechanism defends organisms not just against viruses but also from other mobile genetic elements including plasmids and transposons.

The host-related specificities of CRISPR-Cas contribute to its amazing variety and adaptability. The CRISPR array and cas gene sequences, which exhibit diversity, enable the categorization of these systems into two main classes, which are further subdivided into many groups (Makarova and Koonin 2015). These systems are classified according to the characteristic Cas proteins. As of right now, CRISPR-Cas systems may be divided into two main classes, each of which has several groups (Koonin and Makarova (2019). Because it contains vital promoter sequences for the transcription of CRISPR loci, the leader nucleotide sequence is important. Apart from its function as a promoter, the leader sequence comprises certain signals that are essential for the adaption phase in the first step of CRISPR-Cas activation (Alkhnbashi et al. 2016). Presently, CRISPR-Cas9 is the most extensively utilized

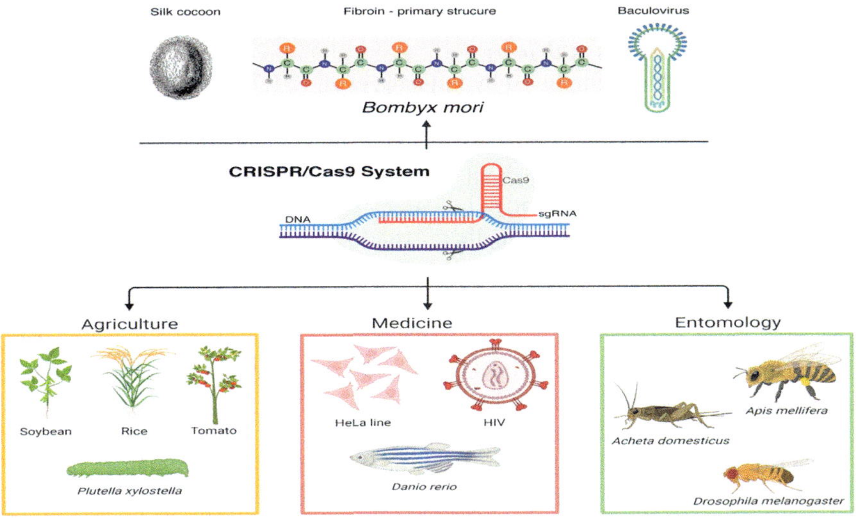

Fig. 4 CRISPR-Cas9 applications in agriculture, medicine, and entomology. On top, a simplified description of CRISPR-Cas9 in silk worm. Source: Baci et al. (2021)

tool in labs around the globe among all the CRISPR-Cas systems (Manghwar et al. 2019). In CRISPR-Cas II systems, double-stranded DNA breaks are induced by the Cas9 nuclease, a hallmark protein (Makarova and Koonin 2015). There are three different ways that the Cas9 endonuclease can be delivered. While the other two ways use a plasmid containing the Cas9 enzyme or a messenger RNA (mRNA) sequence encoding it, the first approach requires directly injecting the enzyme into embryos. Because of its benefits over the other two methods—namely, less off-target activity and minimal immunogenic effects—direct protein administration is regarded as the best option for genome engineering (Li et al. 2021b). A thorough analysis of CRISPR-Cas9's influence on emerging research trends has been done elsewhere (Tyagi et al. 2020). CRISPR-Cas9 is a straightforward yet effective genome editing technique with a wide range of applications.

CRISPR-Cas9 was first successfully used to modify the *B. mori* genome, as reported by Wang et al. 2013 and Baci et al. (2021). The crucial gene BmBlos2, which is orthologous to the human gene Blos2, was the focus of this study's investigation (Fujii et al. 2010). Two 23-bp sgRNAs were created by them in order to cause mutations that would impair the function of the target gene. Each sgRNA-Cas9 combination that was injected during the preblastoderm embryonic stage produced a phenotypic marker in which the larval integument, which is typically opaque, turned transparent when the BmBlos2 gene function was lost.

CRISPR-Cas9 technology's multiplexable potential was highlighted by Liu et al. (2014). Researchers may quickly perform accurate target mutagenesis because of this property, which makes it easier to precisely and simultaneously induce mutations at diverse places (Sakuma et al. 2014). In a different research, the second exon of the BmKu70 gene was targeted using CRISPR-Cas9 (Ma et al. 2014). The findings showed an increased incidence of homologous repair, offering bright futures for basic studies in *B. mori*.

CRISPR-Cas9 was employed by Fujinaga et al. (2017) to examine the function of insulin-like growth factor-like peptide (IGFLP) in *B. mori*. It became clear that IGFLP was important for the development of the genital disc when it was shown to be absent from smaller ovaries and lower egg counts. Additionally, Zhang et al. (2019) used CRISPR-Cas9 in 2019 to introduce spider silk genes into the silkworm genome, improving the silk's qualities. This method showed that improving the mechanical characteristics of silk on an industrial scale was feasible. Using CRISPR-Cas9, Ling et al. (2015) investigated the role of miR-2 in *B. mori* and discovered that it is essential for wing development. Using CRISPR-Cas9, Liu et al. (2020) examined the effect of miR-34 on insect development and found that overexpression had a detrimental influence on wing shape and body size. Park et al. (2018) used CRISPR-Cas9 site-target mutagenesis to target the BmCactus gene in a particular *B. mori* cell line in the context of silkworms' inherent defense systems. This work adds to our understanding of the activation mechanisms involved in antimicrobial peptide genes, notwithstanding the poor survival rate observed. Furthermore, the juvenile hormone esterase producing gene, BmJhe, was depleted by Zhang et al. (2017) using CRISPR-Cas9, demonstrating its effect on larval stages and silk production.

Certain genes within the tasar silkworm genome can be targeted and modified by researchers using certain gRNAs. Scientists can examine the role of particular genes in *A. mylitta*. By monitoring the phenotypic changes that arise when a gene of interest is disrupted, researchers can get additional insight into the role that a gene plays in the biology of an organism. This might be applied to the study of gene function, the induction of particular phenotypes, or even the development of strains with desired properties. CRISPR-Cas9 may be utilized to improve silk production or alter its characteristics if particular genes linked to silk quality or manufacturing are found. This could have implications for the silk industry. *A. mylitta* comprises different ecoraces with variations in traits. CRISPR-Cas9 could aid in understanding and conserving specific ecoraces by studying and potentially modifying relevant genes.

TALENs

Transcription activator-like effector nucleases (TALENs) have been extensively studied and proven to be highly effective chimeric enzymes across a broad spectrum of organisms, including zebrafish, mice, rats, *Xenopus*, *Caenorhabditis elegans*, and multiple insect species. These engineered nucleases are synthesized by fusing the nonspecific cleavage domain of the Type II restriction endonuclease FokI with the DNA-binding domain of bacterial TAL-effector transcription factors, commonly derived from *Xanthomonas* species, where TAL-effector sequences exhibit notable conservation. The operational mechanism of TALENs necessitates their dimerization, as the FokI nuclease domain functions in pairs, targeting DNA sequences that consist of 2 half-sites separated by a spacer whose length can vary from 12 to 30 nucleotides, depending on the architectural design of the TALEN construct employed (Fig. 5). Despite the diversity in the origins of TAL constructs used for mutagenesis, such variations are not anticipated to significantly affect their functionality (Takasu et al. 2014). Enhancements in TALEN activity have been observed when employing various C- and N-terminal truncations of native TAL proteins in conjunction with the FokI nuclease domain, as delineated by Miller et al. (2011).

Building on this foundational knowledge, a study by Li et al. (2022) elucidated the successful application of TALEN-mediated targeted gene integration within the silk gland of *B. mori*, leveraging the Ser1 infusion expression system. This innovative approach harnessed the natural regulatory elements of Ser1 to specifically direct the expression of recombinant proteins in the middle silk gland, overcoming the limitations associated with previous transposon-based transgenesis techniques, which were plagued by issues of genetic variability and instability as noted by Tomita et al. (2003). The infusion expression system, employing 2A peptide-mediated integration at the C-terminal of Ser1, yielded a significant production of EGFP in both the silk glands and cocoon shells while simultaneously preserving the expression of the natural Ser1 gene. This strategy effectively minimized the fitness costs traditionally associated with transgenic modifications. However, the study also noted that the overexpression of EGFP could potentially impact the thickness of the cocoon shell, hinting at possible interference with the native Fibroin gene expression. Furthermore, the research highlighted size-dependent fitness costs, with smaller proteins like hEGF altering Ser1 expression. In sum, the infusion expression system showcased by Li et al. (2022) represents a significant advancement in the field of recombinant protein expression within the silk gland of *B. mori*, offering a promising avenue for efficient and precise genetic engineering in sericulture.

Expanding this concept to *A. mylitta* could open new vistas in the customization and functionalization of non-mulberry silk, potentially enabling the production of silk fibers with novel properties or enhanced functionalities tailored for specific applications. Such advancements could significantly augment the value of *A. mylitta* silk in various industrial and biomedical applications, underscoring the importance of integrating cutting-edge genome editing technologies in the study and manipulation of silk-producing lepidopterans.

ZFN

The modulation of gene expression is intricately facilitated by transcription factors (TFs), which play a pivotal role in the orchestration of cellular processes. These molecular entities are characterized predominantly by their DNA-binding domains, which serve as the quintessential feature facilitating their interaction with genomic DNA. Moreover, a subset of transcription factors is further distinguished by the presence of additional effector domains, which augment their functional repertoire (Fig. 6). The influence of TFs extends across a broad spectrum of biological phenomena, including but not limited to developmental processes (Sommer et al. 1992),

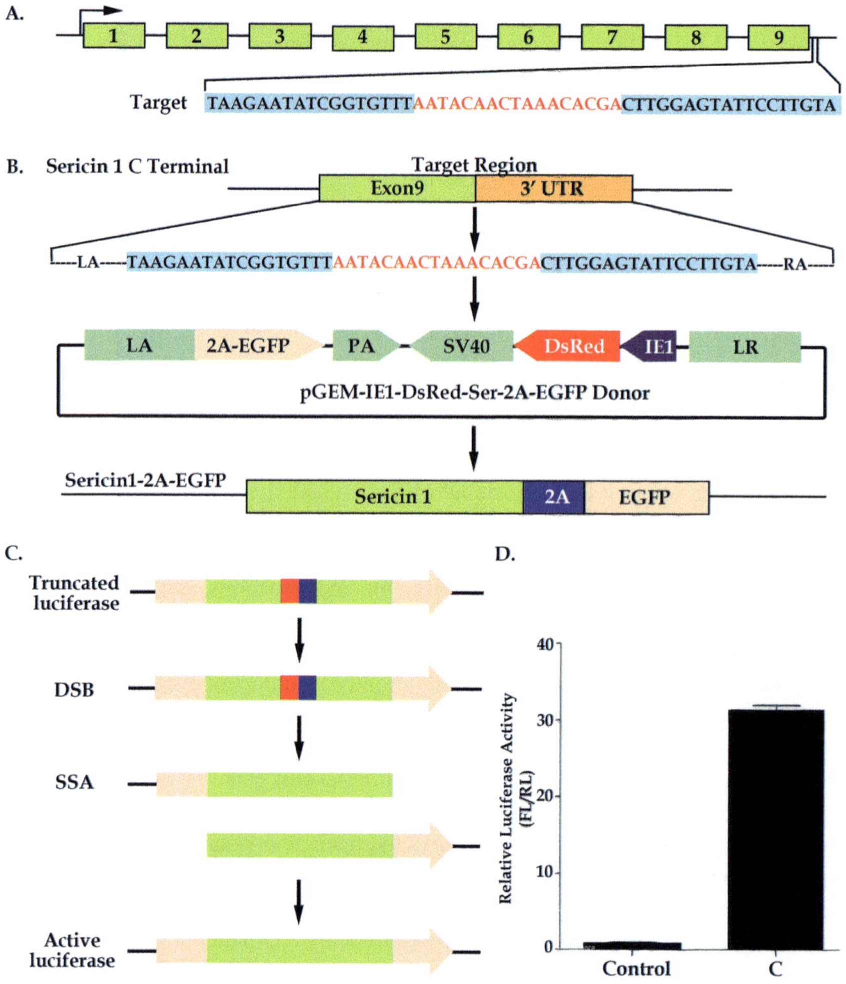

Fig. 5 Ser1 infusion expression system by using TALENs mediated targeted gene expressionin silkworm. Source: Li et al. (2022)

metabolic pathways (Krebs et al. 2014), programmed cell death mechanisms (apoptosis) (Ma et al. 2016), and cellular degradation processes (autophagy) (Chauhan et al. 2013). A significant advancement in our understanding of the zinc finger gene family, specifically the C2H2-ZF proteins, was made through genome-wide analysis conducted by Wu et al. (2019), supplemented by our own investigations employing sequencing and microarray techniques. These studies illuminate the complexity of the C2H2-ZF protein families, suggesting that our findings, in conjunction with those of Wu et al. (2019), could provide vital insights for future research into the functional roles of TFs within *A. mylitta* and, by extension, other members of the Lepidoptera order. The predictive capacity for TF recognition sequences is often

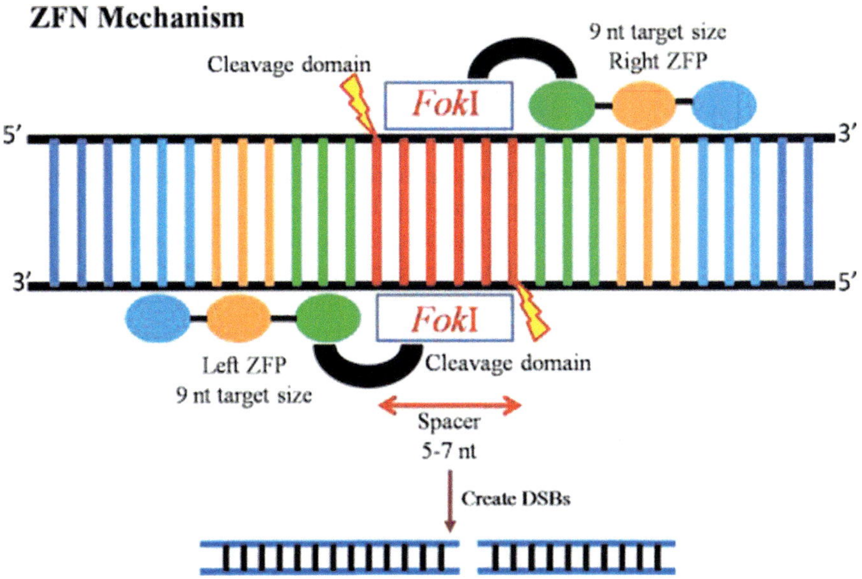

Fig. 6 Mechanism of zinc finger nuclease (ZFN) inducing double-stranded breaks at the target region in the genome. Source: Hillary and Ceasar (2021)

attributed to individual zinc fingers within these proteins, which are commonly organized in tandem arrays. Recent methodological advancements, such as tandem zinc finger protein chromatin immunoprecipitation followed by sequencing (ChIP-seq) and complementary bioinformatic analyses (Lam et al. 2011), have enhanced our ability to decipher the genomic interactions of these transcription factors. Furthermore, the research by Gupta et al. (2014) underscores the extensive species diversity within the zinc finger protein family, highlighting the regulatory significance of zinc finger transcription factors across a myriad of biological processes. Given the complex regulatory roles and vast diversity of the zinc finger protein family, it becomes imperative to delve deeper into the molecular mechanisms and functionalities of zinc fingers to fully appreciate the scope and evolution of TFs in biological systems. This necessitates a concerted effort toward more comprehensive research in this domain, particularly in understanding the nuances of transcription factor roles in *A. mylitta* and related lepidopteran species.

Metagenomic Studies

The microbes that inhabit the guts of insects are essential to their development, growth, and adaption. The identification and molecular study of gut bacteria from insects can help us create new approaches to industrial product creation, efficiently use by-products, and implement the best pest control techniques. 20–30 °C is the perfect temperature for silkworms to go about their daily lives. Raised in tropical climates, certain bivoltine silkworms exhibit thermotolerance. On the other hand, extended exposure to extreme temperatures can be lethal. In this investigation, bivoltine silkworm larvae in their fifth instar were subjected to heat shock for a brief hour each day at 40 ± 2 °C in order to look at alterations in the gut microbiota. Heat shock impacts silkworms' intestinal microbiota, which regulates the activity of related digestive enzymes that influence nutrition intake and digestion, ultimately affecting the growth and development of the larvae and cocoons that are generated (Ismail et al. 2023). In tropical silkworms, alterations in intestinal tissue damage and metabolism are linked to shifts in the makeup and activity of the intestinal microbiota. According to Baig et al. (2023), the gut microbiota plays a crucial role in the development and environmental adaption of the host insect, exerting a considerable influence on several aspects of host physiology, including immunology and gut structure.

Unique environments are created in the stomachs of insects for the colonization of microbes, and gut bacteria in particular may perform a multitude of beneficial tasks for their hosts. The primary source of energy for the host is the breakdown of carbohydrates, which is accomplished by the intestinal microflora's secretion of many digesting enzymes. Research has indicated that gut bacteria can lead to the buildup of fat in the host by modifying the host's energy metabolism and regulating gene expression. If the equilibrium between the gut flora and the host environment is upset, the host's metabolic process becomes abnormal and affects growth and development (Ismail et al. 2023). Using culture-dependent studies, we were able to identify the following microorganisms from the gut sample cultures: *Micrococcus*, *Lactobacillus*, *Bacillus cerus*, *Enterococcus*, *Erwinia*, and *Pseudomonas*. *Proteobacteria* are the primary group of bacteria that are thought to help break down plant biomass when it is ingested (Rajan et al. 2020). The bacterial ecology in the stomach of *A. mylitta* was dominated by *Firmicutes* at the phylum level. *Acinetobacter*, *Desulfomicrobium*, *Sphingomonas*, *Faecalibacterium*, *Staphylococcus*, *Ralstonia*, *Bacillus*, *Azospirillum*, *Candidatus*, and *Kocuria* were the other leading genera after *Turicibacter* (Rajan et al. 2020). *Firmicutes* make up more than 92% of the gut bacterial microbiome, according to culture-independent study, but *Proteobacteria* were also detected by culture-dependent analysis. *Firmicutes* improve an insect's capacity to obtain energy from food and help it endure harsh environments. Consequently, it is advantageous for *A. mylitta* to have a large amount of *Firmicutes* microflora in its intestines in wild populations. Furthermore, *Firmicutes* synthesis of antibiotics and enzyme inhibitors may be the cause of the lack of other microbiota in the *A. mylitta* gut (Pandey et al. 2018,). Since insects have a strong immune system, they can fight off harmful illnesses. The gut microbiota is the defense system, supporting immunity through the innate immune system or the production of antimicrobial peptides. A variety of mechanisms may be used by the innate immune system (Eleftherianos et al. 2013). In ingested bacteria, Miyashita et al. (2015) demonstrated a primed immune response in silkworms that led to systemic infection tolerance. This was accomplished by infusing *P. aeruginosa*, heat-killed *S. marcescens* cells, and heat-killed *C. albicans* cells that show resistance against *P. aeruginosa*, among other heat-killed microorganisms. *Bacillus subtilis*, the other strain, has likewise demonstrated immunity in a number of settings. By raising blood IFN-gamma and salivary SIgA levels, Lefevre et al. (2015) demonstrated the ability of a probiotic *Bacillus subtilis* strain to boost immunity against common infectious diseases in the elderly. Conversely, medium-chain-length polyhydroxyalkanoate (PHAMCL) and alginate oligosaccharides (AO) have been reported to be synthesized by *Pseudomonas mendocina*. Numerous biological processes, including immunological modulation, anti-oxidation, and anticoagulation, are carried out by alginate oligosaccharide (Guo et al. 2011). According to Dubey et al. (2016), a nano-vaccine that was created utilizing *Aeromonas hydrophila*'s outer membrane protein (OmpW) demonstrated dose-dependent immunity in *Labeo rohita* or *Rohu fish*. Recent studies have revealed that *Pseudomonas aeruginosa* secretes type II protease IV, which works as an immunological elicitor in *Arabidopsis* (Cheng 2016).

Epigenetics

In the domain of epigenetics, a sophisticated examination of the silkworm (*B. mori*) methylation landscape has unveiled the pivotal role of DNA methylation, a fundamental epigenetic modification, in the orchestration of gene regulation during embryonic development. This revelation, however, opens the door to intricate inquiries concerning the dynamics of DNA methylation patterning throughout the development of insects. Specifically, the investigation seeks to elucidate the mechanisms through which DNA methylation influences the functional processes governing early embryonic development and the diapause phenomenon in insects. Notably, in *B. mori*, the genomic landscape is characterized by a minimal presence of cytosine methylation within transposable elements, promoters, and ribosomal RNA sequences. Conversely, a significant concentration of methylation is observed within gene bodies, where it exhibits a positive correlation with gene expression levels. This correlation underscores the critical function of DNA methylation in the regulation of gene transcription, as highlighted by the findings of Xia et al. (2009).

Furthermore, the application of CRISPR-Cas technology in *B. mori* has paved the way for innovative modifications in the field of epigenetics. The exploratory work of Liu et al. (2019) delves into the impact of methylation on silkworm development, illuminating the profound significance of DNA methylation within specific genomic regions. Complementary studies by Xing et al. (2020) have harnessed the capabilities of CRISPR-Cas9 technology not only for fundamental research purposes but also for gaining insights into the mechanisms underlying pesticide resistance. These studies employed innovative imaging techniques for labeling endogenous regions within a *B. mori* embryonic cell line, thereby contributing to the expanding body of knowledge on the potential applications of CRISPR-Cas9 in epigenetic research.

Extending this context to *A. mylitta*, an important sericultural insect known for its economic significance in the production of tasar silk, it becomes imperative to explore the epigenetic mechanisms underlying its development and stress responses.

Given the comparative insights drawn from *B. mori*, investigating the role of DNA methylation in *A. mylitta* could unveil novel regulatory mechanisms. Such research could elucidate the epigenetic landscape of *A. mylitta*, shedding light on the specificity of methylation patterns during its embryonic development and diapause states. By leveraging CRISPR-Cas technologies, further studies could explore the modulation of epigenetic markers in *A. mylitta*, potentially offering groundbreaking perspectives on gene regulation, development, and adaptation strategies in this and related lepidopteran species.

Unraveling the Molecular Responses to Thermal Stress in Tasar Silkworms

The exigency for an enriched comprehension of the thermal stress impacts on insects, including *A. mylitta*, has become paramount in light of the anticipated escalation in global mean surface temperature by 2100, which is projected to range between 1.4 and 5.8 °C. *A. mylitta*, distinguished for its polyphagous nature, thrives on a plethora of forest tree species along with other host plants, many of which possess significant economic value. Within the cellular milieu, the mitigation of thermal stress is facilitated by heat shock proteins (HSPs), such as HSP21, which serve as molecular chaperones. These proteins are instrumental in aiding the folding, unfolding, and translocation of other proteins, particularly under conditions of thermal duress.

The expression of heat shock proteins, a cellular defense mechanism against stress-induced protein denaturation and other cellular adversities, is a critical area of study. This is exemplified by the work of Feder and Hofmann (1999), which underscores the inducible nature of HSPs in response to thermal stress. Recent investigations into the genetic responses of *A. mylitta* to heat stress shed light on the vital aspects of the silk industry, focusing on the differential expression patterns of heat shock proteins, notably HSP21 and HSP70, to elucidate the mechanisms enabling tasar silkworms to withstand elevated temperatures.

- Role of HSPs as Molecular Chaperones: The elevation of HSP21 and HSP70 during thermal stress highlights their indispensable role in cellular safeguarding, showcasing their universal function across diverse species as reported by Evgen'ev et al. (1987). This upregulation signifies the critical protective measures undertaken by cells in the presence of thermal stress.
- Genetic Underpinnings of Thermal Resilience: The resilience of *A. mylitta* to thermal stress is modulated by a complex interplay of environmental and genetic factors. The research by Kumar et al. (2010) illuminates the molecular strategies that silkworm strains employ to navigate survival challenges in their habitats. Particularly, the enhanced expression of the hsp21 gene post-heat shock underscores its pivotal role in bestowing thermotolerance.
- Proteomic Insights into HSP Expression Dynamics: Comparative proteomic analyses across different silkworm strains and developmental stages have

revealed distinct expression patterns of HSPs, as demonstrated by Chavadi et al. (2006). This variability accentuates the nuanced nature of the thermal stress response, contingent upon variables such as strain specificity and developmental phase.

- Molecular Markers for Thermotolerance: The introduction of a DNA-based molecular marker, SCAR marker TT-PB1, by Prabhu et al. (2023), represents a significant advancement in selecting thermotolerant lines of *A. mylitta*. This marker's specificity and reproducibility are promising for the analysis of thermotolerance inheritance patterns, heralding the need for comprehensive gene sequence elucidation.

The differential HSP expression patterns observed in tasar silkworms under thermal stress unveil the intricate molecular mechanisms governing their adaptability. The preservation of HSP functions, coupled with genomic discoveries and the development of molecular markers, enriches our understanding of *A. mylitta*'s resilience to heat stress. Ongoing and future research endeavors are poised to impact biotechnological applications, sericulture advancements, and the sustainable exploitation of tasar silkworms for various industrial applications.

Harnessing Silkworms as Bioreactors for Therapeutic Protein Synthesis

The advent of transgenic technology has facilitated innovative methodologies for the biosynthesis of valuable proteins, with transgenic silkworms emerging as viable bioreactors. A seminal study by Tomita (2011) underscored the capability of transgenic *B. mori* silkworms to produce recombinant human type III procollagen within their silk glands, showcasing the potential of leveraging silkworm larvae's natural silk production abilities. This approach offers a scalable and cost-effective alternative to conventional protein production methodologies.

- Silkworm Silk Glands as Protein Synthesis Platforms: The silk glands, constituting a substantial fraction of the silkworm's body mass, are responsible for the production of liquid silk, composed of fibroin and sericins. By incorporating gene integration techniques, researchers have achieved stable and long-term expression of recombinant proteins within these glands. This breakthrough paves the way for the mass production of therapeutic proteins, utilizing the ancient art of silk production while broadening the scope of silkworm silk's applicability beyond traditional uses, as delineated by Zhang et al. (2006).
- Targeted Expression in Silk Gland Regions: By manipulating the expression of recombinant protein genes within specific regions of the silk gland, namely, the posterior (PSG) and middle silk glands (MSG), it is possible to exert precise control over the localization of the synthesized proteins. Proteins expressed in the PSG are incorporated into the fibroin core, whereas those in the MSG become part of the sericin layer. This strategic localization facilitates the straightforward

extraction of recombinant proteins from the cocoon's sericin layer through the use of gentle aqueous solutions, thereby reducing the risk of contamination with native silk proteins.

- Production of Therapeutic Proteins: The utility of transgenic silkworms extends to the production of therapeutic proteins on a large scale. For instance, human platelet-derived growth factor (PDGF-BB) has been successfully produced in transgenic *B. mori*, with high levels of gene expression observed in the MSG, leading to secretion into the cocoon's sericin layer (Zhang et al. 2017). The extracted PDGF-BB retains significant bioactivity, fostering cell growth, proliferation, and migration, which underscores the efficiency of silkworms as bioreactors for therapeutic protein production, with promising implications for applications such as wound healing.
- Neurotrophin-4 Synthesis Innovations: In a groundbreaking study by Zhang et al. (2023), the silk gland bioreactor of silkworms was genetically modified to produce activated human neurotrophin-4 (NT-4) at scale. The NT-4-functionalized silk material exhibited potent bioactivity, enhancing cell proliferation and potentially inducing differentiation. This novel approach offers significant promise for applications in tissue engineering and neural regeneration.

Expanding this paradigm to *A. mylitta*, a species renowned for its role in tasar silk production, could further diversify the range of therapeutic proteins producible through silkworm bioreactors. Given *A. mylitta*'s unique silk properties and its potential for genetic modification, it stands as a compelling candidate for bioengineering efforts aimed at therapeutic protein production. Such endeavors not only contribute to the biotechnology and pharmaceutical fields but also bolster the sericulture industry by adding value to the traditional silk production process through biotechnological innovations.

Additional Therapeutic Applications

The exploration of silkworm bioreactors extends far beyond the production of platelet-derived growth factor (PDGF-BB) and neurotrophin-4 (NT-4). The intrinsic properties of silkworm silk, characterized by its biocompatibility and versatility, present a fertile ground for the biosynthesis of a diverse array of therapeutic proteins. This includes, but is not limited to, the production of growth factors, cytokines, antibodies, and various bioactive molecules utilizing the silk produced by genetically modified silkworms. The attractiveness of silkworm bioreactors to the pharmaceutical and biotechnological industries is underscored by their cost-effectiveness, scalability, and relative simplicity in operation.

The pioneering efforts, such as those by Tomita (2011) and subsequent researchers, have significantly propelled the field of therapeutic protein manufacturing forward. These studies underscore the feasibility of employing silkworm silk glands for the stable and prolonged expression of recombinant proteins. This innovation opens vistas for the mass production of critical therapeutic agents, including neurotrophins and growth factors, with wide-ranging applications in tissue engineering, neural regeneration, and wound healing processes. Silkworm bioreactors offer a unique confluence of biological utility, cost-efficiency, and scalability, making them invaluable assets in the realm of bioproduction.

Moreover, the potential medical applications of silkworm bioreactors, particularly those derived from silkworm, are poised for expansion as research in this domain advances. This evolution holds promising implications for the future of biotechnology and medicine, heralding a new era of therapeutic interventions.

In addition to their role in producing therapeutic proteins, silkworms offer a novel in vivo model for investigating human diseases, such as periodontal diseases. A compelling study by Miyashita et al. (2015) demonstrated the modeling of periodontal disease in silkworms through the injection of the pathogen *Porphyromonas gingivalis*. This led to rapid mortality in silkworms, mirroring the inflammatory response and pathology observed in human periodontal tissues. The study highlighted the induction of a cytokine storm as a pivotal mechanism underlying silkworm mortality, drawing parallels with the pathophysiological processes observed in human periodontal disease-related complications.

The application of silkworms, particularly *A. mylitta*, in modeling human diseases and the production of therapeutic agents underscore their versatility and potential in advancing medical research and treatment modalities. As the field of silkworm biotechnology continues to evolve, the scope for utilizing these organisms in diverse therapeutic applications and disease models is expected to broaden, thereby enriching both biotechnological innovation and medical science.

Gene Pool for Breeding

The entire genetic variety of the *A. mylitta* species that can be employed in selective breeding initiatives is included in the gene pool for breeding. This entails finding and protecting genetic variants linked to desired qualities including the caliber of the silk, the properties of the cocoon, disease resistance, and other factors that are significant to the economy. Tasar silkworm populations are generally improved through the use of breeding procedures that select individuals with favorable genetic markers or characteristics, hence enhancing specific attributes. For a species to be resilient and adaptable to shifting environmental conditions and new challenges, genetic diversity is essential. Keeping a variety of breeding stocks, avoiding inbreeding, and adding new genetic material as needed are all part of maintaining and regulating the gene pool. Molecular tools, such as DNA markers, are often employed to assess genetic diversity and guide breeding decisions for the sustainable development of tasar silkworm populations.

Colored Cocoon

In the pioneering field of genetic engineering, *A. mylitta*, akin to its counterpart *B. mori*, has been the subject of transformative research aimed at producing colored silk cocoons through transgenesis. This innovative approach involves the integration of exogenous genes coding for color-producing proteins into the genome of *A. mylitta*, paving the way for the customization of silk hues beyond the conventional spectrum of white or yellowish tones inherent to natural silk. The initial phases of this process necessitate the meticulous identification and subsequent genetic modification of the target genes responsible for the desired coloration, followed by the application of sophisticated gene integration techniques to engender transgenic *A. mylitta* specimens.

Upon successful integration, these exogenously introduced genes are expressed within the silkworms, culminating in the production of silk cocoons characterized by the predetermined colors. This technological advancement heralds a new era for the textile industry, offering the prospect of fabricating diversified silk products in a palette of colors without the need for postproduction dyeing processes. However, this endeavor is not without its challenges; paramount among these is the imperative to preserve the intrinsic quality of the silk and to ensure consistent color expression across generations of transgenic silkworms.

The realm of transgenic research, particularly within the context of *A. mylitta*, demands rigorous adherence to regulatory and ethical standards to mitigate potential ecological and health impacts. The ethical implications and regulatory compliance associated with genetic modifications are critical considerations that must be navigated with utmost diligence. Emblematic of the strides made in this domain, the works of Xia et al. (2009) and Takasu et al. (2010), while primarily focused on *B. mori*, shed light on the broader implications of genetic interventions aimed at altering pigmentation and other phenotypic traits in silkworms. These studies underscore the complexity and potential of genetic engineering in revolutionizing silk production, setting the stage for further exploration into the genetic manipulation of *A. mylitta* for the production of colored silk cocoons, thereby expanding the horizons of sustainable and versatile silk manufacturing.

Antimicrobial Textile Production

The goal is to create antimicrobial fabrics that are inherently resistant to the growth of microorganisms, providing benefits including improved cleanliness, reduced odor, and extended textile life. A variety of techniques are taken into consideration while producing antimicrobial textiles with *A. mylitta*, depending on the desired textile qualities, safety, and efficacy. Using genetic engineering, one method produces silk with inherent antibacterial qualities by modifying the tasar silkworm's genome to express antimicrobial proteins right in the silk gland. As an alternative, antimicrobial compounds can be produced by microbes or enzymes, which can then be introduced into the silk glands of tasar silkworms to aid in the incorporation of antimicrobial materials into the silk. Another technique involves applying an anti-bacterial coating or treatment to silk fibers, while they are being manufactured into textiles. These approaches represent innovative strategies for creating antimicrobial textiles using tasar silkworms, showcasing the potential for advancing textile technology.

Biomedical Engineering

The burgeoning field of biomedical engineering has identified non-mulberry silk polymers, particularly those derived from *A. mylitta*, as materials with significant potential for a myriad of applications. However, the challenge of dissolving non-mulberry silk fibers has historically hindered their broader utilization due to issues related to their poor processability. Addressing this bottleneck, Parekh et al. (2022) have developed an innovative protocol that significantly enhances the processability of *A. mylitta* silk fibers. This method involves the cryo-milling of silk fibers, a process that effectively reduces the beta-sheet content by over 10%, yielding a silk fibroin (SF) powder. This powder is capable of complete dissolution in common solvents such as trifluoroacetic acid (TFA), forming highly concentrated solutions (~20 wt%) within a matter of hours. These solutions can then be seamlessly integrated into conventional fabrication processes, such as electrospinning, to create three-dimensional scaffolds. For comparative purposes, *B. mori* (BM) silk served as a control within this study.

The versatility of these processed silk solutions was further demonstrated through the fabrication of silk hydrogels. Kaewchuchuen et al. (2023) aimed at expanding the utility of silk-based materials in tissue engineering domains. This was achieved by embedding microfibers of BM and AM (tasar) silk into self-assembling BM silk hydrogels, thereby enhancing the mechanical properties of the bulk hydrogel. This method not only improved the structural integrity of the hydrogels but also allowed for the tuning of their mechanical properties. Microfibers, prepared with lengths ranging between 250 and 500 μm and exhibiting high beta-sheet contents (61% for BM and 63% for tasar), were incorporated into silk solutions at various concentrations. This addition altered the hydrogels' initial stiffness and stress relaxation properties in a concentration-dependent manner, with higher microfiber concentrations yielding more pronounced effects.

Subsequent in vitro and in vivo evaluations focused on cell adhesion, proliferation, and differentiation, alongside a specific investigation into the osteogenic potential of the scaffolds through a 4-week implantation study in a rat calvarial model. These studies conclusively demonstrated that the novel processing techniques employed did not compromise the biocompatibility of the material. Notably, AM silk scaffolds were found to be conducive to bone regeneration. Furthermore, the embedding of silk microfibers into hydrogels significantly influenced the biological responses of human induced pluripotent stem cell-derived mesenchymal stem cells (iPSC-MSCs), with enhanced cell attachment and morphological changes observed in hydrogels containing tasar microfibers.

These findings collectively underscore the efficacy of cryo-milling in enhancing the processability of non-mulberry silk, thereby broadening its applicability in biomedical applications. Additionally, the strategic incorporation of silk microfibers into hydrogels presents a novel approach to modulate the mechanical and structural properties of silk-based biomaterials, offering further insights into the design of advanced materials for tissue engineering and regenerative medicine.

In the realm of maxillofacial reconstruction, the selection of regenerative biomaterials is pivotal, yet the availability of suitable polymeric biomaterials for addressing critical size bony defects remains constrained. There's a pronounced need for 3D printable biomaterials capable of generating patient-specific structures tailored to the defect morphology. Bojedla et al. (2022) introduces a 3D printable composite blend of polycaprolactone (PCL) and silk fibroin microfibers, alongside a refined protocol for crafting complex geometric structures using this composite. Comparative analyses revealed that scaffolds fabricated from the PCL-silk fibroin composite exhibit a superior compressive modulus relative to those made from PCL alone, with PCL-*A. mylitta* (AM) silk constructs outperforming PCL-*B. mori* (BM) silk variants in mechanical robustness at identical compositional ratios.

Further, human gingival mesenchymal stem cells (hGMSCs) were incorporated into these 3D printed scaffolds to assess their osteogenic capability. The *A. mylitta* silk fibroin-based scaffolds displayed enhanced bioactivity, attributed to the presence of arginine-glycine-aspartic acid (RGD) sequences, which are conducive to cell adhesion and proliferation. Notably, the metabolic activity of hGMSCs on PCL-40 AM scaffolds was observed to increase significantly by day 14, underscoring the potential of these scaffolds for bone regeneration. Additionally, leveraging computed tomography (CT) scans for the creation of a patient-specific mandible model demonstrated the feasibility of using this composite material for the precise reconstruction of complex maxillofacial bony defects in a clinical setting, highlighting the translational potential of this research in advancing personalized regenerative therapies.

Conclusion

The exploration of *A. mylitta*, a pivotal species in the sericulture industry, unfolds a narrative of scientific ingenuity and ecological significance. This narrative is enriched by the concerted efforts to decipher the molecular, genetic, and epigenetic landscapes governing the tasar silkworm's existence and its contribution to silk production. The journey from understanding the environmental adaptability and stress responses of *A. mylitta* to leveraging its biological processes for therapeutic protein production encapsulates a blend of traditional sericulture with modern biotechnological advancements. The identification and manipulation of key genes responsible for silk quality, disease resistance, and thermal tolerance, facilitated by cutting-edge genomic tools such as CRISPR-Cas9, TALENs, and ZFNs, signify a leap toward enhancing the economic and ecological value of tasar silk. These technologies not only promise improved strains of *A. mylitta* but also offer a sustainable pathway to address challenges posed by climate change and disease pressures. The production of colored cocoons and antimicrobial textiles, alongside the development of biomedical applications such as 3D printable scaffolds for tissue engineering, exemplifies the multifaceted potential of *A. mylitta* silk beyond the conventional textile industry.

Moreover, the exploration of *A. mylitta*'s gut microbiota through metagenomic studies reveals the intricate symbiosis between the silkworm and its microbial guests, emphasizing the importance of microbial communities in the health and development of the silkworm. These insights pave the way for innovative approaches to enhance silk production and disease resistance through the manipulation of gut microbiota. The journey through the chapter of this narrative underscores the significance of *A. mylitta* in the tapestry of sericulture, biotechnology, and ecological conservation. It highlights the necessity of preserving genetic diversity within *A. mylitta* populations to ensure the resilience and adaptability of this species in the face of environmental and anthropogenic challenges. The confluence of traditional sericultural practices with modern scientific research opens new horizons for the sustainable utilization of tasar silkworms, offering promising prospects for the silk industry, biomedical engineering, and beyond.

As research continues to unravel the molecular intricacies of tasar silkworm biology and its interaction with the environment, it is clear that the future of sericulture lies in harnessing the full genetic potential of this remarkable organism. The fusion of traditional sericulture practices with modern biotechnological advancements holds the promise of elevating the tasar silk industry to new heights, ensuring its economic viability and ecological sustainability.

References

Akkir D, Budak Yıldıran FA, Çakir S (2010) Molecular analysis of three local silkworm breeds (Alaca, Bursa Beyazi and HataySarisi) by RAPD-PCR and SDS-PAGE methods. Kafkas Univ Vet Fak Derg 16:S265–S269

Alkhnbashi OS, Shah SA, Garrett RA, Saunders SJ, Costa F, Backofen R (2016) Characterizing leader sequences of CRISPR loci. Bioinformatics 32(17):i576–i585

Arora V, Ghosh MK, Pal S, Gangopadhyay G (2017) Allele specific CAPS marker development and characterization of chalcone synthase gene in Indian mulberry (*Morus spp.,* family *Moraceae*). PLoS One 12(6):e0179189

Arunkumar KP, Sahu AK, Mohanty AR, Awasthi AK, Pradeep AR, Urs SR, Nagaraju J (2012) Genetic diversity and population structure of Indian golden silkmoth (*Antheraea assama*)

Arunkumar KP, Metta M, Nagaraju J (2014) Molecular phylogeny of silkmoths reveals the origin of domesticated silkmoth, *Bombyx mori* from Chinese *Bombyx mandarina* and paternal inheritance of *Antheraea proylei* mitochondrial DNA. Mol Phylogenet Evol 80:472–485. https://doi.org/10.1016/j.ympev.2014.09.013

Ashok Kumar K, Somasundaram P, Vijaya Bhaskara Rao A, Vara Prasad P, Kamble CK, Smitha S (2011) Genetic diversity and enzymes among selected silkworm races of *Bombyx mori (L.)*. Int J Sci Nat 2:773–777

Baci GM, Cucu AA, Giurgiu AI, Muscă AS, Bagameri L, Moise AR et al (2021) Advances in editing silkworms *(Bombyx mori)* genome by using the CRISPR-cas system. Insects 13(1):28

Baig MM, Singh G, Prabhu DIG, Manjappa SA, Kutala S (2023) Characterization of tasar silkworm *Antheraea mylitta*drury (Saturniidae: Lepidoptera) midgut bacterial symbionts through metagenomic analysis. Int J Trop Insect Sci 3:1–13

Bojedla SSR, Yeleswarapu S, Alwala AM, Nikzad M, Masood SH, Riza S, Pati F (2022) Three-dimensional printing of customized scaffolds with polycaprolactone–silk fibroin composites and integration of gingival tissue-derived stem cells for personalized bone therapy. ACS Appl Bio Mater 5(9):4465–4479

Chakraborty S, Muthulakshmi M, Vardhini D, Jayaprakash P, Nagaraju J, Arunkumar KP (2015) Genetic analysis of Indian tasar silkmoth (*Antheraea mylitta*) populations. Sci Rep 5(1):15728

Chatterjee SN, Pradeep AR (2003) Molecular markers (RAPD) associated with growth, yield, and origin of the silkworm, *Bombyx mori* L. in India. Russ J Genet 39(12):1365–1377

Chatterjee SN, Vijayan K, Roy GC, Nair CV (2004a) ISSR profiling of genetic variability in the ecoraces of *Antheraea mylitta* Drury, the tropical tasar silkworm. Russ J Genet 40:152–159

Chatterjee SN, Vijayan K, Roy GC, Nair CV (2004b) ISSR profiling of genetic variability in the ecoraces of *Antheraeae mylitta* Drury, the tropical tasar silkworm. Genetika 40(2):210–217

Chauhan S, Goodwin JG, Chauhan S, Manyam G, Wang J, Kamat AM, Boyd DD (2013) ZKSCAN3 is a master transcriptional repressor of autophagy. Mol Cell 50(1):16–28

Chavadi VB, Sosalegowda AH, Boregowda MH (2006) Impact of heat shock on heat shock proteins expression, biological and commercial traits of *Bombyx mori*. Insect Sci 13(4):243–250

Cheng Z (2016) A *Pseudomonas aeruginosa*-secreted protease modulates host intrinsic immune responses, but how? BioEssays 38(11):1084–1092

Chowdhury T, Mandal SM, Dutta S, Ghosh AK (2021) Identification of a novel proline-rich antimicrobial protein from the hemolymph of *Antheraea mylitta*. Arch Insect Biochem Physiol 106(3):e21771

Dubey S, Avadhani K, Mutalik S, Sivadasan SM, Maiti B, Paul J et al (2016) *Aeromonas hydrophila* OmpWPLGA nanoparticle oral vaccine shows a dose-dependent protective immunity in rohu (Labeorohita). Vaccines 4(2):21

Dubey H, Pradeep AR, Neog K, Debnath R, Aneesha PJ, Shah SK et al (2023) Genome sequencing and assembly of Indian Golden Silkmoth, Antheraea Assamensis Helfer (Saturniidae, Lepidoptera). Genomics 116(3):110841

Dutta S, Mohapatra J, Ghosh AK (2020) Molecular characterization of *Antheraea mylitta*arylphorin gene and its encoded protein. Arch Biochem Biophys 692:108540

Eleftherianos I, Atri J, Accetta J, Castillo JC (2013) Endosymbiotic bacteria in insects: guardians of the immune system? Front Physiol 4:46

Evgen'ev MB, Sheinker VS, Levin AV, Braude-Zolotareva TY, Titarenko EA, Shuppe NG, … Ul'masov KA (1987) Molecular mechanisms of adaptation to hyperthermia in higher organisms. I. Synthesis of heat-shock proteins in cell cultures of different species of silkworms and in caterpillars

Faure G, Shmakov SA, Yan WX, Cheng DR, Scott DA, Peters JE et al (2019) CRISPR–Cas in mobile genetic elements: counter-defence and beyond. Nat Rev Microbiol 17(8):513–525

Feder ME, Hofmann GE (1999) Heat-shock proteins, molecular chaperones, and the stress response: evolutionary and ecological physiology. Annu Rev Physiol 61(1):243–282

Fujii T, Daimon T, Uchino K, Banno Y, Katsuma S, Sezutsu H et al (2010) Transgenic analysis of the BmBLOS2 gene that governs the translucency of the larval integument of the silkworm, *Bombyx mori*. Insect Mol Biol 19(5):659–667

Fujinaga D, Kohmura Y, Okamoto N, Kataoka H, Mizoguchi A (2017) Insulin-like growth factor (IGF)-like peptide and 20-hydroxyecdysone regulate the growth and development of the male genital disk through different mechanisms in the silkmoth, *Bombyx mori*. Insect Biochem Mol Biol 87:35–44

Gargi RK, Singh MK, Prasad BC (2010) Field screening of *Terminalia arjuna* Bedd and *T. tomentosa* W. & A. for leaf spot and black nodal girdling diseases. Indian Forester 136(8):1129–1132

Gargi YH, Deka M, Kumar R, Sahay A (2015) Propagation technique and rearing performance on *Lagerstroemia speciosa:* anew food plant of Tasar silkworm *Antheraea mylitta* D. Int J Curr Res 7(10):22054–22057

Gattu R, Vodithala S (2024) Next-generation sequencing (NGS) in populations of Indian Tropical Tasar Silkworm, *Antheraea mylitta*

Ghosh AK, Datta A, Mahendran B, Kundu SC (2005) Molecular characterization of a Pao-like long terminal repeat retrotransposon, Tamy in saturniid silkworm *Antheraea mylitta*. Curr Sci 89:539–543

Guan RB, Li HC, Fan YJ, Hu SR, Christiaens O, Smagghe G, Miao XX (2018) A nuclease specific to lepidopteran insects suppresses RNAi. J Biol Chem 293(16):6011–6021

Guha TK, Wai A, Hausner G (2017) Programmable genome editing tools and their regulation for efficient genome engineering. Comput Struct Biotechnol J 15:146–160

Guo W, Wang Y, Song C, Yang C, Li Q, Li B, ... Wang S (2011) Complete genome of Pseudomonas mendocina NK-01, which synthesizes medium-chain-length polyhydroxyalkanoates and alginate oligosaccharides

Gupta A, Christensen RG, Bell HA, Goodwin M, Patel RY, Pandey M et al (2014) An improved predictive recognition model for Cys2-His2 zinc finger proteins. Nucleic Acids Res 42(8):4800–4812

Haque E, Taniguchi H, Hassan MM, Bhowmik P, Karim MR, Śmiech M et al (2018) Application of CRISPR/Cas9 genome editing technology for the improvement of crops cultivated in tropical climates: recent progress, prospects, and challenges. Front Plant Sci 9:617

Hillary VE, Ceasar SA (2021) Genome engineering in insects for the control of vector borne diseases. Prog Mol Biol Transl Sci 179:197–223

Ismail KS, Kumar CS, Aneesha U, Syama PS, Sajini KP (2023) Comparative analysis of gut bacteria of silkworm *Bombyx mori L.* on exposure to temperature through 16S rRNA high throughput metagenomic sequencing. J Invertebr Pathol 201:107992

Kaewchuchuen J, Roamcharern N, Phuagkhaopong S, Bimbo LM, Seib FP (2023) Microfibre-functionalised silk hydrogels. Cells 13(1):10

Kar PK, Vijayan K, Mohandas TP, Nair CV, Saratchandra B, Thangavelu K (2005) Genetic variability and genetic structure of wild and semi-domestic populations of tasar silkworm (*Antheraea mylitta*) ecorace Daba as revealed through ISSR markers. Genetica 125(2–3):173–183

Khurana P, Checker VG (2011) The advent of genomics in mulberry and perspectives for productivity enhancement. Plant Cell Rep 30:825–838

Koonin EV, Makarova KS (2019) Origins and evolution of CRISPR-Cas systems. Philos Trans R Soc B 374(1772):20180087

Krebs CJ, Zhang D, Yin L, Robins DM (2014) The KRAB zinc finger protein RSL1 modulates sex-biased gene expression in liver and adipose tissue to maintain metabolic homeostasis. Mol Cell Biol 34(2):221–232

Krishnan RR, Naik VG, Ramesh SR, Qadri SMH (2014) Microsatellite marker analysis reveals the events of the introduction and spread of cultivated mulberry in the Indian subcontinent. Plant Genet Resources 12(1):129–139

Kumar R, Alpana A, Richa S, Vijayprakash NB (2009) Morphological characterization of *Terminalia tomentosa*-primary food plant of tasar silkworm, *Antheraea mylitta D.* Indian Forester 135(12):1677–1685

Kumar R, Anupam A, Supriya R, Singh MK, Vijayprakash NB (2010) Genetic divergence and gene source studies in *Terminalia arjuna* Bedd. Indian J Plant Genet Resources 23(03):306–309

Lam KN, van Bakel H, Cote AG, van der Ven A, Hughes TR (2011) Sequence specificity is obtained from the majority of modular C2H2 zinc-finger arrays. Nucleic Acids Res 39(11):4680–4690

Lefevre M, Racedo SM, Ripert G, Housez B, Cazaubiel M, Maudet C, ... & Urdaci MC (2015) Probiotic strain Bacillus subtilis CU1 stimulates immune system of elderly during common infectious disease period: a randomized, double-blind placebo-controlled study. Immun Ageing 12(1):1–11

Li YuPing LY, Xia RunXi XR, Wang Huan WH, Li XiSheng LX, Liu YanQun LY, Wei ZhaoJun WZ et al (2009) Construction of a full-length cDNA library from Chinese oak silkworm pupa and identification of a KK-42-binding protein gene in relation to pupa-diapause termination

Li M, Hou C, Miao X, Xu A, Huang Y (2007) Analyzing genetic relationships in *Bombyx mori* using intersimple sequence repeat amplification. J Econ Entomol 100(1):202–208

Li H, Yang Y, Hong W, Huang M, Wu M, Zhao X (2020) Applications of genome editing technology in the targeted therapy of human diseases: mechanisms, advances and prospects. Signal Transduct Target Ther 5(1):1

Li JJ, Shi Y, Wu JN, Li H, Smagghe G, Liu TX (2021a) CRISPR/Cas9 in lepidopteran insects: progress, application and prospects. J Insect Physiol 135:104325

Li P, Wang L, Yang J, Di LJ, Li J (2021b) Applications of the CRISPR-Cas system for infectious disease diagnostics. Expert Rev Mol Diagn 21(7):723–732

Li Z, You L, Zhang Q, Yu Y, Tan A (2022) A targeted in-fusion expression system for recombinant protein production in *Bombyx mori*. Front Genet 12:816075

Lichun F, Guangwei Y, Maode Y, Yifu K, Chengjun J, Zhonghuai X (1996) Studies on the genetic identities and relationships of mulberry cultivated species (Morus L.) via a random amplified polymorphic DNA assay. Can ye kexue= CanyeKexue= Acta Sericologica Sinica 22(3):135–139

Ling L, Ge X, Li Z, Zeng B, Xu J, Chen X et al (2015) MiR-2 family targets awd and fng to regulate wing morphogenesis in *Bombyx mori*. RNA Biol 12(7):742–748

Liu Y, Ma S, Wang X, Chang J, Gao J, Shi R et al (2014) Highly efficient multiplex targeted mutagenesis and genomic structure variation in *Bombyx mori* cells using CRISPR/Cas9. Insect Biochem Mol Biol 49:35–42

Liu Y, Shen H, Wu H, Wang Z, Wang X, Xing D et al (2018) A PCR-based method for detecting the nuclear polyhedrosis virus in silkworm larvae. Insects 9(4):144. https://doi.org/10.3390/insects9040144

Liu Y, Ma S, Chang J, Zhang T, Chen X, Liang Y, Xia Q (2019) Programmable targeted epigenetic editing using CRISPR system in *Bombyx mori*. Insect Biochem Mol Biol 110:105–111

Liu Z, Xu J, Ling L, Luo X, Yang D, Yang X et al (2020) miR-34 regulates larval growth and wing morphogenesis by directly modulating ecdysone signalling and cuticle protein in *Bombyx mori*. RNA Biol 17(9):1342–1351

Lou CF, Zhang YZ, Zhou JM (1998) Polymorphisms of genomic DNA in parents and their resulting hybrids in mulberry Morus. Sericologia 38(3):437–449

Lusser M, Parisi C, Plan D, Rodríguez-Cerezo E (2012) Deployment of new biotechnologies in plant breeding. Nat Biotechnol 30(3):231–239

Ma S, Chang J, Wang X, Liu Y, Zhang J, Lu W et al (2014) CRISPR/Cas9 mediated multiplex genome editing and heritable mutagenesis of BmKu70 in *Bombyx mori*. Sci Rep 4(1):4489

Ma X, Huang M, Wang Z, Liu B, Zhu Z, Li C (2016) ZHX1 inhibits gastric cancer cell growth through inducing cell-cycle arrest and apoptosis. J Cancer 7(1):60

Mahendran B, Ghosh SK, Kundu SC (2006a) Molecular phylogeny of silk-producing insects based on 16S ribosomal RNA and cytochrome oxidase subunit I genes. J Genet 85:31–38

Mahendran B, Padhi BK, Ghosh SK, Kundu SC (2006b) Genetic variation in ecoraces of tropical tasar silkworm, *Antheraea mylitta* D. using RFLF technique. Curr Sci 90:100–103

Makarova KS, Koonin EV (2015) Annotation and classification of CRISPR-Cas systems. In: CRISPR: methods and protocols, pp 47–75

Manghwar H, Lindsey K, Zhang X, Jin S (2019) CRISPR/Cas system: recent advances and future prospects for genome editing. Trends Plant Sci 24(12):1102–1125

Miller JC, Tan S, Qiao G, Barlow KA, Wang J, Xia DF et al (2011) A TALE nuclease architecture for efficient genome editing. Nat Biotechnol 29(2):143–148

Minagawa S, Nakaso Y, Tomita M, Igarashi T, Miura Y, Yasuda H, Sekiguchi S (2018) Novel recombinant feline interferon carrying N-glycans with reduced allergy risk produced by a transgenic silkworm system. BMC Vet Res 14:1–9

Miyashita A, Takahashi S, Ishii K, Sekimizu K, Kaito C (2015) Primed immune responses triggered by ingested bacteria lead to systemic infection tolerance in silkworms. PLoS One 10(6):e0130486

Nagaraju J, Kathirvel M, Subbaiah EV, Muthulakshmi M, Kumar LD (2002) FISSR-PCR: a simple and sensitive assay for highthroughput genotyping and genetic mapping. Mol Cell Probes 16(1):67–72

Niranjan K, Prabhu IG, Manjappa AKS, Sahay A (2018) Development of RAPD-SCAR markers for the identification of *Antheraea mylitta* with high shell weight. J Entomol Zool Stud 6:828–833

Pandey JP, Jena K, Singh GP, Gupta VP, Baig MM, Sinha AK (2018) Why biotechnological exploration and endeavor are requisite for Tasar silk industry? In: Bioprospecting in life sciences, Chapter-8, pp 1–12

Parekh N, Bijosh CK, Kane K, Panicker A, Nisal A, Wangikar P, Agawane S (2022) Superior processability of *Antheraea mylitta* silk with cryo-milling: performance in bone tissue regeneration. Int J Biol Macromol 213:155–165

Park JW, Yu JH, Kim SW, Kweon HY, Choi KH, Kim SR (2018) Enhancement of antimicrobial peptide genes expression in cactus mutated *Bombyx mori* cells by CRISPR/Cas9. Int J Ind Entomol 37(1):21–28

Ponnuvel KM, de Miranda JR, Terenius O, Li W, Ito K, Khajje D et al (2022) Genetic characterisation of an Iflavirus associated with a vomiting disease in the Indian tropical tasar silkworm, *Antheraea mylitta*. Virus Res 311:198703

Prabhu IG, Baig MM, Kumar N, Sinha AK, Kutala S (2023) Molecular cloning and development of RAPD-SCAR markers for the selection of thermo-tolerant line of tropical tasar silkworm. J Environ Biol 44:464–471

Prasad MD, Nagaraju J (2003) A comparative phylogenetic analysis of full-length mariner elements isolated from the Indian tasar silkmoth, *Antheraea mylitta* (Lepidoptera: saturniidae). J Biosci 28:443–453

Qian Q, You Z, Ye L, Che J, Wang Y, Wang S, Zhong B (2018) High-efficiency production of human serum albumin in the posterior silk glands of transgenic silkworms, *Bombyx mori* L. PLoS One 13(1):e0191507

Rajan R, Chanda SD, Rani A, Gattu R, Vodithala S, Mamillapalli A (2020) Bacterial gut symbionts of *Antheraea mylitta* (Lepidoptera: Saturniidae). J Entomol Sci 55(1):137–140

Renuka G, Shamitha G (2016) Genetic variation in ecoraces of tropical tasar silkworm, *Antheraea mylitta* using SSR markers. J Genet 95:777–785

Renuka G, Natra N, Shamita G (2018) Development of EST-derived SSR markers for tasar ecoraces and their application in genetic diversity analysis. Nat Environ Pollut Technol 17(4):1315–1324

Rohela GK, Jogam P, Mir MY, Shabnam AA, Shukla P, Abbagani S, Kamili AN (2020) Indirect regeneration and genetic fidelity analysis of acclimated plantlets through SCoT and ISSR markers in Morus alba L. cv. Chinese white. Biotechnol Rep 25:e00417

Rudd S (2003) Expressed sequence tags: alternative or complement to whole genome sequences? Trends Plant Sci 8(7):321–329

Saha M, Kundu SC (2006) Molecular identification of tropical tasar silkworm *(Antheraea mylitta)* ecoraces with RAPD and SCAR markers. Biochem Genet 44:72–85

Saha M, Mahendran B, Kundu SC (2008) Development of random amplified polymorphic DNA markers for tropical tasar silkworm *Antheraea mylitta*. J Econ Entomol 101(4):1176–1182

Sakuma T, Nishikawa A, Kume S, Chayama K, Yamamoto T (2014) Multiplex genome engineering in human cells using all-in-one CRISPR/Cas9 vector system. Sci Rep 4(1):5400

Schuijff M, De Jong MD, Dijkstra AM (2021) AQ methodology study on divergent perspectives on CRISPR-Cas9 in The Netherlands. BMC Med Ethics 22(1):1–13

Sharma A, Niphadkar MP, Kathirvel P, Nagaraju J, Singh L (1999) DNA fingerprint variability within and among the silkworm *Bombyx mori* varieties and estimation of their genetic relatedness using Bkm-derived probe. J Hered 90(2):315–319

Sinha AK (1994) A review on the breeding and genetic aspects of tropical tasar silkworm *Antheraea mylitta*. In: Poc Conf on Cyto & Genet, vol 4, pp 7–16

Sommer RJ, Retzlaff M, Goerlich K, Sander K, Tautz D (1992) Evolutionary conservation pattern of zinc-finger domains of Drosophila segmentation genes. Proc Natl Acad Sci 89(22):10782–10786

Sreekumar S, Ashwath SK, Slathia M, Kumar SN, Qadri SM (2011) Detection of a single nucleotide polymorphism (SNP) DNA marker linked to cocoon traits in the mulberry silkworm, *Bombyx mori* (Lepidoptera: Bombycidae). Eur J Entomol 108(3):347

Staub JE, Serquen FC, Gupta M (1996) Genetic markers, map construction, and their application in plant breeding. HortScience 31(5):729–741

Suetsugu Y, Futahashi R, Kanamori H, Kadono-Okuda K, Sasanuma SI, Narukawa J et al (2013) Large scale full-length cDNA sequencing reveals a unique genomic landscape in a lepidopteran model insect, *Bombyx mori*. G3 Genes Genom Genet 3(9):1481–1492

Takasu Y, Kobayashi I, Beumer K, Uchino K, Sezutsu H, Sajwan S et al (2010) Targeted mutagenesis in the silkworm *Bombyx mori* using zinc finger nuclease mRNA injection. Insect Biochem Mol Biol 40(10):759–765

Takasu Y, Tamura T, Sajwan S, Kobayashi I, Zurovec M (2014) The use of TALENs for nonhomologous end joining mutagenesis in silkworm and fruitfly. Methods 69(1):46–57

Tamura T, Thibert C, Royer C, Kanda T, Eappen A, Kamba M et al (2000) Germline transformation of the silkworm *Bombyx mori* L. using a piggyBac transposon-derived vector. Nat Biotechnol 18(1):81–84

Thangavelu K (1992) Population ecology of *Antheraea mylitta* Drury (Lepidoptera Saturnidae). Wild Silkmoths 92:99–104

Tomita M (2011) Transgenic silkworms that weave recombinant proteins into silk cocoons. Biotechnol Lett 33(4):645–654

Tomita M, Munetsuna H, Sato T, Adachi T, Hino R, Hayashi M et al (2003) Transgenic silkworms produce recombinant human type III procollagen in cocoons. Nat Biotechnol 21(1):52–56

Tyagi S, Kumar R, Das A, Won SY, Shukla P (2020) CRISPR-Cas9 system: a genome-editing tool with endless possibilities. J Biotechnol 319:36–53

Velu D, Ponnuvel KM, Muthulakshmi M, Sinha RK, Qadri SM (2008) Analysis of genetic relationship in mutant silkworm strains of *Bombyx mori* using inter simple sequence repeat (ISSR) markers. J Genet Genomics 35(5):291–297

Vestergaard LK, Oliveira DNP, Høgdall CK (2021) Next generation sequencing technology in the clinic and its challenges. Cancers 13:1–18

Vijayan K, Nair CV, Kar PK, Mohandas TP, Saratchandra B, Raje US (2005) Genetic variability within and among three ecoraces of the tasar silkworm *Antheraea mylitta Drury*, as revealed by ISSR and RAPD markers. Int J Ind Entomol 10(1):51–59

Wang Y, Li Z, Xu J, Zeng B, Ling L, You L et al (2013) The CRISPR/Cas system mediates efficient genome engineering in *Bombyx mori*. Cell Res 23(12):1414–1416

Wu S, Tong X, Li C, Lu K, Tan D, Hu H et al (2019) Genome-wide identification and expression profiling of the C2H2-type zinc finger protein genes in the silkworm *Bombyx mori*. PeerJ 7:e7222

Xia Q, Guo Y, Zhang Z, Li D, Xuan Z, Li Z et al (2009) Complete resequencing of 40 genomes reveals domestication events and genes in silkworm (Bombyx). Science 326(5951):433–436

Xing WQ, Ma SY, Liu YY, Xia QY (2020) CRISPR/dCas9-mediated imaging of endogenous genomic loci in living *Bombyx mori* cells. Insect Sci 27(6):1360–1364

Xu J, Wang XF, Chen P, Liu FT, Zheng SC, Ye H, Mo MH (2016) RNA interference in moths: mechanisms, applications, and progress. Genes 7(10):88

Xuan Y, Wu Y, Li P, Liu R, Luo Y, Yuan J et al (2019) Molecular phylogeny of mulberries reconstructed from ITS and two cpDNA sequences. PeerJ 7:e8158

Yamamoto K, Nohata J, Kadono-Okuda K, Narukawa J, Sasanuma M, Sasanuma SI et al (2008) A BAC-based integrated linkage map of the silkworm *Bombyx mori*. Genome Biol 9(1):1–14

Zhang P, Aso Y, Yamamoto K, Banno Y, Wang Y, Tsuchida K et al (2006) Proteome analysis of silk gland proteins from the silkworm, *Bombyx mori*. Proteomics 6(8):2586–2599

Zhang Z, Liu X, Shiotsuki T, Wang Z, Xu X, Huang Y et al (2017) Depletion of juvenile hormone esterase extends larval growth in *Bombyx mori*. Insect Biochem Mol Biol 81:72–79

Zhang X, Xia L, Day BA, Harris TI, Oliveira P, Knittel C et al (2019) CRISPR/Cas9 initiated transgenic silkworms as a natural spinner of spider silk. Biomacromolecules 20(6):2252–2264

Zhang D, Hussain A, Manghwar H, Xie K, Xie S, Zhao S et al (2020) Genome editing with the CRISPR-Cas system: an art, ethics and global regulatory perspective. Plant Biotechnol J 18(8):1651–1669
Zhang W, Li Z, Lan W, Guo H, Chen F, Wang F et al (2023) Bioengineered silkworm model for expressing human neurotrophin-4 with potential biomedical application. Front Physiol 13:1104929

Silkworm Databases and Research Tools: A Comprehensive Guide for Advancing Sericulture Research

Megha Murthy, V. S. Raviraj, Anu Sonowal, and Jula S. Nair

Abstract

Farmers in several countries rely on sericulture as an important source of subsistence. Due to the economic significance of silkworms, there has been extensive research in sericulture, resulting in a vast amount of biological data. In recent years, research in silkworms, pathogens, and hosts has made tremendous progress. High-throughput analysis methods used to study genomes generate large amounts of data, leading to the creation of several databases to organize information. Databases such as KAIKObase, SilkDB, and SilkBase provide genetic maps, genomic and transcriptomic annotations, functional annotations, and information on cDNAs, EST clusters, TEs, mutants, and SNPs. This chapter offers an up-to-date review of the silkworm and associated databases, exploring the content, features, tools, and scope of each database. It aims to provide a comprehensive guide to researchers, enabling them to leverage the databases and accelerate research.

Keywords

Silkworm genome · Silkworm database · KAIKObase · SilkDB · Silkbase

M. Murthy (✉)
Queen Square Brain Bank for Neurological Disorders, UCL Queen Square Institute of Neurology, London, UK
e-mail: m.murthy@ucl.ac.uk

V. S. Raviraj · A. Sonowal · J. S. Nair
Silkworm Breeding and Genetics Lab, Central Sericultural Research and Training Institute, Central Silk Board, Ministry of Textiles, Government of India, Berhampore, West Bengal, India

R. V Suresh et al. (eds.), *Biotechnology for Silkworm Crop Enhancement*,
https://doi.org/10.1007/978-981-97-5061-0_13

Introduction

The silkworm holds significant economic and scientific importance, playing a central role in sericulture and important roles as model organism in genetics, molecular biology, and biotechnology (Li et al. 2023). Silkworm rearing becomes imperative for large-scale silk production and the maintenance of the silk quality; the most common species used for silk production, *Bombyx mori*, is a result of domestication of its wild counterpart *B. mandarina*. Silkworms have therefore been extensively studied, with various genotypes being maintained and different genetic stocks of silkworm strains being preserved (Cappellozza et al. 2022).Moreover, the scientific applications of silkworm extend to diverse fields such as human disease modeling, drug screening, environmental monitoring, and pest control. This has propelled substantial advancements in silkworm research, particularly in the fields of genetics, genomics, and functional mechanisms (Goldsmith et al. 2005; Meng et al. 2017).

Rapid advancements in silkworm research have resulted in the accumulation of tremendous amounts of data, with a wealth of information encompassing genomic landscapes and molecular markers across the different silkworm strains. This has triggered a need for organization of the generated data, and as a result several databases have been created. A comprehensive examination of these databases is vital to ensure optimal utilization of the existing resources. It also enables research advancements and facilitates the development of innovative breeding strategies ultimately increasing silk quality, disease resistance, and productivity, thus enhancing sericulture practices. The last comprehensive review of the available resources related to silkworm was conducted nearly a decade ago (Singh et al. 2016) and warrants a subsequent review to identify new databases and integrate new genetic information and updated research findings within the existing databases. This chapter, therefore, focuses on detailing the information and utilities of the existing databases related to silkworm that are currently functional.

Silkworm Databases

Silkworm databases serve as repositories of genetic, molecular, and functional information related to the domesticated silkworm *B. mori* and related species. Several silkworm databases exist that provide a comprehensive resource encompassing genomic, transcriptomic, proteomic, and other molecular data, and the information contained in each database and their utilities will be discussed in detail (Table 1).

SilkBase

Originally established in 1999 as an expressed sequence tag (EST) database, SilkBase (https://silkbase.ab.a.u-tokyo.ac.jp/cgi-bin/index.cgi) has evolved into a comprehensive database containing genomic, transcriptomic, proteomic, and

Table 1 List of silkworm databases

Database	Reference	Data available
SilkBase	https://silkbase.ab.a.u-tokyo.ac.jp/cgi-bin/index.cgi	Central repository for silkworm genome and gene sequence data for basic sequence searches and downloads
SilkDB	https://silkdb.bioinfotoolkits.net/main/species-info/-1	Silkworm genome, transcriptome, Hi-C data, pangenome, mutant database, orthologs, and synteny
SGID (Silkworm Genome Informatics Database)	http://sgid.popgenetics.net/	Silkworm gene and protein information, including functional annotations, expression data, and links to external resources
KAIKObase	https://kaikobase.dna.affrc.go.jp/	Genomic sequences, cDNA libraries with the corresponding ESTs, genetic and physical maps of the silkworm genome, and genome browsers
KAIKO2DDB	http://kaiko2ddb.dna.affrc.go.jp/cgi-bin/search_2DDB.cgi	2D polyacrylamide gel electrophoresis images of silkworm proteins
KAIKOGAAS	http://kaikogaas.dna.affrc.go.jp/	Annotation of genome scaffolds generated from the silkworm genome sequencing efforts of Japan and China
Bombyx trap database	https://sgp.dna.affrc.go.jp/ETDB/	Transposon insertion lines for various developmental stages, organs, and tissues
KAIKOcDNA	https://sgp.dna.affrc.go.jp/EST/	Silkworm EST (cDNA) sequences
KAIKO Full-length cDNA Database	https://sgp.dna.affrc.go.jp/FLcDNA//	Full-length cDNA (FLcDNA) sequences of the silkworm genome
KAIKO Linkage Map	https://sgp.dna.affrc.go.jp/LinkageMap/cgi-bin/index.cgi	Chromosome-wise linkage map
SilkwormBase	https://shigen.nig.ac.jp/silkwormbase/top.jsp	Genetic stock resources, genes, linkage maps, mutant genetics, distribution and ordering different silkworm strains at various developmental stages
SilkOrganPDB	https://silkorgan.biodb.org/	Searching, browsing, downloading, and analyzing silkworm proteins, homology search (BLAST, HMMER), organelle proteins, protein locations, sequences, gene ontology, and phylogeny
InsectBase	http://www.insect-genome.com/	Integrated genome and transcriptomic resource for insects
SilkSatDB	http://www.cdfd.org.in/SILKSAT/index.php	Microsatellite repeats of *Bombyx mori* extracted from WGS and EST sequences; primers developed and validated by the group; primer-designing program (Autoprimer)
NCBI	https://www.ncbi.nlm.nih.gov/datasets/taxonomy/7090/	Silkworm genome, genes, mRNA, and protein sequences, conserved sequences, and 3D structures

epigenomic data of the domesticated silkworm *B. mori* and related species (Kawamoto et al. 2022). SilkBase also provides several functional genomic tools making it an integrated database for silkworm research (Fig. 1).

The centerpiece of SilkBase is an updated genome assembly of *B. mori* (ver. 2016), obtained by performing hybrid assembly based on 140x deep sequencing of long (PacBio) and short (Illumina) reads followed by closure of the remaining gaps using BAC and Fosmid sequences, to yield a 460.3-Mb-long assembly (Kawamoto et al. 2019). In addition, it also contains a hypothetically reconstructed genome of *B. mandarina*, raw reads of the genome of 2 *B. mandarina* strains, and 155 scaffolds of the *Samia ricini* genome assembly (Lee et al. 2021). The database also contains 16,880 gene models, which have been annotated with transcript levels, GO terms, BLAST results against nucleotide, and InterProScan results.

SilkBase further houses transcriptomic data, with de novo assemblies of RNA-seq reads and RNA-seq raw reads for *B. mori* and *B. mandarina*, and putative ORFs predicted from stage- and tissue-specific RNA-seq assemblies for *B. mori*, *B. mandarina*, *Trilocha varians*, *Ernolatia moorei*, and *S. ricini*. The predicted ORFs have been annotated with GO term prediction and descriptions against nucleotide and transcript levels for certain species. RNA-sequencing data is available for multiple tissue regions including early embryo, fat body, midgut, epidermis, brain, internal genitalia, anterior silk gland, and middle silk gland.

SilkBase also contain spiRNA libraries that span multiple tissues and stages including the ovary (OV) and testis (TE) of day 4 wild-type male and female pupae, W chromosome of *B. mandarina* (MW), W chromosome-fragment attached Z chromosome (WF) of day 4 pupal testes of *B. mori* strain without Fem, Limited Yellow (LY) containing a truncated W chromosome constructed from day 4 pupal ovaries

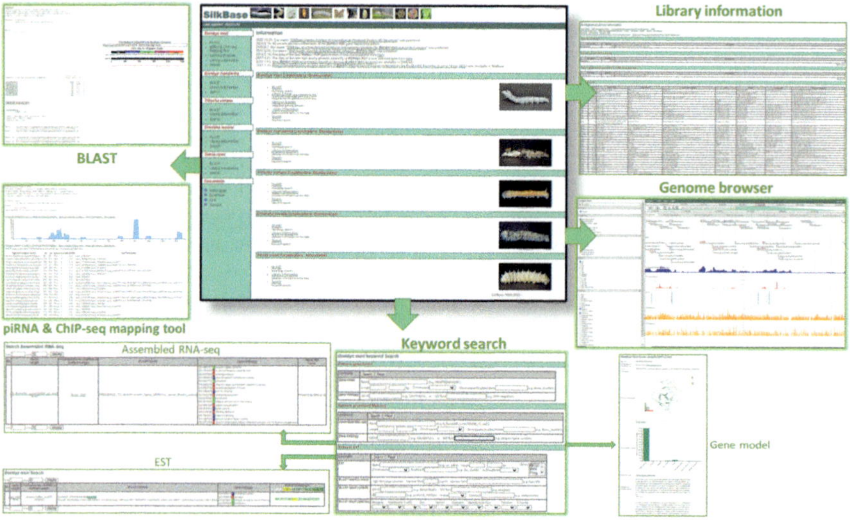

Fig. 1 An overview of the analysis tools in SilkBase

of LY female of *B. mori* strain, Siwi and BmAgo3 strains, BmN4 cultured cells, and embryos at various stages post fertilization (0–1 h (0 h egg), 6 h (6 h egg), 12 h (12 h egg), 24 h (24 h egg), and 40 h (40 h diapause egg)) (Kawaoka et al. 2009, 2011, 2012). The available ChIP-seq data covers major histone modifications that are known euchromatin or heterochromatin markers and include H3K4me2, H3K4me3, H3K9ac, H3K9me2, and H3K9me3 data, among others (Kawaoka et al. 2013).

The *B. mori* genome browser acts as a powerful platform that integrates genomic, transcriptomic, and epigenomic data to provide a detailed view of tissue- and stage-specific gene expression, alternative splicing, piRNA production, and histone modifications in addition to the data download features. Furthermore, SilkBase equips researchers with functional features such as BLAST, for interspecies homology searches, and the piRNA and ChIP-seq mapping tool, which can be used to visualize piRNA production status, identify protein-binding sites, and pinpoint the abundance, location, and sequences of relevant piRNAs and ChIP-seq reads within the *B. mori* genome, empowering comprehensive analyses of piRNA and chromatin dynamics.

SilkDB

Launched in 2005, SilkDB (https://silkdb.bioinfotoolkits.net/main/species-info/-1) emerged in the wake of the 2004 completion of silkworm genome drafts parallelly by Chinese and Japanese teams, positioning itself as a pivotal database for functional genomics in silkworms and other insects (Lu et al. 2020). Updates to the database were released once in 2010 (SilkDB 2.0) and again in 2020 (SilkDB 3.0), with the latest update showcasing heightened performance, augmented data, additional tools, and an enhanced visual interface. The reference *B. mori* genome in SilkDB has been constructed by error-correcting and assembling the Pacbio long reads generated by the group that created SilkBase (Kawamoto et al. 2019), followed by the construction of chromosome-scale scaffolds using HiC data and categorizing and ordering the assemblies. The protein coding genes (~16,069) were annotated using KEGG Ontology (KO), Gene Ontology (GO), Pfam, and KEGG enzyme databases, ensuring a robust foundation for functional genomics research.

SilkDB User Interface

The user interface of the database incorporates a search panel that accepts gene identifiers (IDs) or gene names as input presenting queried genes beneath the search panel. The platform also facilitates input of nucleotide or protein sequence of interest into the BLAST function to identify genes with sequences similar to the input within the database. SilkDB houses 14 modules, which include a silkworm mutant database (SilwormMD) and modules on gene information, tissue-specific expression, cellular localization and expression, coexpression network, 3D protein structure, gene family, pangenome, an interactive genome browser, chromosome viewer,

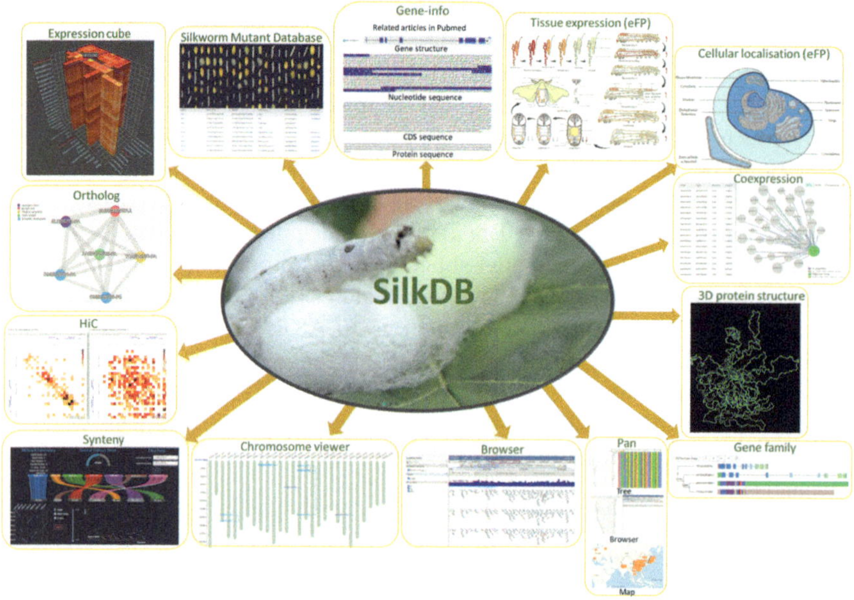

Fig. 2 Overview of the search engines and analysis tools in SilkDB

expression cube and heatmap, synteny viewer, Hi-C viewer, and an ortholog viewer. These modules empower users to explore silkworm genes at diverse molecular levels and compare them with other related insect genomes (Fig. 2).

Functionalities of the different modules are detailed below:

SilwormMD This module features phenotypic and genotypic data from 300 CRISPR mutant lines, enabling rapid discovery of unknown gene functions and facilitating functional studies. SilwormMD also provides information on plasmid libraries, comprising 92,917 sgRNA-expressing vectors and 1726 sgRNA transgenic lines.

Gene-Info The Gene-Info module offers information related to gene ID, synonyms, gene structure, protein domain description, gene distribution, and gene location, in addition to genomic, CDS, protein sequences, and related research articles in PubMed. It also allows the user to highlight nucleotide sequences corresponding to elements, such as 3′ UTRs, within the gene structure, by clicking on the element. The user can also perform functional annotations and design primers for sequences of interest.

eFP Gene expression levels for the selected genes can be visualized in 19 tissues (hemolymph, epidermis, head, posterior silk gland, testis, ovary, Malpighian tubule, trachea, anterior silk gland, midgut, fat body, middle silk gland, thorax, antenna, legs, and wing) during various growth stages (instar, larval, wandering, pupal, and moth) of the silkworm using this module.

Cell Displaying the predicted localization of a gene within a cell, the Cell eFP module provides a confidence-scored color gradient for subcellular compartment localization.

Coexpression This module provides interactive gene coexpression networks for the selected genes based on gene coexpression relationships generated using transcriptome data and provides the Pearson correlation coefficient between genes.

3D This module allows the visualization of 3D models of a given protein, generated using Phyre2.

Gene family This module displays the gene structure and a similarity dendrogram of comparisons of the gene of interest with all the annotated genes available for different insect genomes such as *Aedes aegypti*, *Drosophila melanogaster*, *Spodoptera litura*, *Tribolium castaneum*, *Trichoplusia ni*, and *B. mori*, to aid phylogenetic analyses and to help in understanding evolutionary changes in gene structure.

Pan The pangenome module showcases variations such as SNPs and indels, identified by comparing 163 different geographically representative samples (Xia et al. 2009; Xiang et al. 2018) to the reference silkworm assembly. To aid the visualization of these variations, the module features a multiple sequence alignment (MSA) viewer that displays the SNPs, aJBrowse interface (Buels et al. 2016) to view the indels, and a map of the geographic location of each sample.

JBrowse This interactive genome browser can be used to visualize gene distributions in silkworm sequences, including exons, introns, CDS, and reference sequences.

Chr The chromosome viewer maps multiple genes on chromosomes. In addition, the tool also displays the location and duplications of other genes in the gene family of the selected gene.

Exp-Cube The expression cube and heatmap viewer allow customizable plotting and provide the expression profile of the selected gene in different tissues at different development stages. It can also plot expression profiles of other genes that are highly correlated with the gene of interest. Alternatively, it can also be used to plot the expression profiles of several genes of interest, for a comparative analysis.

Synteny Displaying the selected gene's location on chromosomes, this module illustrates collinearity between silkworms and other insect species.

Hi-C Utilizing HiGlass (Kerpedjiev et al. 2018), the Hi-C module enables interactive browsing of Hi-C heatmaps to identify chromatin interactions in a three-dimensional space and allows simultaneous comparison of two heatmaps from different developmental stages.

Ortholog The ortholog viewer displays ortholog clusters of silkworm and other related insects containing the gene of interest.

Silkworm Genome Informatics Database (SGID)

Established in 2019, the SGID (http://sgid.popgenetics.net/) is another comprehensive database that features a compilation of silkworm genes meticulously annotated for function, protein structure, homologs, transcription factors (TFs), repeat elements, population statistic tests, epigenomic data, and an array of functional tools (Zhu et al. 2019). Similar to SilkDB, SGID also houses the *B. mori* reference genome assembly generated by SilkBase creators (Kawamoto et al. 2019) but distinguishes itself by encompassing a more extensive gene annotation than SilkBase.

Navigation of the SGID interface is facilitated by a user-friendly search bar accepting gene names, symbols, IDs, and gene lists as input, which retrieves gene information, ontology and pathway, transcriptional data, epigenomic data, protein structure, and population genetic information while also providing an interactive view of the gene in a genome browser. The interface extends its utility with individual links to diverse query and analysis tools, such as BLAT/BLAST, chromosomal segment or gene-based querying, gene ontology, transcriptome, epigenomics, protein structure, transcription factors, clusters, population genetics, transposable elements and repeat sequences, and subcellular localization. This design empowers users to conduct targeted searches and perform in-depth functional analyses (Fig. 3).

The results obtained from individual gene search include the following sections. The gene information tab encompasses vital details such as gene name, location, CDS, and protein sequences, along with a summary section featuring links to UniProt, PubMed, EMBL, Proteomes, PRIDE, ProteinModelPortal, and PDB databases. Additionally, this section includes KEGG and GO ontologies, topology, population genetic test statistics, multiple alignment of orthologs, and a gene tree. The genome browser tab offers an interactive view of the gene structure with information on repeat elements and multiple alignments of homologous proteins. The gene ontology and pathway tab provides links to GO, KEGG, and ENTREZ annotations. In the transcriptional analysis tab, one can explore expression levels for the epidermis and fat body of wild-type *B. mori*, as well as "Tanaka's mottled translucent" (otm) mutant. The epigenomic data tab displays peak for IgG-R, H3K4me2, H3K4me3, H3K9Ac, H3K9me2, H3K9me3, IgG-M, and PolII data derived from various projects. The protein structure tab provides a link to PDB, where in-depth information about the protein structure can be accessed. The tab dedicated to population genetics displays various metrics, including Pi, theta, Tajima's D estimates, and composite likelihood ratio (CLR). These metrics collectively portray polymorphism levels, nucleotide and genetic diversity, selection processes, and directional selection detected from DNA sequence data.

The abovementioned tools can also be accessed individually within the database and are complemented by additional functionalities offered through the interface. The BLAT/BLAST tool enables the users to perform sequence alignments against genes and genomes. The "List of segments" and "List of genes" tools provide insights into the specifics of a given list of segments or genes, respectively. The TF tool displays transcription factors for a particular query as predicted by AnimalTFDB (http://bioinfo.life.hust.edu.cn/AnimalTFDB4/#/) (Shen et al. 2023). The Cluster tool facilitates access to clusters of silkworm genes, while the "TE and repeat sequence" tool allows exploration of transposable elements and repeat sequences within the silkworm genome. Additionally, the "Subcellular localization" tool provides information on the subcellular localization of genes.

Fig. 3 Homepage and overview of the tools in Silkworm Genome Informatics database (SGID)

KAIKObase

KAIKObase (https://kaikobase.dna.affrc.go.jp) is an integrated silkworm genome database created in 2009 comprising genomic sequences, cDNA libraries with the corresponding ESTs, and genetic and physical maps of the silkworm genome, in addition to four browsers (PGmap, UnifiedMap, GBrowse, and UTGB) and two databases (KAIKO2DDB and *Bombyx* trap database) (Shimomura et al. 2009). KAIKObase is a core database set up by the Silkworm Genome Research Program (SGP) (https://sgp.dna.affrc.go.jp/index.html) to integrate all silkworm genome data including ESTs, chromosome linkage maps, and genome sequence data.

The KAIKObase user interface includes the "Genome browser (p50T)" tab containing the improved genome assembly generated using PacBio long reads and Illumina short reads similar to SilkBase and SGID [33,645,624, 30,802,494] and a "Genome browser (Nichi01)" tab containing the genome assembly of an improved Japanese strain Nichi01 (Waizumi et al. 2023). In addition to the abovementioned genome browsers, the KAIKObase user interface also provides the following informational and functional tools: The "Chromosome view" tab allows queries related to the linkage and physical map and contains genome contigs, BAC-ends, SNP markers, and fingerprint contigs (FPC); the "Keyword and position search" tab allows targeted queries pertaining to scaffold, contig, FPC-contig, SNP marker, trait marker, gene model, EST/cDNA, BAC, BAC-end, fosmid-end, and position on scaffold, to retrieve comprehensive information comprising the chromosomal map of the position of the queried element and corresponding links to GBrowse, gene description page, and UnifiedMap; the "BLAST Search" tab allows nucleotide and protein searches against silkworm genes and genesets; and the "Curated genes" tab provides a list of manually curated genes from different categories (Fig. 4).

KAIKO2DDB (KAIKO Proteome Database)

KAIKO2DDB (http://kaiko2ddb.dna.affrc.go.jp/cgi-bin/search_2DDB.cgi) is the proteome database of KAIKO base and contains two-dimensional polyacrylamide gel electrophoresis images with molecular weight, isoelectric point, gene models, and corresponding ESTs of silkworm proteins from various tissues (midgut, fat body, middle silk gland, posterior silk gland, Malpighian tubule, ovary, and hemolymph) at different developmental stages (fourth and fifth larval instars, spinning, and pupation stages). The database has a "Search by" section that allows the user to query using several search criteria such as by accession number, description, ID, gene name, UniProtKB/Swiss-Prot keywords (KW), author, spot id/serial number, identification methods, and pI/Mw range or by using a combination of fields. In addition, the database has a "Maps" section that provides links to experimental information, protein list, and a graphical viewer (Fig. 5).

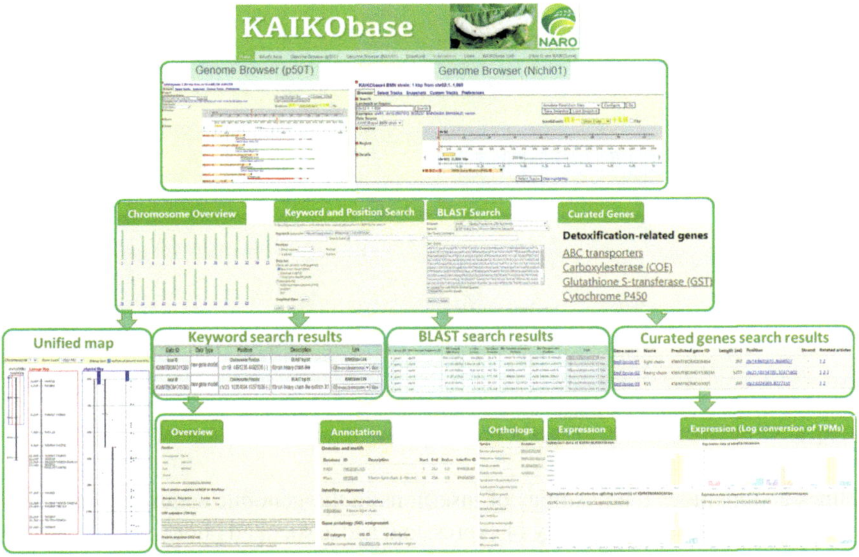

Fig. 4 Functionalities and tools within KAIKObase

KAIKOGAAS

The KAIKOGAAS (http://kaikogaas.dna.affrc.go.jp/) database integrates genome annotation of scaffolds generated from the silkworm genome sequencing efforts of Silkworm Genome Research Program (SGP) in Japan and the Silkworm Genome Project in China. The database provides detailed annotation of 192 scaffolds anchored to 28 silkworm chromosomes (total length: 417.7 Mb) and 4615 unmapped scaffolds which are >1 kb long, in addition to the annotations of 55 BAC clones used to check the quality of the WGS assembly. The annotation system includes coding region prediction programs (GENSCAN, FGENESH, MZEF), splice site prediction programs (SplicePredictor), DNA sequence homology search analysis programs (Blast, HMMER, ProfileScan, MOTIF), tRNA gene prediction program (tRNAscan-SE), repetitive DNA analysis programs (RepeatMasker, Printrepeats), protein localization site prediction program (PSORT), and program of classification and membrane protein classification and secondary structure prediction program (SOSUI) and automatically assigns a unique function for the predicted gene based on the protein homology using GFSelectorK.

Bombyx Trap Database

The *Bombyx* trap database (https://sgp.dna.affrc.go.jp/ETDB/) is a database of transposon insertion lines for various developmental stages, organs, and tissues. Information is available for enhancer-trap lines (Uchino et al. 2008) and gene-trap

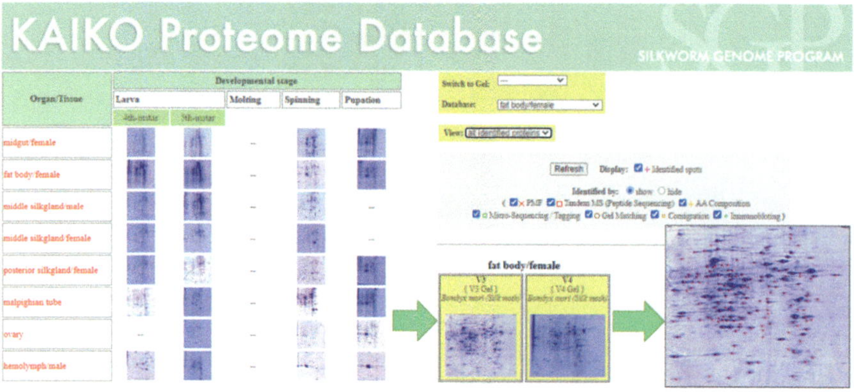

Fig. 5 Overview of the information available in KAIKO proteome database

lines and includes the positions of insertions in the genomic sequence, expression profiles of genes with intensity of expression, and associated photos. User can query the database using either the "Word Search Option" using keywords, search position of the chromosome or scaffold, fluorescent sites of different tissues, or photo images or through the "Pictorial Search Option" that contains images from different developmental stages that are apparent by ordinary, fluorescence, or fluorescent organs/tissues (Fig. 6).

Other related databases provided by SGP include the KAIKO cDNA database (https://sgp.dna.affrc.go.jp/EST/), KAIKO full-length cDNA database (https://sgp.dna.affrc.go.jp/FLcDNA//), KAIKOBLAST (http://kaikoblast.dna.affrc.go.jp/), and KAIKO Linkage Map (https://sgp.dna.affrc.go.jp/LinkageMap/cgi-bin/index.cgi).

SilkwormBase

SilkwormBase (https://shigen.nig.ac.jp/silkwormbase/top.jsp) is an integrated database of silkworm resources jointly developed by the resource center for the National BioResource Project-Silkworm (Graduate School of Agriculture, Kyushu University) and the information center (National Institute for Genetics). The database contains information pertaining to genetic stock resources, mutant genetics, distribution and order, and links to related sites. The database user interface has six main sections, namely, strains, *Bombyx* mandarina, genes, references, distribution request, and related links (Fig. 7). The database can also be accessed through the Silkworm National BioResource Project (NBRP) (https://silkworm.nbrp.jp/index_en.html).

The "Strains" section contains several links including a list of silkworm strains classified by their phenotype, strain list, images of the various strains at different developmental stages, a larva period time list, feeding ability of artificial diets, and a field search link which allows users to search based on selected keywords. The "*Bombyxmandarina*" section has links to place order requests for eggs, larvae,

THIS NOTE IS NOT LEGAL TENDER FOR DEBTS BOTH PUBLIC AND PRIVATE
1M
SERIES 2019 D
SEEN BY MILLIONS
FULLYFREEFILMS.COM
NEEDGOD.COM
MILLION
SERIES OF 2019
Department of Eternal Affairs
1,000,000
FRANKLIN
THE UNITED STATES OF AMERICA
RC S 72
D4
© 2019 LIVING WATERS
1,000,000
JS 2102COR135G
LIVINGWATERS.COM
ONE MILLION DOLLARS
1,000,000

1,000,000 IN GOD WE TRUST 1,000,000

THE MILLION-DOLLAR QUESTION: Will you go to Heaven when you die? Have you lied, stolen, used God's name in vain, or lusted (which Jesus said was adultery, Mt. 5:28)? If so, God sees you as a liar, thief, blasphemer, and adulterer at heart. If you die in your sins, you will end up in a terrible place called Hell. But there's good news. Though we broke God's Law, Jesus paid the fine by dying on the cross: "God so loved the world that He gave His only begotten Son, that whoever believes in Him should not perish but have everlasting life" (Jn. 3:16). Then Jesus rose from the dead and was seen by hundreds (it's no fairytale). He fulfilled all the prophecies of the promised Savior. Please, today, repent and trust Jesus, and God will forgive you and grant you the gift of eternal life (Eph. 2:8,9). Then, to show your gratitude, read the Bible daily and obey it, join a Christian church, and be baptized. Visit NeedGod.com and LivingWaters.com.

ONE MILLION THE UNITED STATES OF AMERICA ONE MILLION

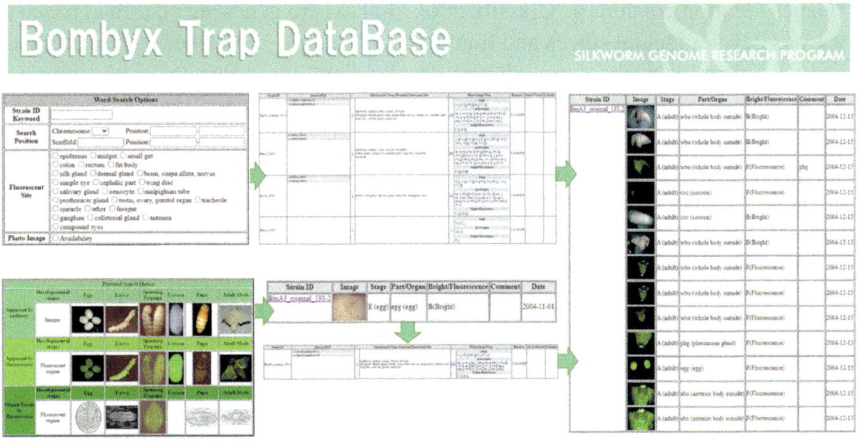

Fig. 6 Overview of the query and results from the *Bombyx* trap database

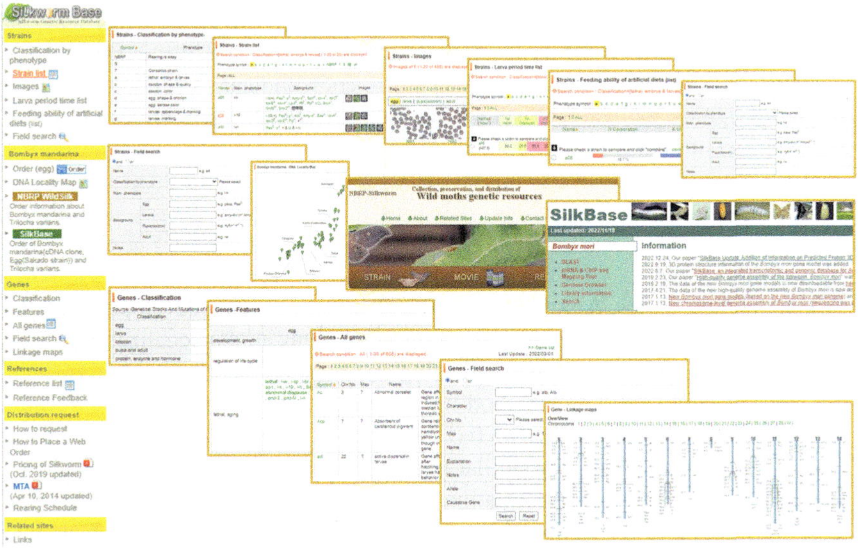

Fig. 7 Summary of the tools available in SilkwormBase

pupae, cocoon, or adult silk moths of different strains and provides a DNA locality map through which users can check availability of the different strains and order DNA samples from selected strains. The "*Bombyxmandarina*" section also has a link to the NBRP-Silkworm database dedicated to the collection, preservation, and distribution of wild moth genetic resource database (https://shigen.nig.ac.jp/wildmoth/) where users can order different silkworm strains including *B. mandarina* and *T. varians* at various developmental stages and a link to SilkBase where the user can order cDNA clones and eggs (Sakado strain) of *B. mandarina* and *T. varians*. The

"Genes" section contains a "Classification" link where genes are classified based on various developmental stages and functions; a "Feature" link where genes are listed based on features such as growth and development, morphogenesis, physiological traits, behavior, and reproduction; an "All genes" link where all genes are listed in alphabetical order; a "Field search" link which allows keyword based searches for the different genes and their features; and a "Linkage maps" link, which provides the chromosome-wise distribution of the different genes and images for the related strains. The "Reference" section contains citations to all resources, and the "Distribution request" section contains information on requesting, ordering, pricing of silkworm, MTA, and the rearing schedule.

SilkSatDB

The silk moth microsatellite database SilkSatDb (http://www.cdfd.org.in/SILKSAT/index.php) catalogs information on microsatellites extracted from whole genome sequencing (WGS) and EST sequences of silkworm genome (Prasad et al. 2005). The database houses informative figures and polymorphism status of the microsatellites along with the protocols used. The database also houses "PrimerBase," which contains a list of experimentally validated primers with their PCR conditions and a primer-designing tool called "Autoprimer" that facilitates users to design primers for their loci of interest.

SilkOrganPDB

SilkOrganPDB (https://silkorgan.biodb.org/) is a recently developed database (launched in 2021) and provides information on subcellular localization of silkworm organelle proteomes identified using an organelle-targeted method, which is an experimentally validated method unlike predictions made using bioinformatic tools that is employed in SilkDB 3.0 (Li et al. 2021; Lu et al. 2020). The user interface of SilkOrganPDB is very similar to that of SilkDB, with a keyword input section at the left corner and information provided in different modules in the right. User can access information either through the keyword section, which takes sequence identifier, description, or annotations as input, followed by selecting the organism and available organelles before search, or through the "Browse" module that contains a list of available *B. mori* genomes, organelles, and proteins. Both searches direct the user to the "Analysis" module that displays results for the queried protein. The "Analysis" module comprises functional tools such as BLAST and HMMER that facilitate homology searches. In addition, the module also provides links to other tabs including the "Organelle Proteome" tab, which lists all organelle proteins identified by the APEX-tagging method; the "eFP" tab, which displays the subcellular localization map; the "Sequences" tab, which displays nucleotide and protein sequences; and the "GeneOntology," "Homologs," and "Phylogeny" tabs, which provide the gene ontologies, homologs, and phylogeny trees, respectively, for the selected proteins. Other modules in the database include the “Download” and “Links” modules, which allow the user to download genomic and protein sequences and provide links to other relevant databases, respectively.

Other Databases with Information Related to Silkworms

National Center for Biotechnology Information (NCBI)

The NCBI database was created in 1988 within the National Library of Medicine as a molecular biology resource (Sayers et al. 2020). NCBI contains extensive resources including nucleotide sequences, reference genomes, genes, transcripts, and proteins and is an excellent tool for mining resources pertaining to several species of the silkworm (https://www.ncbi.nlm.nih.gov/search/all/?term=bombyx). As of 2023, the NCBI Assembly database has complete genome sequences of three silkworm species, *B. mori*, *B. huttoni*, and *B. mandarina*, in addition to other wild species of silk moths. NCBI also houses nucleotide, mRNA, and protein sequences, 3D protein structures, and information regarding conserved domains for multiple strains, and the resource goes beyond static data, with functional tools such as homology searches and phylogenetic studies that facilitate functional characterization and comparative studies. The Gene Expression Omnibus (GEO) contains raw and processed data for expression profiles of different organs and developmental stages, as well as ATAC-seq data, RNA m6A methylation, noncoding RNA, microRNA, and single-cell expression profiles (Fig. 8).

Within the NCBI browser, ongoing research projects related to silkworms can be tracked using BioProject, which provides links to associated datasets and relevant publications, while data pertaining to the individual samples can be accessed through BioSamples. The taxonomy browser provides information on species and related organisms, and tools such as BLAST and NCBI sequence viewer can be used for sequence alignment for comparisons with other organisms and to interactively visualize and explore gene and protein sequences, respectively. NCBI, therefore, stands as a powerful and versatile molecular toolkit for silkworm research, with vast data resources and cutting-edge tools, making it an indispensable resource for silkworm researchers.

InsectBase

InsectBase (http://www.insect-genome.com/) is a comprehensive database established in 2015 and houses genomes of 138 insects and provides information on genomic sequences, transcriptomes, gene sets and families, miRNAs, piRNAs, untranslated regions (UTRs), and pathways for various insects including *B. mori* (Yin et al. 2016). The *B. mori* data included in this database comprises genomic, mRNA, and protein information (http://www.insect-genome.com/data/detail.php?id=5), and the search toolbar allows users to access details of individual genes

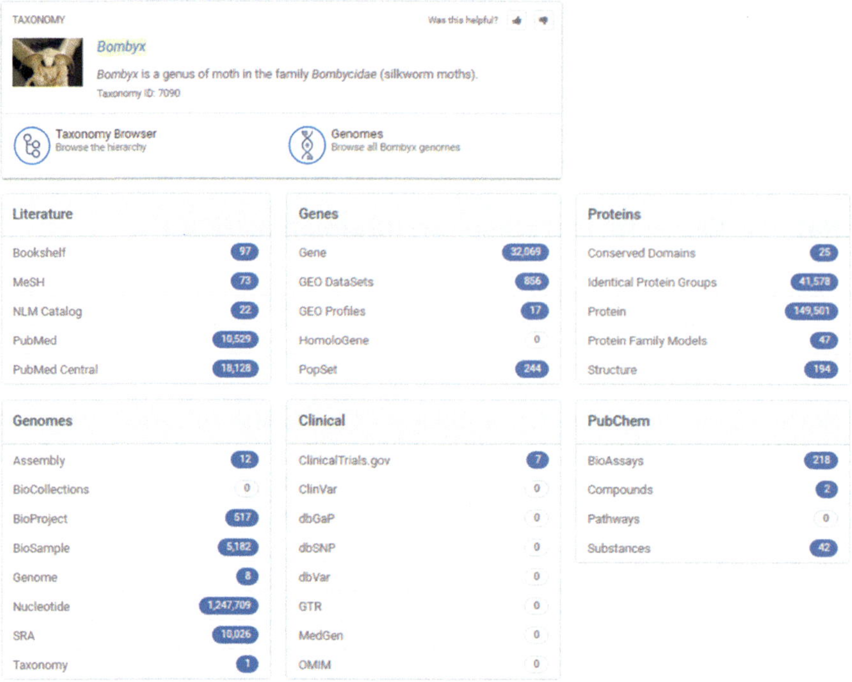

Fig. 8 NCBI webpage showing data available for the genus *Bombyx* as of December 2023

of interest using gene names and IDs, KEGG orthologs, Pfam, superfamily, and prosite annotations. The database also has BLAST and Gbrowse toolbars; the iPathway toolbar from which the user can download iPathCons result for *B. mori* and has a phylogenetic tree for 52 insect species along with raw data used in iPathDB for the 2 tables; the Ortholog toolbar that contains information ortholog groups and phylogenetic trees for 7 insects including *B. mori*, *Danaus plexippus*, *Linepithema humile*, *Nasonia vitripennis*, *Tribolium castaneum*, *Aedes aegypti*, and *Pediculus humanus*; the Gene Family toolbar containing gene family information obtained using homology searches from all insect gene data; the ncRNA toolbar housing conserved miRNAs of *B. mori* obtained by homology searches using RNA-seq data; the Transposon toolbar containing 1308 silkworm transposons obtained from the BmTEdb database; and the UTR and CDS tab containing UTR sequences of *B. mori* obtained from UTRdb and CDS sequences predicted from transcriptomic data.

Silkworm Host Plant, Pathogen, and Other Related Databases

In addition to silkworm related resources, information on other resources related to sericulture also provide immense utility in advancing silkworm research. These resources include information regarding silkworm host plants and pathogens and are crucial to sericulture as they directly impact the quality and quantity of silk generated. Silkworms primarily feed on mulberry leaves, and understanding the relationship between silkworms and their host plants and identifying nutritional requirements, growth conditions, and potential alternatives for silkworm host plants help in enhancing silk yield and ensuring the overall health and productivity of silkworms. Similarly, silkworms are vulnerable to various pathogens, including bacteria, viruses, and fungi, and diseases caused by these pathogens can result in lower yield and reduced quality of silk. Several databases dedicated to silkworm host plants exist, such as the mulberry genome database MorusDB (https://morus.biodb.org/index) (Li et al. 2014) and the *Morus notabilis* transposable element database (MnTEdb) (https://mntedb.biodb.org/) (Ma et al. 2015). These databases facilitate the understanding of the molecular and genetic aspects of silkworm-host and silkworm-pathogen interactions allowing researchers to provide optimal nutrition and breed disease-resistant silkworm varieties ultimately advancing the silk industry.

Conclusion

Silkworm and related databases form an indispensable resource for researchers in the field of sericulture and for those using the silkworm as a model organism, with a wide range of databases such as KAIKObase, SilkDB, and SilkBase offering vast amounts of information on sequences, annotations, functions, evolutionary and comparative analyses, and other molecular aspects. This comprehensive review of silkworm databases aims to provide researchers interested in studying silkworm with the necessary information to accelerate research and unravel the molecular structure and mechanisms in the silkworm, understand their nutritional requirements, unravel silkworm-pathogen interactions, and facilitate groundbreaking discoveries.

References

Buels R, Yao E, Diesh CM, Hayes RD, Munoz-Torres M, Helt G, Goodstein DM, Elsik CG, Lewis SE, Stein L, Holmes IH (2016) JBrowse: a dynamic web platform for genome visualization and analysis. Genome Biol 17:66. https://doi.org/10.1186/s13059-016-0924-1

Cappellozza S, Casartelli M, Sandrelli F, Saviane A, Tettamanti G (2022) Silkworm and silk: traditional and innovative applications. Insects 13(11):1016. https://doi.org/10.3390/insects13111016

Goldsmith MR, Shimada T, Abe H (2005) The genetics and genomics of the silkworm, Bombyx mori. Annu Rev Entomol 50:71–100. https://doi.org/10.1146/annurev.ento.50.071803.130456

Kawamoto M, Jouraku A, Toyoda A, Yokoi K, Minakuchi Y, Katsuma S, Fujiyama A, Kiuchi T, Yamamoto K, Shimada T (2019) High-quality genome assembly of the silkworm, Bombyx mori. Insect Biochem Mol Biol 107:53–62. https://doi.org/10.1016/j.ibmb.2019.02.002

Kawamoto M, Kiuchi T, Katsuma S (2022) SilkBase: an integrated transcriptomic and genomic database for Bombyx mori and related species. Database 2022:baac040. https://doi.org/10.1093/database/baac040

Kawaoka S, Hayashi N, Suzuki Y, Abe H, Sugano S, Tomari Y, Shimada T, Katsuma S (2009) The Bombyx ovary-derived cell line endogenously expresses PIWI/PIWI-interacting RNA complexes. RNA 15(7):1258–1264. https://doi.org/10.1261/rna.1452209

Kawaoka S, Kadota K, Arai Y, Suzuki Y, Fujii T, Abe H, Yasukochi Y, Mita K, Sugano S, Shimizu K, Tomari Y, Shimada T, Katsuma S (2011) The silkworm W chromosome is a source of female-enriched piRNAs. RNA 17(12):2144–2151. https://doi.org/10.1261/rna.027565.111

Kawaoka S, Mitsutake H, Kiuchi T, Kobayashi M, Yoshikawa M, Suzuki Y, Sugano S, Shimada T, Kobayashi J, Tomari Y, Katsuma S (2012) A role for transcription from a piRNA cluster in de novo piRNA production. RNA 18(2):265–273. https://doi.org/10.1261/rna.029777.111

Kawaoka S, Hara K, Shoji K, Kobayashi M, Shimada T, Sugano S, Tomari Y, Suzuki Y, Katsuma S (2013) The comprehensive epigenome map of piRNA clusters. Nucleic Acids Res 41(3):1581–1590. https://doi.org/10.1093/nar/gks1275

Kerpedjiev P, Abdennur N, Lekschas F, McCallum C, Dinkla K, Strobelt H, Luber JM, Ouellette SB, Azhir A, Kumar N, Hwang J, Lee S, Alver BH, Pfister H, Mirny LA, Park PJ, Gehlenborg N (2018) HiGlass: web-based visual exploration and analysis of genome interaction maps. Genome Biol 19(1):125. https://doi.org/10.1186/s13059-018-1486-1

Lee J, Nishiyama T, Shigenobu S, Yamaguchi K, Suzuki Y, Shimada T, Katsuma S, Kiuchi T (2021) The genome sequence of Samia ricini, a new model species of lepidopteran insect. Mol Ecol Resour 21(1):327–339. https://doi.org/10.1111/1755-0998.13259

Li T, Qi X, Zeng Q, Xiang Z, He N (2014) MorusDB: a resource for mulberry genomics and genome biology. Database 2014:bau054. https://doi.org/10.1093/database/bau054

Li T, Xu C, Xu J, Luo J, Yu B, Meng X, Li C, Pan G, Zhou Z (2021) Proteomic identification of Bombyx mori organelles using the engineered ascorbate peroxidase APEX and development of silkworm organelle proteome database (SilkOrganPDB). Int J Mol Sci 22(9):5051. https://doi.org/10.3390/ijms22095051

Li K, Dong Z, Pan M (2023) Common strategies in silkworm disease resistance breeding research. Pest Manag Sci 79(7):2287–2298. https://doi.org/10.1002/ps.7454

Lu F, Wei Z, Luo Y, Guo H, Zhang G, Xia Q, Wang Y (2020) SilkDB 3.0: visualizing and exploring multiple levels of data for silkworm. Nucleic Acids Res 48(D1):D749–D755. https://doi.org/10.1093/nar/gkz919

Ma B, Li T, Xiang Z, He N (2015) MnTEdb, a collective resource for mulberry transposable elements. Database 2015:bav004. https://doi.org/10.1093/database/bav004

Meng X, Zhu F, Chen K (2017) Silkworm: a promising model organism in life science. J Insect Sci 17(5):97. https://doi.org/10.1093/jisesa/iex064

Prasad MD, Muthulakshmi M, Arunkumar KP, Madhu M, Sreenu VB, Pavithra V, Bose B, Nagarajaram HA, Mita K, Shimada T, Nagaraju J (2005) SilkSatDb: a microsatellite database of the silkworm, Bombyx mori. Nucleic Acids Res 33(Database issue):D403–D406. https://doi.org/10.1093/nar/gki099

Sayers EW, Beck J, Brister JR, Bolton EE, Canese K, Comeau DC, Funk K, Ketter A, Kim S, Kimchi A, Kitts PA, Kuznetsov A, Lathrop S, Lu Z, McGarvey K, Madden TL, Murphy TD, O'Leary N, Phan L, Ostell J (2020) Database resources of the National Center for Biotechnology Information. Nucleic Acids Res 48(D1):D9–D16. https://doi.org/10.1093/nar/gkz899

Shen WK, Chen SY, Gan ZQ, Zhang YZ, Yue T, Chen MM, Xue Y, Hu H, Guo AY (2023) AnimalTFDB 4.0: a comprehensive animal transcription factor database updated with variation and expression annotations. Nucleic Acids Res 51(D1):D39–D45. https://doi.org/10.1093/nar/gkac907

Shimomura M, Minami H, Suetsugu Y, Ohyanagi H, Satoh C, Antonio B, Nagamura Y, Kadono-Okuda K, Kajiwara H, Sezutsu H, Nagaraju J, Goldsmith MR, Xia Q, Yamamoto K, Mita K (2009) KAIKObase: an integrated silkworm genome database and data mining tool. BMC Genomics 10:486. https://doi.org/10.1186/1471-2164-10-486

Singh D, Chetia H, Kabiraj D, Sharma S, Kumar A, Sharma P, Deka M, Bora U (2016) A comprehensive view of the web-resources related to sericulture. Database 2016:baw086. https://doi.org/10.1093/database/baw086

Uchino K, Sezutsu H, Imamura M, Kobayashi I, Tatematsu K, Iizuka T, Yonemura N, Mita K, Tamura T (2008) Construction of a piggyBac-based enhancer trap system for the analysis of gene function in silkworm Bombyx mori. Insect Biochem Mol Biol 38(12):1165–1173. https://doi.org/10.1016/j.ibmb.2008.09.009

Waizumi R, Tsubota T, Jouraku A, Kuwazaki S, Yokoi K, Iizuka T, Yamamoto K, Sezutsu H (2023) Highly accurate genome assembly of an improved high-yielding silkworm strain, Nichi01. G3 13(4):jkad044. https://doi.org/10.1093/g3journal/jkad044

Xia Q, Guo Y, Zhang Z, Li D, Xuan Z, Li Z, Dai F, Li Y, Cheng D, Li R, Cheng T, Jiang T, Becquet C, Xu X, Liu C, Zha X, Fan W, Lin Y, Shen Y, Wang J (2009) Complete resequencing of 40 genomes reveals domestication events and genes in silkworm (Bombyx). Science 326(5951):433–436. https://doi.org/10.1126/science.1176620

Xiang H, Liu X, Li M, Zhu Y, Wang L, Cui Y, Liu L, Fang G, Qian H, Xu A, Wang W, Zhan S (2018) The evolutionary road from wild moth to domestic silkworm. Nat Ecol Evol 2(8):1268–1279. https://doi.org/10.1038/s41559-018-0593-4

Yin C, Shen G, Guo D, Wang S, Ma X, Xiao H, Liu J, Zhang Z, Liu Y, Zhang Y, Yu K, Huang S, Li F (2016) InsectBase: a resource for insect genomes and transcriptomes. Nucleic Acids Res 44(D1):D801–D807. https://doi.org/10.1093/nar/gkv1204

Zhu Z, Guan Z, Liu G, Wang Y, Zhang Z (2019) SGID: a comprehensive and interactive database of the silkworm. Database 2019:baz134. https://doi.org/10.1093/database/baz134

The Application of Biostatistical Techniques in Silkworm Breeding and Improvement

Rahul Banerjee, Manjunatha Gyarehalli Rangappa, Ritwika Das, Tauqueer Ahmad, Pradip Kumar Sahu, P. A. Sangannavar, S. Manthira Moorthy, and V. Sivaprasad

Abstract

Statistical techniques hold a pivotal position in the realm of silkworm breeding and enhancement, encompassing both traditional and molecular methodologies. Despite the prevalence of various biostatistical tools, their utilization tends to be fragmented across different sources. The objective of this chapter is to elucidate the application of biostatistical methods, bridging theory and practice, with pertinent illustrations employing statistical software like Excel or R. Encompassing a diverse array of subjects, it addresses the conduct of experiments (design of experiments), assessment of data variability, identification of optimal genetic resources/germplasm, selection of requisite parental stock, assurance of breed/hybrid adaptability, determination of the most suitable breed/hybrid for commercialization, and the significance of statistical software in silkworm genomics. This chapter furnishes a lucid and thorough overview of employing statistical tools (univariate/multivariate and parametric/nonparametric) in the domain of silkworm breeding and enhancement.

Keywords

Biostatistics · Descriptive statistics · Inferential statistics · Statistical software · Silkworm genetic resources

R. Banerjee · R. Das · T. Ahmad
ICAR-Indian Agricultural Statistics Research Institute, New Delhi, India

M. G. Rangappa (✉) · P. A. Sangannavar · S. M. Moorthy · V. Sivaprasad
Central Silk Board, Bengaluru, Karnataka, India

P. K. Sahu
Bidhan Chandra Krishi Viswavidyalaya, Mohanpur, West Bengal, India

R. V Suresh et al. (eds.), *Biotechnology for Silkworm Crop Enhancement*,
https://doi.org/10.1007/978-981-97-5061-0_14

Introduction

Biostatistics is a branch of statistics that involves the application of statistical methods to analyze and interpret data related to biological, health, and medical phenomena (Pagano et al. 2022). It plays a crucial role in scientific research, particularly in fields such as epidemiology, medicine, genetics, environmental health, and public health. The primary objective of biostatistics is to extract meaningful information from data to inform decision-making, draw conclusions, and make inferences about biological processes or health-related outcomes. The scope of biostatistics broadly involves the following areas:

- *Study Design:* Biostatisticians are involved in designing experiments and observational studies, helping researchers determine the most appropriate methods for collecting data to answer specific research questions.
- *Data Collection and Management:* Biostatisticians contribute to the development of data collection protocols, ensuring that data is collected in a systematic and unbiased manner. They are also involved in data cleaning and management to ensure data accuracy.
- *Descriptive Statistics:* Biostatistics involves summarizing and describing data through measures such as mean, median, mode, and variability, providing a clear understanding of the characteristics of the data.
- *Inferential Statistics:* Biostatisticians use inferential statistical techniques to make predictions and inferences about populations based on sample data. This includes hypothesis testing, confidence intervals and regression analysis.
- *Epidemiological Studies:* Biostatistics is fundamental in epidemiology, where it is used to analyze patterns of disease occurrence, identify risk factors, and assess the impact of interventions on public health.
- *Clinical Trials:* Biostatisticians play a crucial role in the design and analysis of clinical trials, ensuring that the results are statistically valid and reliable. This is essential for the evaluation of new drugs, treatments, or medical procedures.
- *Genetic Studies:* In genetics, biostatistics is used to analyze and interpret data related to inheritance patterns, gene mapping, and population genetics.
- *Public Health Research:* Biostatistics contributes to research in public health by analyzing data related to health disparities, health behaviors, and the evaluation of public health programs.
- *Environmental Health Studies:* Biostatisticians may be involved in analyzing data related to environmental exposures and their impact on human health.
- *Policy and Decision-Making:* Biostatistical analyses often inform policy decisions in health and medicine, providing evidence-based recommendations for public health interventions.

The scope of biostatistics encompasses a wide range of applications and all aimed at extracting meaningful information from data in the biological and heal. Biostatistics plays a crucial role in biological research by providing a systematic and quantitative framework for collecting, analyzing, and interpreting data. Its importance is evident across various aspects of biological research, contributing to the advancement of knowledge, decision-making, and the development of effective interventions. Biostatistics is a fundamental tool in biological research that enhances the rigor, validity and reliability of studies. It provides researchers with the means to draw meaningful conclusions, make informed decisions, and contribute to advancements in understanding the complexities of biological systems.

Role of Biostatistics in Silkworm Breeding

The role of biostatistics in silkworm breeding is essential for optimizing breeding programs, analyzing genetic data, and making informed decisions to enhance silk production and quality. Biostatistics helps in the design of breeding experiments, including the selection of appropriate sample sizes, randomization techniques, and control groups. Proper experimental design is crucial for obtaining reliable and statistically valid results in silkworm breeding studies. Biostatistical methods are applied to collect and analyze data related to silkworm traits such as cocoon weight, silk quality and disease resistance. Statistical analyses help identify significant differences, trends and patterns in the data. Biostatistical techniques, particularly QTL mapping, are used to identify regions of the silkworm genome associated with specific traits (Xu et al. 2011). This information is valuable for understanding the genetic basis of desirable characteristics and facilitating selective breeding. Biostatistics plays a key role in the implementation of MAS in silkworm breeding. It helps identify molecular markers associated with important traits, allowing for the selection of individuals with desirable genetic profiles, thereby accelerating the breeding process (Trochez et al. 2019). Biostatistical methods are employed to assess the genetic diversity within silkworm populations. This information is crucial for maintaining a healthy breeding population, avoiding inbreeding depression, and preserving a diverse gene pool (Furdui et al. 2014). It helps in studying the distribution of genetic variation within and among silkworm populations. Understanding population genetics contributes to effective breeding strategies and the conservation of genetic resources. Biostatistical methods are employed for assessing the heritability of particular traits in silkworms, which quantifies the extent to which genetic factors contribute to the observed variation in a trait. This aids breeders in forecasting the efficacy of selective breeding initiatives (Reza et al. 2016). These techniques also aid in the visual representation of complex genetic data through graphs and charts. This facilitates the interpretation of breeding outcomes and helps in communicating findings to stakeholders. Biostatistical tools contribute to monitoring and maintaining the quality of silkworm breeding data. This includes identifying outliers, checking for consistency, and ensuring the accuracy of recorded observations (Kwak and Kim 2017). Biostatistics provides breeders with quantitative insights that inform decision-making in selecting breeding pairs, determining optimal breeding conditions, and assessing the success of breeding programs. The significance of employing biostatistical methodologies in silkworm breeding cannot be overstated, as they play a pivotal role in systematically designing experiments, analyzing genetic data, and making strategic decisions. Through the application of statistical techniques, researchers and breeders can optimize the efficiency and outcomes of silkworm breeding initiatives, thereby enhancing both silk production and quality. This chapter will provide a concise overview of various biostatistical techniques, highlighting their critical role in sericultural research and development.

Descriptive Statistics

An initial and crucial phase in any scientific endeavor is the descriptive phase. Before delving into the analysis of causes, it's essential to accurately describe phenomena. The inquiry of "What?" precedes that of "How?" (Sokal and Rohlf 2009). Descriptive statistics are a set of techniques used to summarize and describe the main features of a dataset. These statistics provide a simple and effective way to analyze and interpret data. The primary goal of descriptive statistics is to provide a clear and concise summary of the essential characteristics of a dataset, allowing researchers and analysts to gain insights into the data without making inferences about a larger population. Some common descriptive statistics are as follows:

Measures of Central Tendency

Measures of central tendency are statistical measures that describe the center or average of a distribution of values in a dataset. The three main measures of central tendency are the mean, median and mode.

- *Mean:* the average of all values in a dataset
- *Median:* the middle value when data is arranged in ascending or descending order
- *Mode:* the most frequently occurring value in a dataset

The measures of central tendency can be conveniently calculated using any statistical software. Following is a sample of the R code in Fig. 1 that can be used to calculate mean, median and mode for an ungrouped data. By replacing "*data*" in the script with one's dataset, this script will output the mean, median, and mode of the dataset.

Measures of Dispersion or Spread

Measures of dispersion, also known as measures of spread, quantify the extent to which the values in a dataset vary or spread out from a central point. These measures provide insights into the distribution of the data and its variability. Some common measures of dispersion include the following:

- *Range:* the difference between the maximum and minimum values in a dataset.

```r
# Sample data
data <- c(10, 20, 30, 40, 50, 50, 60, 70, 80, 90)

# Mean calculation
mean_value <- mean(data)
print(paste("Mean:", mean_value))

# Median calculation
median_value <- median(data)
print(paste("Median:", median_value))

# Mode calculation
mode_value <- as.numeric(names(sort(table(data), decreasing = TRUE)[1])
print(paste("Mode:", mode_value))
```

Fig. 1 Sample R code for calculation of measures of central tendency

```r
# Sample data
data <- c(10, 20, 30, 40, 50, 50, 60, 70, 80, 90)

# Range calculation
range_value <- max(data) - min(data)
print(paste("Range:", range_value))

# Variance calculation
variance_value <- var(data)
print(paste("Variance:", variance_value))

# Standard deviation calculation
std_deviation <- sd(data)
print(paste("Standard Deviation:", std_deviation))
```

Fig. 2 Sample R code for calculation of measures of dispersion

- *Variance:* a measure of how spread out the values in a dataset are from the mean.
- *Standard deviation:* the square root of the variance, providing a more interpretable measure of spread.

```
# Load the e1071 package
library(e1071)

# Sample data
data <- c(10, 20, 30, 40, 50, 50, 60, 70, 80, 90)

# Calculate skewness
skewness_value <- skewness(data)
print(paste("Skewness:", skewness_value))

# Calculate kurtosis
kurtosis_value <- kurtosis(data)
print(paste("Kurtosis:", kurtosis_value))
```

Fig. 3 Sample R code for calculation of measures of shape

The measures of dispersion can be conveniently calculated using any statistical software. Following is a sample of the R code in Fig. 2 that can be used to calculate range, variance and standard deviation for an ungrouped data. By replacing "**data**" in the script with one's dataset, this script will output the range, variance and standard deviation of the dataset.

Measures of Shape

Measures of shape describe the characteristics of the distribution of data points in a dataset. Two common measures of shape are skewness and kurtosis.

- *Skewness:* a measure of the asymmetry of a distribution. Positive skewness indicates a tail on the right, while negative skewness indicates a tail on the left.
- *Kurtosis:* a measure of the "tailedness" of a distribution. It describes the thickness of the tails relative to the normal distribution.

One can calculate skewness and kurtosis in R using the "**e1071**" package. Then, one can use the **skewness()** and **kurtosis()** functions from the package to calculate skewness and kurtosis, respectively, as shown in Fig. 3. By replacing "data" in the script with one's dataset, this script will output the skewness and kurtosis of the dataset.

```
# Sample data
data <- c(1, 2, 3, 4, 1, 2, 3, 4, 5, 1, 2, 3)

# Creating frequency distribution
freq <- table(data)

# Displaying frequency distribution
print(freq)
```

Fig. 4 Sample R code for calculation of frequency distribution

Frequency Distribution

Frequency distribution is a table that shows the number of occurrences of each value or range of values in a dataset. It organizes data into categories or intervals and records how many times each value or category appears. The purpose of creating a frequency distribution is to summarize and understand the distribution or pattern of the data, making it easier to analyze and interpret large datasets. Typically, frequency distributions are presented in tabular or graphical form, such as histograms or bar charts, to visualize the distribution of values or categories within the dataset. In R, you can create a frequency distribution using various functions and packages. One common way is to use the **table()** function, which calculates the frequency of each unique value in a vector. An example of the same is illustrated in Fig. 4.

Additionally, one can visualize the frequency distribution using various plotting functions like **hist()** for histograms or **barplot()** for bar charts. However, these are just basic examples, and depending on the data and specific needs, one might need to explore other packages or functions for more advanced analyses and visualizations.

Percentiles and Quartiles

- Percentiles divide a dataset into hundredths (percent) or 100 equal parts and indicate the relative standing of a particular value.
- Quartiles divide a dataset into 4 equal parts, with the first quartile (Q_1) representing the 25th percentile, the second quartile (Q_2) representing the median and the third quartile (Q_3) representing the 75th percentile.

Descriptive statistics are crucial for understanding the main characteristics of a dataset, identifying patterns and making informed decisions. However, it's important to note that they do not allow for making predictions or inferences about a larger population that's the realm of inferential statistics.

Inferential Statistics

Inferential statistics is a branch of statistics that involves using data from a sample to make inferences or draw conclusions about a population. The primary goal of inferential statistics is to make predictions, generalizations, or hypotheses about a larger group (population) based on a subset of that group (sample). This process involves analyzing and interpreting data, and it relies on probability theory. Inferential statistics helps researchers and analysts make predictions or inferences about populations based on data from samples. It provides a framework for drawing conclusions and making decisions in the presence of uncertainty. The crux of inferential statistics is hypothesis testing. Hypothesis testing is a statistical method used to make inferences about a population based on a sample of data. It involves formulating a hypothesis, collecting and analyzing data, and then using statistical techniques to assess the validity of the hypothesis. The process typically includes the steps illustrated through a flowchart in Fig. 5:

In silkworm breeding, hypothesis testing and confidence intervals can be applied to various aspects of research and decision-making, including assessing the effectiveness of breeding strategies, evaluating the impact of environmental factors, and estimating genetic parameters. Hypothesis testing can be used to compare the performance of different breeding strategies or treatments. For example, researchers may hypothesize that a new breeding technique improves silk yield compared to traditional methods (Banerjee et al. 2009). Confidence intervals can provide estimates of the potential improvement in silk yield associated with the new breeding technique. This information can guide breeders in making decisions about whether to adopt the new strategy. Hypothesis testing can be used to investigate the impact of environmental factors, such as temperature, humidity, or feed quality, on

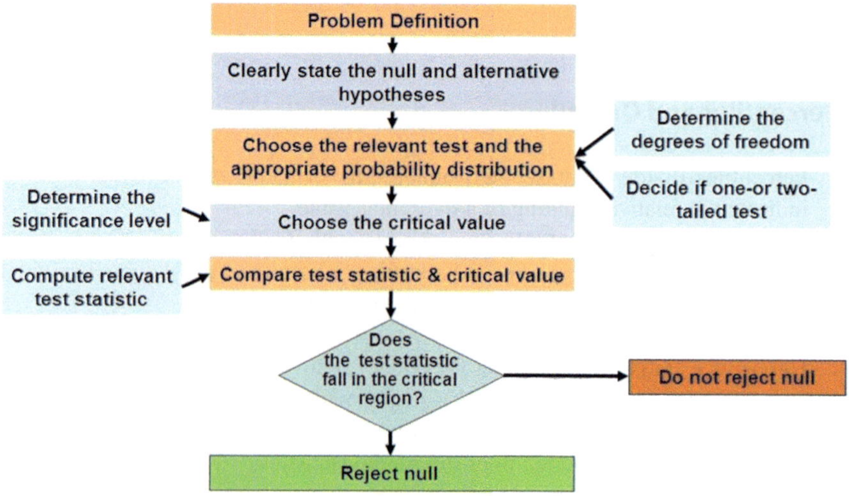

Fig. 5 Flowchart representing the steps in hypothesis testing

silkworm performance. Researchers may hypothesize that certain environmental conditions affect silk production or quality. Confidence intervals can provide estimates of the effect size of environmental factors on silkworm traits, along with the level of uncertainty. This information can inform management practices to optimize environmental conditions for silk production. Hypothesis testing and confidence intervals play crucial roles in silkworm breeding by enabling researchers to make evidence-based decisions about breeding strategies, genetic parameters, environmental factors, and disease resistance. These statistical techniques provide valuable insights into the effectiveness of breeding programs and help optimize silk production and quality.

Regression Analysis

Regression analysis is a statistical method used to examine the relationship between one or more independent variables (predictors) and a dependent variable (response or target). It aims to model the relationship between the variables and make predictions based on that model. Regression analysis is widely used in various fields, including economics, finance, social sciences and biology. Regression analysis is used for various purposes, including prediction, forecasting, hypothesis testing, and understanding the relationships between variables. It provides valuable insights into how changes in one or more independent variables are associated with changes in the dependent variable, allowing researchers to make informed decisions and draw conclusions based on data.

Linear Regression

Linear regression is a statistical method used to model the relationship between a dependent variable (target) and one or more independent variables (predictors) by fitting a linear equation to the observed data. The linear equation takes the following form:

$$y = \beta_0 + \beta_1 x_1 + \beta_2 x_2 + \ldots + \beta_n x_n + \varepsilon$$

where y is the dependent variable (target) and $x_1, x_2, \ldots, x_n$ are the independent variables (predictors). $\beta_0, \ldots, \beta_n$ are the coefficients (parameters) representing the intercept and slopes of the linear relationship between the variables. ε represents the error term, which accounts for the variability in the data not explained by the linear relationship. The goal of linear regression is to estimate the coefficients (parameters) that best fit the observed data, minimizing the sum of squared differences between the observed values and the values predicted by the linear equation. Once the model is fitted, it can be used for various purposes, including the following:

- *Prediction:* Given values of the independent variables, predict the value of the dependent variable.

- *Inference:* Understand the relationship between the independent and dependent variables, and make inferences about the population parameters.
- *Variable selection:* Identify which independent variables have a significant impact on the dependent variable.

Linear regression can be implemented using various techniques, including the ordinary least squares (OLS) method, gradient descent, and matrix algebra. It is widely used in fields such as economics, finance, social sciences, engineering, and machine learning for modeling and prediction tasks. Linear regression techniques have been used to study relationship between six biochemical parameters and four yield attributes in the mulberry silkworm, *Bombyx mori* (Chatterjee et al. 1993).

Logistic Regression

Logistic regression is a statistical method used for modeling the relationship between a binary dependent variable and one or more independent variables. Unlike linear regression, which predicts a continuous outcome, logistic regression predicts the probability that the dependent variable belongs to a particular category or class. In logistic regression, the dependent variable y is categorical and typically takes on two values, often encoded as 0 and 1. The logistic regression model applies a logistic (or sigmoid) function to the linear combination of the independent variables, resulting in predicted probabilities between 0 and 1. The logistic function is given by the following:

$$P(y = 1|x) = \frac{1}{1 + e^{-z}}$$

where $P(y = 1|x)$ is the probability that the dependent variable y equals 1 given the values of the independent variables x. z is the linear combination of the independent variables and their coefficients, defined as $z = \beta_0 + \beta_1 x_1 + \beta_2 x_2 + \ldots + \beta_n x_n$. e is the base of the natural logarithm (approximately equal to 2.71828). The coefficients are estimated using maximum likelihood estimation or other optimization techniques, aiming to maximize the likelihood of observing the data given the model parameters. Once the model is trained, it can be used to predict the probability that an observation belongs to the positive class (usually denoted as 1). A decision threshold can be set to classify observations into the positive class (1) or negative class (0) based on these predicted probabilities. Logistic regression is widely used in various fields, including healthcare (e.g., predicting disease risk), finance (e.g., credit scoring), marketing (e.g., customer churn prediction), and machine learning (e.g., binary classification tasks). Logistic regression techniques have been successfully used in several aspects of sericulture (Manjunatha et al. 2019). Logistic regression ensemble has been used for the classification of male and female cocoons with the help of X-ray images without destructing the cocoon (Thomas and Thomas 2022).

Experimental Designs

Experimental designs play a vital role in scientific inquiry, enabling the exploration of cause-and-effect relationships. They encompass various types, including preexperimental, true experimental, and quasi-experimental designs. Pre-experimental designs, such as the one-shot case study and the one-group pretest-posttest design, exhibit limited control and lack randomization. True experimental designs, exemplified by the randomized controlled trial (RCT), stand as the gold standard owing to their random allocation of participants to groups and deliberate manipulation of the independent variable. Quasi-experimental designs, such as the nonequivalent control group design, step in when random assignment is impractical. These design variations afford researchers varying degrees of control and facilitate the investigation of causal relationships across diverse contexts (Monette et al. 2011). Experimental designs are a crucial component of scientific research, allowing researchers to investigate causal relationships between variables. There are several types of experimental designs, each with its own strengths and limitations. There exist several categories of experimental designs, encompassing pre-experimental, quasi-experimental, true-experimental, factorial, and single-subject designs. Pre-experimental designs, the simplest among them, include the one-shot case study, one-group pretest-posttest design, and static-group comparison. Quasi-experimental designs entail independent variable manipulation but lack random assignment. True experimental designs entail random participant assignment to various conditions, offering the highest control over extraneous factors. Factorial designs manipulate two or more independent variables concurrently, whereas single-subject designs observe individual behavior longitudinally. When choosing an experimental design, researchers should meticulously evaluate the research question and available resources. Experimental designs in silkworm breeding focus on optimizing silkworm production, improving silk quality, and exploring the biological mechanisms underlying silkworm development and physiology. Due to the long history of silkworm domestication and breeding, natural variations exist among wild, local, and improved silkworm strains. By applying experimental designs, scientists aim to understand these variations and develop strategies for enhancing silkworm performance. Creating standardized laboratory settings allows for consistent monitoring of silkworm development and reproduction across generation. Artificial selection of desirable traits, such as increased silk yield, faster growth, or enhanced disease resistance, leads to improvements in silkworm stocks (Gao et al. 2020). Tools like CRISPR-Cas9 have been used to introduce desired genetic alterations in silkworm genomes. Genetic basis of complex traits has been identified through linkage analyses and association studies. Experimental designs have been used for analyzing differences in gene expression, epigenetic regulation, and physiological processes between wild and domesticated silkworms (Meng et al. 2017). These experimental approaches help elucidate the molecular and genetic bases of silkworm adaptability and productivity, ultimately contributing to the improvement of silkworm breeding programs worldwide.

Applications in Silkworm Breeding

Marker-Assisted Selection (MAS)

The field of molecular genetics and its related technologies have made significant progress in comprehending the underlying genetics of the qualities that are sought through breeding. Selection in plant and animal breeding typically entails testing a breeding population for one or more qualities using chemical tests (such as grain quality) or field or glasshouse trials (such as agronomic attributes, disease resistance, or abiotic stress tolerance). The aim of breeding is to create a new elite line with more favorable gene combinations. Agronomic qualities and stress tolerance are evaluated visually when choosing better plants or animals, and other traits are evaluated in the lab. It takes a lot of time and money to complete the process. Recently, advanced molecular biotechnology tools have been integrated into the traditional breeding scheme to enhance the breeding efficiency, and a new technique, namely, marker-assisted selection (MAS), has been introduced. It is an indirect selection process where a trait of interest is selected based on a marker (morphological, biochemical, or DNA/RNA variation) linked to a trait of interest (e.g., productivity, disease resistance, abiotic stress tolerance, and quality), rather than on the trait itself. For instance, rather than focusing on the degree of disease resistance, MAS is used to select individuals who have disease resistance by identifying a marker allele that is associated with disease resistance. It is assumed that the marker's high frequency of association with the gene or quantitative trait locus (QTL) of interest results from genetic linkage, which is the close chromosomal proximity between the marker locus and the locus that determines disease resistance. When selecting for qualities that are expensive or difficult to quantify, have little heredity, or manifest late in development, MAS might be helpful. Over the last 10 years, numerous mapping experiments have been conducted with the goal of establishing marker-assisted selection due to the increasing capacity to transfer specific genomic areas using DNA markers.

Types of Markers

Overall, the markers used in MAS can be categorized into four broad groups:

(a) *Morphological:* These were the first markers to facilitate indirect selection of an interest trait. These markers are frequently visible to the human eye through simple observation. The presence or absence of an awn, the color of the leaf sheath, height, rice grain color, scent, cocoon color, etc. are a few examples of this marker.
(b) *Biochemical:* These are protein based markers like isozymes, storage proteins, heat shock proteins (HSPs), etc.
(c) *Cytological:* These are chromosomal features identifiable by microscopy. Presence or absence of a particular band on such markers is correlated with the desirable trait.

(d) *DNA-based:* Nowadays, various types of DNA-based markers are used in most MAS researches. Selection of appropriate DNA marker is dependent on five key factors (Collard and Mackill 2008), including cost, polymorphism level, assay process, quantity and quality of DNA required, and reliability.
(e) *Reliability:* Markers should be highly linked to the target gene, and the genetic distance between them should be <5 cM. The reliability of the markers to predict phenotype will be significantly increased by the use of intragenic or flanking markers.
   - DNA quantity and quality: Large amount of high-quality DNA is required for MAS.
   - Technical procedure: High throughput, simple, and quick method is preferrable.
   - Polymorphism: Selected marker should be highly polymorphic.
   - Cost: Marker assay should be cost-effective.

Among all the markers, simple sequence repeats (SSRs) are highly preferred for marker-assisted selection as they are codominant, highly reproducible, simple, highly polymorphic, and very cheap to use. SSRs have the drawbacks of usually requiring polyacrylamide gel electrophoresis and typically providing information on just one locus per experiment. These issues have frequently been resolved by multiplexing numerous SSR markers in a single reaction and using SSR markers with size differences significant enough for agarose gel detection. Other than SSRs, sequence tagged site (STS), sequence characterized amplified region (SCAR), single nucleotide polymorphism (SNP) markers, restriction fragment length polymorphisms (RFLPs), etc. are also incredibly helpful for MAS. Currently, the most potent method of marker-assisted selection for efficiently using DNA markers is the transfer of single alleles. However, marker-assisted selection has proven to be less effective for polygenic traits. More studies are required to improve the effectiveness of marker-assisted selection for quantitative traits by properly characterizing the target genes. New technology is needed for the efficient and successful use of marker-assisted selection for polygenic trait improvement.

Quantitative Trait Locus (QTL) Mapping

Quantitative trait locus (QTL) is a genetic region that is correlated with variations in a quantitative trait in an organism population's phenotype. These QTLs are identified by QTL mapping method which is genetic linkage analysis based on the principle of genetic recombination. In this approach, linkage maps are prepared using genetic markers for a particular population. Backcross (BC) populations, F2, F3, and other segregating populations are widely employed. Recombinant inbreds and doubled haploids are two examples of populations which are preferred as they are useful for repeated and repeatable tests. Genes or QTLs can be identified in connection to a linkage map using statistical techniques like single-marker analysis or interval mapping to find relationships between DNA markers and phenotypic data.

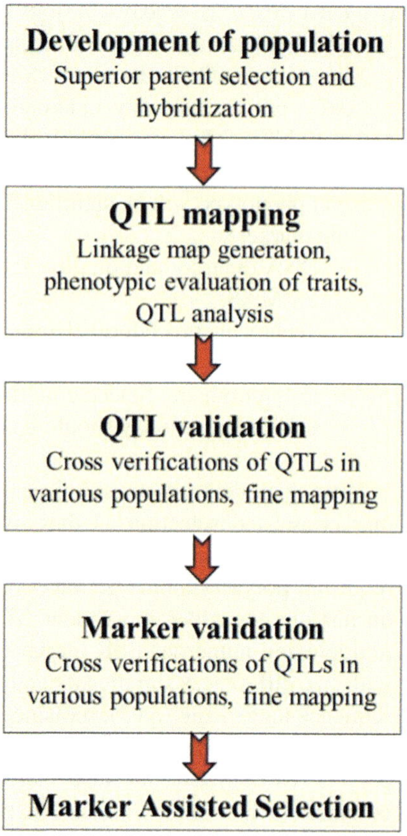

Fig. 6 QTL mapping and marker-assisted selection

It was previously believed that the majority of markers linked to QTLs in early mapping studies had direct application in MAS. However, it is observed that QTL confirmation, QTL validation, and fine (or high resolution) mapping are also highly necessary (Collard and Mackill 2008) (Fig. 6).

While some highly accurate preliminary QTL mapping data have been obtained from earlier QTL mapping studies, preferably a confirmation phase is preferred to avoid the sampling bias and other factors which might lead to erroneous QTL placements and effects. QTL validation step is required to assess QTL's effectiveness across several genetic backgrounds. Further steps in marker testing could include finding a set of markers inside a 10 cM genetic distance that surrounds and spans a QTL and transforming markers into a format that permits easier detection of them. Once tightly linked markers that reliably predict a quantitative trait phenotype have been identified, they may be used for MAS (Fig. 6). MAS and QTL mapping are highly advantageous over conventional phenotype screening methods as they are time-, resource-, and laborsaving and can be used to select the desired individuals at a very early stage of development.

Application of QTL Mapping and MAS in Silkworm Breeding

In the last decade, application of QTL mapping approach as well as MAS in silkworm breeding has gained vast popularity. Although some abiotic stress tolerant silkworm host varieties have been developed, the hosts' highly heterozygous genetic makeup complicates understanding the inheritance and manifestation of these quantitative features. Hence, genome assisted breeding for silkworm host species for the abiotic and biotic stress management purposes is emerging (Duo et al. 2023). Similarly, molecular breeding/marker-assisted breeding procedures are being used in mulberry cultivation for genetic improvement toward the generation of productive and climate-resilient cultivars for the horizontal and vertical spread of high-quality silk production across the globe (Sarkar et al. 2021). *Bombyx mori bidensovirus* (BmBDV) causes "flacherie" disease, resulting in severe economic losses to the sericulture sector, but it can be controlled by the expression of the recessive gene nsd-2. In a recent study, two parental thermotolerant *Bombyx mori* breeds abundant in Northern India, namely, SK6 and SK7, are made homozygous for nsd-2 gene through marker-assisted selection (Gundi et al. 2023). Multi-viral (viz., densonucleosis (BmDV), infectious flacherie (BmIFV), and nuclear polyhedrosis (BmNPV)) disease-tolerant bivoltine silkworm breeds of *Bombyx mori* (Lepidoptera: Bombycidae) have been developed using marker-assisted breeding (Satish et al. 2023). MAS approach has also been employed to produce silkworm strains with high tolerance to fluoride, resistance to densonucleosis virus, scaleless wings, and high silk production. QTL mapping approaches have been applied to identify 11 QTLs enhancing cocoon yield of silkworm (Fang et al. 2020). In another study (Lingaiah et al. 2023), genetic diversities across bivoltine and multivoltine silkworm genotypes have been analyzed using inter-simple sequence repeat (ISSR) and simple sequence repeat (SSR) markers.

Population Genetics

Population genetics is a subfield of genetics in which one can study the genetic makeup of populations and how those compositions vary over time. It sheds light on the distribution of genetic variation within a population and the ways in which it changes in response to different evolutionary pressures. Analyzing gene frequencies and comprehending population dynamics are two essential components of population genetics.

Analyzing Gene Frequencies

Gene frequencies refer to the proportion of alleles of a particular gene within a population. Analyzing gene frequencies provides valuable information about the genetic diversity and structure of a population. The Hardy-Weinberg equilibrium (HWE) is a widely used technique for analyzing gene frequencies. It states that in a large, random mating population, the gene and genotyping frequency remain constant generation after generation in the absence of systematic forces like mutation, migration, and selection. A divergence from the Hardy-Weinberg equilibrium may signify the existence of natural selection, genetic drift, migration, or mutation, among other evolutionary processes. Researchers can also examine genetic inheritance patterns, gene flow across populations, and the possibility of genetic adaptability by analyzing gene frequencies.

Understanding Population Dynamics

Population dynamics is the study of how the number and makeup of a population fluctuate over time due to a variety of causes, such as immigration, emigration, births and deaths. For the purpose of forecasting population growth and evaluating population resistance to shocks such as disease outbreaks and environmental changes, it is essential to comprehend population dynamics. Age structure, carrying capacity, and population growth rate are important notions in population dynamics. The pace at which a population grows or shrinks over a certain time period, impacted by birth and death rates, is known as the population growth rate. The carrying capacity of an environment refers to the maximum population number that it can sustainably maintain, taking into account environmental limits and resource availability. The distribution of people within a population across various age groups is referred to as its age structure, and it can have an impact on the dynamics and rates of population increase. The logistic growth model and the Lotka-Volterra equations are two examples of the mathematical models and theoretical frameworks for studying population dynamics that are provided by population genetics. These models aid in the understanding of how populations react to alterations in the environment, rivalry for resources, and other ecological elements. To sum up, population genetics is essential in understanding population dynamics and gene frequency analysis. Population genetics offers important insights into evolutionary processes, biodiversity conservation, and the management of natural resources by examining the distribution of genetic variation within populations and the mechanisms influencing changes in population size and composition. By means of multidisciplinary methodologies that integrate genetics, ecology, and mathematics, scientists can enhance their understanding of the intricate dynamics of natural populations and devise tactics for their preservation and sustainable administration.

Ethical Considerations

Ethical concerns in silkworm breeding primarily revolve around the treatment of silkworms, which are living organisms used in the production of silk. Some of the key ethical considerations include the following:

- *Treatment of silkworms:* During breeding and silk production, silkworms are frequently put through a variety of procedures and manipulations, including handling, feeding, and harvesting silk. In order to minimize stress, discomfort, and harm, breeding procedures should guarantee that silkworms are treated with care throughout their lives.

- *Genetic manipulations:* Genetic modifications are carried out in silkworms to introduce particular traits. Techniques for genetic alteration should be applied carefully, considering any wider ecological effects as well as possible welfare implications for the silkworms.
- *Conservation of genetic diversity:* Preserving the diversity of genes within silkworm populations needs to be the top priority for breeding initiatives. Keeping a variety of genetic resources helps protect silkworm health and resilience against environmental changes, disease outbreaks, and other hazards.
- *Environmental impact:* Breeding silkworms and producing silk can have an impact on the environment in terms of waste production, resource consumption, and land use. Sustainability issues should be taken into account while developing ethical standards, encouraging eco-friendly behavior, and reducing adverse effects on ecosystems and biodiversity.
- *Alternative silk production methods:* Investigating non-silkworm-based alternatives to conventional silk manufacturing techniques may also be spurred by ethical concerns. Research on cruelty-free and environmentally friendly methods of producing silk, including bio-fabrication or plant-based substitutes, should be encouraged.

Conclusion

Biostatistics serves as a vital tool in biological inquiry, particularly within domains like epidemiology, medicine, genetics, environmental health, and public health. It furnishes a structured and quantitative approach for gathering, scrutinizing, and interpreting data, thereby fortifying the integrity, credibility, and consistency of investigations. Within the realm of silkworm breeding, biostatistical methodologies are indispensable for refining breeding programs, scrutinizing genetic information, and executing well-informed choices to augment silk output and quality. Descriptive statistics are fundamental for grasping the principal attributes of a dataset, discerning trends, and making enlightened determinations. Inferential statistics, a subset of statistical analysis, entails leveraging data from a sample to deduce or infer conclusions about a broader population, thereby furnishing a framework for decision-making amid uncertainty. Regression analysis, another statistical technique, is employed to scrutinize the correlation between one or more independent variables and a dependent variable, empowering researchers to derive insights and make decisions grounded in data. In summary, biostatistics assumes a pivotal role in methodically designing experiments, dissecting genetic data, and formulating strategic choices in the domain of silkworm breeding and beyond in biological exploration.

References

Banerjee A, Chitnis UB, Jadhav SL, Bhawalkar JS, Chaudhury S (2009) Hypothesis testing, type I and type II errors. Ind Psychiatry J 18(2):127–131. https://doi.org/10.4103/0972-6748.62274

Chatterjee SN, Rao CG, Chatterjee GK, Ashwath SK, Patnaik AK (1993) Correlation between yield and biochemical parameters in the mulberry silkworm, *Bombyx mori* L. Theor Appl Genet 87(3):385–391. https://doi.org/10.1007/BF01184928

Collard BC, Mackill DJ (2008) Marker-assisted selection: an approach for precision plant breeding in the twenty-first century. Philos Trans R Soc B Biol Sci 363(1491):557–572. https://doi.org/10.1098/rstb.2007.2170

Duo H, Dorjee L, Raising LP, Zhipao RR (2023) Genomic assisted breeding and holistic management of abiotic and biotic stress in silkworm host cultivation: a review. Indian J Agric Sci 93(7):691–698. https://doi.org/10.56093/ijas.v93i7.138159

Fang SM, Zhou QZ, Yu QY, Zhang Z (2020) Genetic and genomic analysis for cocoon yield traits in silkworm. Sci Rep 10(1):5682. https://doi.org/10.1038/s41598-020-62507-9

Furdui EM, Mărghitaş LA, Dezmirean DS, Paşca I, Pop IF, Erler S, Schlüns EA (2014) Genetic characterization of Bombyx mori (Lepidoptera: Bombycidae) breeding and hybrid lines with different geographic origins. J Insect Sci 1(14):211. https://doi.org/10.1093/jisesa/ieu073

Gao R, Li CL, Tong XL, Han MJ, Lu KP, Liang SB, Hu H, Luan Y, Zhang BL, Liu YY, Dai FY (2020) Identification, expression, and artificial selection of silkworm epigenetic modification enzymes. BMC Genomics 21(1):740. https://doi.org/10.1186/s12864-020-07155-z

Gundi R, Vanitha C, Tulsi KSN, Velusamy L, Ramesha A, Ponnuvel KM, Rabha M, Sivaprasad V, Pradeep AR, Mishra RK (2023) Molecular marker assisted breeding and development of bidensovirus resistant and thermo tolerant silkworm (Bombyx mori) hybrids suitable for tropical climatic conditions. Agric Res 12(4):428–438. https://doi.org/10.1007/s40003-023-00662-x

Kwak SK, Kim JH (2017) Statistical data preparation: management of missing values and outliers. Korean J Anesthesiol 70(4):407–411. https://doi.org/10.4097/kjae.2017.70.4.407

Lingaiah K, Lokanath S, Iyengar P, Gowda H, Shanmugam MM, Kudupaje BC, Vankadara S (2023) Analysis of genetic variability among bivoltine and multivoltine silkworm genotypes using inter simple sequence repeat and simple sequence repeat markers. Nucleus:1–10. https://doi.org/10.1007/s13237-023-00436-4

Manjunatha GR, Hunmily E, Patil KKR, Afroz S, Parmeshwarnaik J, Pandit D, Sivaprasad V (2019) Prognostication of mulberry silk cocoon prices in Kaliachak (West Bengal) market. J Crop Weed 15(3):48–53. https://doi.org/10.22271/09746315.2019.v15.i3.1236

Meng X, Zhu F, Chen K (2017) Silkworm: a promising model organism in life science. J Insect Sci 17(5):97. https://doi.org/10.1093/jisesa/iex064

Monette DR, Sullivan TJ, DeJong CR (2011) Applied social research: a tool for the human services, 8th edn. Cengage Learning

Pagano M, Gauvreau K, Mattie H (2022) Principles of biostatistics, 3rd edn. CRC Press, Taylor & Francis Group

Reza NH, Alireza S, Shahabodin G (2016) A review on correlation, heritability and selection in silkworm breeding. J Appl Anim Res 44(1):9–23. https://doi.org/10.1080/09712119.2014.987289

Sarkar T, Doss SG, Sivaprasad V, Teotia RS (2021) Stress tolerant traits in mulberry (Morus spp.) resilient to climate change. In: Mulberry, 1st edn. CRC Press

Satish L, Kusuma L, Shery AVMJ, Moorthy SM, Manjunatha GR, Sivaprasad V (2023) Development of productive multi-viral disease-tolerant bivoltine silkworm breeds of Bombyx mori (Lepidoptera: Bombycidae). Appl Entomol Zool 58(1):61–71. https://doi.org/10.1007/s13355-022-00803-8

Sokal RR, Rohlf FJ (2009) Introduction to biostatistics, 2nd edn. Dover Publications, Inc, Mineola

Thomas S, Thomas J (2022) Non-destructive silkworm pupa gender classification with X-ray images using ensemble learning. Artif Intell Agric 6:100–110. https://doi.org/10.1016/j.aiia.2022.08.001

Trochez SJD, Ruiz EX, Almanza PM, Zambrano GG (2019) Role of microsatellites in genetic analysis of Bombyx mori silkworm: a review. F1000Research 13(8):1424. https://doi.org/10.12688/f1000research.20052.1. PMID: 32148760; PMCID: PMC7043130

Xu HM, Wei CS, Tang YT, Zhu ZH, Sima YF, Lou XY (2011) A new mapping method for quantitative trait loci of silkworm. BMC Genet 12:19. https://doi.org/10.1186/1471-2156-12-19

Dipteran Parasitoid-Silkworm Interaction: Application of Genomic and Proteomic Tools in Host-Parasitoid Communication

Pooja Makwana, Jula S. Nair,
and Appukuttan Nair R. Pradeep

Abstract

The dipteran endoparasitoid *Exorista bombycis* (Tachinidae) infest the silkworm *Bombyx mori* (Lepidoptera: Bombycidae). The female fly lays eggs on host larval cuticle. *E. bombycis* eggs have proteinase containing a gluey substance on ventral surface that aids to attach egg on larval surface as well as in partial digestion of the larval cuticle for easy invasion of the newly hatched parasitoid larvae into hemocoel. On invasion, host hemocytes in circulating hemolymph recognize the foreign body that induces the accumulation of exclusive proteins in hemocytes and other tissues. These proteins/conjugates are identified by SDS-PAGE analysis or by 2D electrophoresis. Few of the exclusive protein bands/spots are analyzed by mass spectrometry coupled with de novo peptide sequencing.

On *E. bombycis* infestation, detoxification mechanisms are activated in host hemocytes. Detoxification enzyme level, reactive oxygen species (hydrogen peroxide; H_2O_2), and lipid peroxidation activity are increased after the infestation. The antioxidative enzymes, viz., oxidase, superoxide dismutase, thioredoxin peroxidase, catalase, glutathione-S-transferase (GST), and peroxidases, showed enhanced presence in hemolymph plasma, hemocytes, and fat body.

The infestation deactivated host defense by downregulated expression of several defense and immune genes as revealed by genome wide gene expression using hemocyte microarray and substantiated by quantitative PCR. Moreover, microarray showed inhibition of several genes encoding interacting proteins revealed by protein-protein interaction network. The review emphasized on the communication between dipteran parasitoid *E. bombycis* and the host larvae of *B. mori* as well as the supporting data from hymenopteran-lepidopteran interaction.

P. Makwana · J. S. Nair · A. N. R. Pradeep (✉)
Central Sericultural Research & Training Institute, Berhampore, West Bengal, India

R. V Suresh et al. (eds.), *Biotechnology for Silkworm Crop Enhancement*,
https://doi.org/10.1007/978-981-97-5061-0_15

Keywords

Bombyx mori · *Exorista bombycis* · Dipteran parasitoid biology · Host-parasitoid interaction · Hymenopteran-lepidopteran interaction

Introduction

Though insect immune system activation against pathogen infection has been well studied, reports on defense activities against infestation by eukaryotic endoparasitoids as well as strategies and counter interactions of hosts and parasitoids are scanty (Amaya et al. 2005; Etebari et al. 2011; Nighat et al. 2023; Schmidt 2008; Vinson 1977; Wago 1995; Wertheim 2022). Most of the studies are associated with hymenopteran parasitoid infestation in *Drosophila* and lepidopteran hosts. Host immune responses against parasitoids and microbes showed similarities; however, the parasitoids require suitable host internal milieu for successful parasitism. It is well known that hymenopteran parasitoids inject virus, calyx fluid, and venom to the host hemocoel during oviposition (Asgari and Rivers 2011) to generate an environment suitable for parasitoid growth and to protect the parasitoid from the host defense (Burke and Strand 2012; Teng et al. 2016). Injection of wasp venom suppressed prophenoloxidase (PPO) activity and subsequent melanization in hemocytes of *Galleria mellonella* (Dubovskiy et al. 2016).

The uzi fly, *Exorista bombycis* (Diptera: Tachinidae), is an endoparasitoid that infests nearly 95 species of lepidopterans including the commercially important mulberry silkworm, *Bombyx mori*. This fly has been reported in Australian, Asian, and African countries (Kumar et al. 1990; Narayanaswamy and Devaiah 1998; O'Hara et al. 2020), and the economic loss has been recorded as 10–15% (Jadhav et al. 2014). However, the Indian wild silkworm, *Antheraea mylitta*, is infested by another dipteran *Blepharipa zebina* (Tachinidae) (Rath and Sinha 2005). Selection of host by dipteran parasitoids and its direct or indirect oviposition habits are crucial strategies for successful parasitism (Dindo and Nakamura 2018). After infestation by *E. bombycis*, variations induced in integumental epithelial cells are observed by transmission electron microscopy (TEM). Signs of apoptosis are revealed by DNA fragmentation test, and modulated expression of apoptosis-associated genes is reported by qPCR from integumental epithelium of *B. mori* larvae (Anitha et al. 2014; Pradeep et al. 2013). However, host attack and defense responses are being avoided by dipteran parasitoids (Caron et al. 2008). The infestation affected host physiology and development to synchronize with parasitoid development (Hegazi et al. 1977). Infestation by *E. japonica* shortened larval duration in fifth instar larvae of *B. mori*, probably due to decreased juvenile hormone titer and increased ecdysone titer (Dai et al. 2022, 2024; Wang et al. 2023). In addition, nonspecific melanization reactions are activated that end in formation of melanized hemocyte capsule around the invaded parasitoids (Anitha et al. 2014; Meister and Govind 2006; Shambhavi et al. 2023; Sorrentino et al. 2002; Williams 2007).

The habits and origin of dipteran parasitoids vary widely and are considered as models to analyze character convergence and adaptation (Feener and Brown 1997).

The physiological mechanism by which dipteran parasitoids surmount the host defense is not known. Hence to investigate the insect host-dipteran parasitoid interactions, genomic, proteomic, and molecular biology tools have to be utilized as in host-pathogen interaction analyses (Biron et al. 2011).

Application of Genomic Tools

Genome wide gene expression by microarray (Agilent platform, USA) of hemocytes collected from *B. mori* larvae at 6 h after infestation with *E. bombycis* showed downregulation of 587 genes associated with host defense and immunity, thereby hampering the immune reactions (Makwana et al. 2021). Gene ontology analysis by WEGO plot identified three main ontologies, viz., cellular component, molecular function, and biological process. Moreover, genes encoding LP30K, small heat shock proteins, Toll receptors, Notch, apolipoprotein, cell adhesion proteins, recognition proteins, cytoskeletal proteins, and cell regulatory laminin also showed downregulation leading to suppression of different immune events. Interacting partners of downregulated defense genes also showed downregulation and downregulated biological processes, viz., Wnt signaling, DNA repair, transcription, catabolism, glutamine metabolism, and components of cytoskeletal structure and transport that are associated with immune processes, signaling, metabolism, etc. Genes encoding superoxide dismutase, catalase, thiol peroxiredoxin, and thioredoxin reductase showed enhanced expression after the endoparasitoid infestation as shown by the microarray analysis (Makwana et al. 2021). Quantitative PCR of some of these defense genes also confirmed downregulated gene expression after the infestation.

Invasion of hymenopteran parasitoids and its impact on gene modulations have been analyzed in several systems. Activation of prophenoloxidase subunits 1 and 2 is part of the melanization cascade against entry of a foreign body in insects as noticed in the lepidopteran *Spodoptera frugiperda* after infestation by the ichneumonid wasp, *Hyposoter didymator* (Barat-Houari et al. 2006). Infection by malaria parasite, *Plasmodium berghei* NK-65 strain, induced activation of genes that encode protease inhibitors, serine proteases, and regulatory molecules (Oduol et al. 2000), whereas *Plasmodium yoelii* infection upregulated serine protease genes AdSp1 and AdSp3 in the hemocytes (Xu et al. 2006). The defense genes encoding gloverin, moricin-like peptide, serine protease inhibitors, and prophenoloxidase activating protease (PAP) were upregulated in the diamondback moth *Plutella xylostella* after parasitization by the hymenopteran *Diadegma semiclausum* (Huang et al. 2005). Encapsulation of the eggs of the parasitoid *Asobara tabida* requires activation of JAK/STAT and Toll pathways in *Drosophila melanogaster* (Wertheim et al. 2005). whereas host protection is mediated by Toll/IMD pathways in B. mori against Exorista sp infestation (Yang et al 2024). The defense genes are downregulated in the flour beetle *Tribolium confusum* after infection by the tapeworm *Hymenolepis diminuta* (Hitchen et al. 2009).

De novo transcriptome analysis of the sugarcane borer *Diatraea saccharalis* after infestation by the larval endoparasitoid *Cotesia flavipes* showed upregulation of 1432 transcripts and downregulation of 1027 transcripts after the infestation. The differentially expressed genes comprised those associated with Ca^{+2} transduction signaling pathway, glycolysis/gluconeogenesis, chitin metabolism, hormone biosynthesis and degradation, as well as the immune system (Merlin and Cônsoli 2019). Availability of genome and proteome information by whole genome sequencing (Mita et al. 2004; Xia et al. 2004, 2007), transcriptional and EST analyses (Goldsmith and Marec 2010; Mita et al. 2003; Oh et al. 2006; Zhang et al. 2007), as well as development of genome database KAIKOBASE (Shimomura et al. 2009) from *B. mori* made similarity analysis by NCBI-BLAST easy, and it became an integral part to identify defense genes/proteins in response to the infestations. Infestation by *E. bombycis* activated pro- and antioxidative reactions in *B. mori* by upregulation of encoding genes and increased activity of antioxidative enzymes, showing the antioxidative mechanism as a means of defense in the early stage of parasitization.

Polydnavirus (PDV) gene from *Cotesia rubecula CrV1–4* expresses in hemocytes of *Pieris rapae* (Zhang et al. 2004), whereas that associated with *C. kariyai* expresses immunoevasive protein (IEP) in *Pseudaletia separata* (Tanaka et al. 2003).

Application of Proteomic Tools

Host-pathogen interactions are explained by proteomic tools (Biron et al. 2011). Molecular strategies involved in immune responses of the host and in pathogen response for anti-defense activity are analyzed by proteomic tools (Barrett et al. 2000; Schmid-Hempel 2008). Variations in protein profile in hemolymph, hemocytes, and fat body of the host larvae of *B. mori* after infestation by *E. bombycis* are revealed by SDS-PAGE. Protein conjugates exclusive to infested larval tissue are resolved by PAGE gel, performed *in-gel* trypsin digestion and was analysed by mass spectrometry (MALDI-TOF-MS; Rosenfeld et al. 1992) with minor modifications (Sathisha et al. 2008) using an Ultraflex TOF/TOF (Bruker Daltonics, Bremen, Germany) mass spectrometer at the Proteomics Facility of the Indian institute of Science, Bangalore, India. The data was analyzed by MASCOT mass fingerprinting program (Matrix Science) using a fixed modification of carbamidomethyl, a variable modification of methionine oxidation, and a peptide tolerance of 0.6 Da and allowing one missed cleavage. Protein sequences that showed similarity with those available in databases are identified by BLASTp.

Antioxidation proteins identified from hemolymph, hemocytes, and fat body by mass spectrometry included oxidase, esterase, flavin-dependent monooxygenase, thioredoxin-dependent peroxidase, carboxyl/cholinesterase, UDP-glucuronosyl transferase, CuZn-superoxide dismutase, carboxyl esterase clade H, cytochrome P450, glycosyl hydrolase family 7, peroxisomal membrane anchor protein, and peroxisome proliferator-activated receptor gamma coactivator 1 alpha. Activity of most of the enzymes was either suppressed or decreased after the infestation (Makwana et al. 2021; Makwana et al. 2017). On infestation by dipteran parasitoid *E. bombycis*, oxidative reactions are activated in hemocytes of *B. mori* resulting in enhanced activity of oxidase, superoxide dismutase (SOD), thioredoxin peroxidase, catalase (CAT), glutathione-S-transferase (GST), and peroxidases (Makwana et al. 2017). Increased activity of reactive oxygen species and associated increase in hydrogen peroxide and lipid peroxidation are also reported (Makwana et al. 2017). In addition, lipoprotein 30 K (LP30K) showed downregulation after the parasitic microsporidian and endoparasitoid infestation (Makwana et al. 2021; Shambhavi et al. 2023) indicative of parasitic regulation of physiological and immunological roles of LP30K which is immunogenic as well as a storage protein (Ruilin et al. 2019; Ujita et al. 2005).

Hymenopteran venom is complex and consisted of high and low molecular weight proteins that induce cytolysis in the host tissues (Kaeslin et al. 2010). Venom from the hymenopteran *Pimpla hypochondriaca* consisted of trehalase, laccase, putative serine protease, phospholipase B, cathepsin, phenoloxidase (I, II, III), pimplin, VPr3, and cysteine-rich venom protein (Parkinson et al. 2001, 2002, 2003, 2004). From the Ichneumon wasp *P. hypochondriaca*, reprolysin, chitinase, thiol reductase, cathepsin, VG3, VG8, VG10, tetraspanin, and icarapin are reported (Parkinson et al. 2002). Moreover, the venom components possess cytolytic and paralytic activity, possibly through a calcium-dependent mode of action (Ergin et al. 2006; Rivers et al. 2009).

Venom from *P. turionellae* was characterized by a combination of proteomics and transcriptomics (proteome analysis combined with RNA Seq; Aili et al 2020; Dashevsky et al 2023) analysis of the data collected from body tissue and venom gland. It contributed to identify the venom components such as laccase and phenoloxidase, several proteinase inhibitors, metalloproteinase M12B, carboxylesterase, peptidase S1 variants, and a paralyzing factor named pimplin2 (Özbek et al. 2019; Dos-Santos Pinto et al 2018; Price et al 2009). Among the hymenopterans, ferritin from *Microctonus hyperodae* and *M. aethiopoides* (Crawford et al. 2008; Krishnan et al. 1994; Parkinson et al. 2002); aspartylglucosaminidase-like from *Asobara tabida* (Moreau et al. 2004); γ-glutamyl transpeptidase from *Aphidius ervi* (Falabella et al. 2007); acid phosphatase and Vn.11 from *Pteromalus puparum* (Wu et al. 2008; Zhu et al. 2008); Vn4.6, calreticulin, Vn50, and Vn1.5 from *Cotesia rubecula* (Asgari et al. 2003; Zhang et al. 2004, 2006); virulence protein P4 from *Leptopilina boulardi* (Labrosse et al. 2005); and phospholipase B from *P. turionellae* (Uçkan et al. 2006) are also reported as venom constituents.

From the dipteran parasitoids, venom or any substance associated with host-parasitoid interaction has not been reported till recently. Our experiments using combined analysis of HPLC (ultraflow liquid chromatography), in vitro bioassay, and nano LC-MS/MS (facility at C-CAMP, Bangalore, India) of parasitoid tissue from *E. bombycis* larva showed presence of anti-hemocyte components, pyridoxamine phosphate oxidase (PNPO) that releases the reactive oxygen species (hydrogen peroxide), hydrolase, lipase, and mucin in the parasitoid tissue (Makwana et al. 2022) showing the immunosuppressive mechanism by dipteran parasitoids.

Conclusion

Most of the investigations on the impact of hymenopteran parasitoids on host responses are analyzed by molecular biology tools, whereas only scanty information is available on effects of dipteran parasitoids on insect hosts. The review revealed that insect host-dipteran parasitoid communication could also be revealed by genomics and proteomic tools. In our studies cited here, we used the tachinid parasitoid *E. bombycis-B. mori* (parasitoid-host model) system to unravel the host responses that disclose the impact of dipteran parasitic biology on lepidopteran hosts.

Acknowledgments The authors thank the Central Silk Board, Bangalore, for the facilities and the Department of Biotechnology (Government of India), New Delhi, for financial support for the work cited from our lab under a research project to A. R. Pradeep (BT/PR6355/PBD/19/236/2012 dated 08/01/2013).

References

Aili SR, Touchard A, Hayward R, Robinson SD, Pineda SS, Lalagüe H, Vetter MI, Undheim EAB, Kini RM, Escoubas P, Padula MP, Myers GSA, Nicholson GM (2020) An integrated proteomic and transcriptomic analysis reveals the venom complexity of the bullet ant *Paraponera clavata*. Toxins 2020:12. https://doi.org/10.3390/toxins12050324

Amaya KE, Asgari S, Jung R, Hongskulab M, Beckage NE (2005) Parasitization of *Manduca sexta* larvae by the parasitoid wasp *Cotesia congregata* induces an impaired host immune response. J Insect Physiol 51:505–512

Anitha J, Pradeep AR, Sivaprasad V (2014) Upregulation of Atg5 and AIF gene expression in synchronization with programmed cellular death events in integumental epithelium of *Bombyx mori* induced by a dipteran parasitoid infection. Bull Entomol Res 104:794–800. https://doi.org/10.1017/S0007485314000686

Asgari S, Rivers DB (2011) Venom proteins from endoparasitoid wasps and their role in host-parasite interactions. Annu Rev Entomol 56:313–335

Asgari S, Zareie R, Zhang G, Schmidt O (2003) Isolation and characterization of a novel venom protein from an endoparasitoid, *Cotesia rubecula* (Hym: Braconidae). Arch Insect Biochem Physiol 53:92–100

Barat-Houari M, Hilliou F, Jousset FX, Sofer L, Deleury E, Rocher J, Ravallec M, Galibert L, Delobel P, Feyereisen R, Fournier P, Volkoff AN (2006) Gene expression profiling of *Spodoptera frugiperda* hemocytes and fat body using cDNA microarray reveals polydnavirus-associated variations in lepidopteran host genes transcript levels. BMC Genomics 7:160

Barrett J, Jefferies JR, Brophy PM (2000) Parasite proteomics. Parasitol Today 16:400–403. https://doi.org/10.1016/s0169-4758(00)01739-7

Biron DG, Nedelkov D, Missé D, Holzmuller P (2011) Proteomics and host-pathogen interactions: a bright future? Genetics and evolution of infectious diseases, pp 263–303. https://doi.org/10.1016/B978-0-12-384890-1.00011-X

Burke GR, Strand MR (2012) Deep sequencing identifies viral and wasp genes with potential roles in replication of *Microplitis demolitor* Bracovirus. J Virol 86:3293–3306

Caron V, Janmaat AF, Ericsson JD, Myers JH (2008) Avoidance of the host immune response by a generalist parasitoid, *Compsilura concinnata* Meigen. Ecol Entomol 33:517–522

Crawford AM, Brauning R, Smolenski G, Ferguson C, Barton D et al (2008) The constituents of *Microctonus* sp. parasitoid venoms. Insect Mol Biol 17:313–324

Dai M, Jiang Z, Li F, Wei J, Li B (2024) A parasitoid regulates 20E synthesis and antibacterial activity of the host for development by inducing host nitric oxide production. Insect Mol Biol. https://doi.org/10.1111/imb.12890

Dai M-L, Ye W-T, Jiang X-J, Feng P, Zhu Q-Y, Sun H-N, Li F-C, Wei J, Li B (2022) Effect of Tachinid parasitoid *Exorista japonica* on the larval development and pupation of the host silkworm *Bombyx mori*. Front Physiol 13:824203. https://doi.org/10.3389/fphys.2022.824203

Dashevsky D, Baumann K, Undheim EAB, Nouwens A, Ikonomopoulou MP, Schmidt JO, Ge L, Kwok HF, Rodriguez J, Fry BG (2023) Functional and proteomic insights into Aculeata venoms. Toxins 15(3):224. https://doi.org/10.3390/toxins15030224

Dindo ML, Nakamura S (2018) Oviposition strategies of tachinid parasitoids: two *Exorista* species as case studies. Int J Insect Sci 10:1–6. https://doi.org/10.1177/1179543318757491

dos Santos-Pinto JRA, Perez-Riverol A, Lasa AM, Palma MS (2018) Diversity of peptidic and proteinaceous toxins from social hymenoptera venoms. Toxicon 148:172–196

Dubovskiy IM, Kryukova NA, Glupov VV, Ratcliffe NA (2016) Encapsulation and nodulation in insects. Invertebr Surviv J 13:229–246

Ergin E, Uçkan F, Riversm DB, Sak O (2006) In vivo and in vitro activity of venom from the endoparasitic wasp *Pimpla turionellae* (L.) (Hymenoptera: Ichneumonidae). Arch Insect Biochem Physiol 61:87–97. https://doi.org/10.1002/arch.20100

Etebari K, Palfreyman RW, Schlipalius D, Nielsenm LK, Glatz RV, Asgari S (2011) Deep sequencing-based transcriptome analysis of *Plutella xylostella* larvae parasitized by *Diadegma semiclausum*. BMC Genom 12:446

Falabella P, Riviello L, Caccialupim P, Rossodivita T, Valente MT et al (2007) A g-glutamyl transpeptidase of *Aphidius ervi* venom induces apoptosis in the ovaries of host aphids. Insect Biochem Mol Biol 37:453–465

Feener DH Jr, Brown BV. Diptera as parasitoids. Annu Rev Entomol. 1997;42:73–97. https://doi.org/10.1146/annurev.ento.42.1.73.

Goldsmith MR, Marec F (2010) Molecular biology and the genetics of the lepidoptera. In: Contemporary topics in entomology series. CRC Press, Boca Raton

Guido-Patiño JC, Plisson F (2022) Profiling hymenopteran venom toxins: protein families, structural landscape, biological activities, and pharmacological benefits. Toxicon X 14:100119. https://doi.org/10.1016/j.toxcx.2022.100119

Hegazi EM, El-Minshawy AM, Hammad SM (1977) Suitability of *Spodoptera littoralis* larvae for development of *Microplitis rufiventris*. J Agric Sci 29:659–662

Hitchen SJ, Shostak AW, Belosevic M (2009) *Hymenolepis diminuta* (Cestoda) induces changes in expression of select genes of *Tribolium confusum* (Coleoptera). Parasitol Res 105:875–879. https://doi.org/10.1016/j.cois.2022.100896

Huang CY, Chou SY, Bartholomay LC, Christensen BM, Chen CC (2005) The use of gene silencing to study the role of dopa decarboxylase in mosquito melanization reactions. Insect Mol Biol 14:237–244

Jadhav AD, Desai AS, Sathe TV (2014) Distribution and economic status of Uzi fly *Exorista bombycis* Louis: a parasitoid of mulberry silkworm *Bombyx mori* L. Global J Res Anal 3:3–5

Jensen T, Walker AA, Nguyen SH, Jin AH, Deuis JR, Vetter I, King GF, Schmidt JO, Robinson SD (2021) Venom chemistry underlying the painful stings of velvet ants (hymenoptera: mutillidae). Cell Mol Life Sci 78:5163–5177. https://doi.org/10.1007/s00018-021-03847-1

Kaeslin M, Reinhard M, Bühler D, Roth T, Pfister-Wilhelm R, Lanzrein B (2010) Venom of the egg-larval parasitoid *Chelonus inanitus* is a complex mixture and has multiple biological effects. J Insect Physiol 56:686–694

Krishnan A, Nair PN, Jones D (1994) Isolation, cloning and characterization of new chitinase stored in active form in chitin-lined venom reservoir. J Biol Chem 269:20971–20976

Kumar P, Kishore R, Sengupta K (1990) Studies on the alternate host of uzi fly *Exorista sorbillans* (Diptera: Tachinidae). Indian J Seric 29:193–199

Labrosse C, Stasiak K, Lesobre J, Grangeia A, Huguet E et al (2005) A RhoGAP protein as a main immune suppressive factor in the *Leptopilina boulardi* (hymenoptera, Figitidae)- *Drosophila melanogaster* interaction. Insect Biochem Mol Biol 35:93–103

Makwana P, Dubey H, Pradeep ANR, Sivaprasad V, Ponnuvel KM (2021) Dipteran endoparasitoid infestation actively suppressed host defense components in hemocytes of silkworm *Bombyx mori* for successful parasitism. Anim Gene 22:200118. https://doi.org/10.1016/j.angen.2021.200118

Makwana P, Hungund SP, Pradeep ANR (2022) Dipteran endoparasitoid *Exorista bombycis* utilizes antihemocyte components against host defense of silkworm *Bombyx mori*. Arch Insect Biochem Physiol 112:e21976. https://doi.org/10.1002/arch.21976

Makwana P, Pradeep ANR, Hungund SP, Ponnuvel KM, Trivedy K (2017) The dipteran parasitoid *Exorista bombycis* induces pro- and anti-oxidative reactions in the silkworm *Bombyx mori*: enzymatic and genetic analysis. Arch Insect Biochem Physiol 97:e21373. https://doi.org/10.1002/arch.21373

Makwana P, Pradeep ANR, Hungund SP, Sagar C, Ponnuvel KM, Awasthi AK, Trivedy K (2017) Oxidative stress and cytotoxicity elicited lipid peroxidation in hemocytes of *Bombyx mori* larva infested with dipteran parasitoid, *Exorista bombycis*. Acta Parasitol 62:717–727. https://doi.org/10.1515/ap-2017-0086

Meister M, Govind S (2006) Hematopoietic development in *Drosophila*: a parallel with vertebrates. In: Hematopoietic stem cell development. Medical Intelligence Unit, Springer, Boston. https://doi.org/10.1007/978-0-387-33535-3_10

Merlin BL, Cônsoli FL (2019) Regulation of the larval transcriptome of *Diatraea saccharalis* (Lepidoptera: Crambidae) by maternal and other actors of the parasitoid *Cotesia flavipes* (hymenoptera: Braconidae). Front Physiol Sec Invert Physiol 10:446542. https://doi.org/10.3389/fphys.2019.01106

Mita K, Kasahara M, Sasaki S, Nagayasu Y, Yamada T, Kanamori H et al (2004) The genome sequence of silkworm, *Bombyx mori*. DNA Res 11:27–35. https://doi.org/10.1093/dnares/11.1.27

Mita K, Morimyo M, Okano K et al (2003) The construction of an EST database for *Bombyx mori* and its application. Proc Natl Acad Sci USA 100:14121–14126. https://doi.org/10.1073/pnas.2234984100

Moreau SJM, Cherqui A, Doury G, Dubois F, Fourdrain Y et al (2004) Identification of an aspartylglucosaminidase-like protein in the venom of the parasitic wasp *Asobara tabida* (hymenoptera: Braconidae). Insect Biochem Mol Biol 34:485–492

Narayanaswamy KC, Devaiah MC (1998) Silkworm uzi fly. Zen Publishers, Bangalore

Nighat P, Khalid M, Bin MS, Tean Z, Nayla M, Bojan G, Andre SO, Uday K, Lee WA (2023) Host-pathogen interaction in arthropod vectors: lessons from viral infections. Front Immunol 14:1061899. https://doi.org/10.3389/fimmu.2023.1061899

O'Hara, J. E., Henderson, S. J. & Wood, D. M. (2020). Preliminary Checklist of the Tachinidae (Diptera) of the World. Version 2.1, 5th March 2020, p. 1039. http://www.nadsdiptera.org/Tach/WorldTachs/Checklist/Tachchlist_ver2.1.pdf

Oduol F, Xu J, Niare O, Natarajan R, Vernick KD (2000) Genes identified by an expression screen of the vector mosquito *Anopheles gambiae* display differential molecular immune response to malaria parasites and bacteria. Proc Natl Acad Sci USA 97:11397–11402

Oh JH, Jeon YJ, Jeong SY, Hong SM, Lee JS, Nho SK, Kang SW, Kim NS (2006) Gene expression profiling between embryonic and larval stages of the silkworm, *Bombyx mori*. Biochem Biophys Res Commun 343:864–872

Özbek R, Wielsch N, Vogel H, Lochnit G, Foerster F, Vilcinskas A, von Reumont BM (2019) Proteo-transcriptomic characterization of the venom from the endoparasitoid wasp *Pimpla turionellae* with aspects on its biology and evolution. Toxins 11(12):721. https://doi.org/10.3390/toxins11120721

Parkinson NM, Conyers C, Keen J, MacNicoll A, Smith I et al (2004) Towards a comprehensive view of the primary structure of venom proteins from the parasitoid wasp *Pimpla hypochondriaca*. Insect Biochem Mol Biol 34:565–571

Parkinson NM, Conyers C, Keen JN, MacNicoll AD, Weaver ISR (2003) cDNAs encoding large venom proteins from the parasitoid wasp *Pimpla hypochondriaca* identified by random sequence analysis. Comp Biochem Physiol C 134:513–520

Parkinson N, Richards EH, Conyers C, Smith I, Edwards JP (2002) Analysis of venom constituents from the parasitoid wasp *Pimpla hypochondriaca* and cloning of a cDNA encoding a venom protein. Insect Biochem Mol Biol 32:729–735. https://doi.org/10.1016/S0965-1748(01)00155-2

Parkinson N, Smith I, Weaver R, Edwards JP (2001) A new form of arthropod phenoloxidase is abundant in venom of the parasitoid wasp *Pimpla hypochondriaca*. Insect Biochem Mol Biol 31:57–63

Pradeep AR, Anitha J, Awasthi AK, Babu MA, Geetha MN, Arun HK, Chandrashekhar S, Rao GC, Vijayaprakash NB (2013) Activation of autophagic programmed cell death and innate immune gene expression reveals immunocompetence of integumental epithelium in *Bombyx mori* infected by a dipteran parasitoid. Cell Tissue Res 352:371–385. https://doi.org/10.1007/s00441-012-1520-7

Price DRG, Bell HA, Hinchliffe G, Fitches E, Weaver R, Gatehouse JA (2009) A venom metalloproteinase from the parasitic wasp *Eulophus pennicornis* is toxic towards its host, tomato moth (*Lacanobia oleracae*). Insect Mol Biol 18:195–202. https://doi.org/10.1111/j.1365-2583.2009.00864.x

Rath SS, Sinha BR (2005) Parasitization of fifth instar tasar silkworm, *Antheraea mylitta*, by the uzi fly, *Blepharipa zebina,* a host-parasitoid interaction and its effect on host's nutritional parameters and parasitoid development. J Invertebr Pathol 88:70–78. https://doi.org/10.1016/j.jip.2004.09.006

Rivers DB, Dani MP, Richards EH (2009) The mode of action of venom from the endoparasitic wasp *Pimpla hypochondriaca* (hymenoptera: Ichneumonidae) involves Ca+2- dependent cell death pathways. Arch Insect Biochem Physiol 71:173–190. https://doi.org/10.1002/arch.20314

Robinson SD, Mueller A, Clayton D, Starobova H, Hamilton BR, Payne RJ, Vetter I, King GF, Undheim EAB (2018) A comprehensive portrait of the venom of the giant red bull ant, *Myrmecia gulosa*, reveals a hyperdiverse hymenopteran toxin gene family. Sci Adv 4:1–13. https://doi.org/10.1126/sciadv.aau4640

Rosenfeld J, Capdevielle J, Guillemot C, Ferrara P (1992) In-gel digestion for internal sequence analysis after one- or two dimensional gel electrophoresis. Anal Biochem 203:173–179

Ruilin L, Hu C, Shi Y, Geng T, Dingding L, Gao K, Hou C, Guo X (2019) Silkworm storage protein Bm30K-19G1 has a certain antifungal effects on *Beauveria bassiana*. J Invertebr Pathol 163:34–42

Sathisha GJ, Prakash YKS, Chachadi VB, Nagaraja NN, Inamdar SR, Leonidas DD, Savithri HS, Swamy BM (2008) X-ray sequence ambiguities of *Sclerotium rolfsii* lectin resolved by mass spectrometry. Amino Acids 35:309–320

Schmid-Hempel P (2008) Parasite immune evasion: a momentous molecular war. Trends Ecol Evol 23:318–326

Schmidt O (2008) Insect immune recognition and suppression. In: Beckage NE (ed) Insect immunology. Elsevier/Academic Press, San Diego, pp 271–294

Shambhavi HP, Makwana P, Pradeep ANR (2023) LP30K protein manifested in hemocytes of *Bombyx mori* larva on *Nosema bombycis* infection and showed functional evolution based on glucose- binding domain. 3Biotech 13:264. https://doi.org/10.1007/s13205-023-03685-x

Shimomura M, Minami H, Suetsugu Y, Ohyanagi H, Satoh C, Antonio B, Nagamura Y, Kadono-Okuda K, Kajiwara H, Sezutsu H, Nagaraju J, Goldsmith MR, Xia Q, Yamamoto K, Mita K (2009) KAIKObase: an integrated silkworm genome database and data mining tool. BMC Genomics 10:486. https://doi.org/10.1186/1471-2164-10-486

Sorrentino RP, Carton Y, Govind S (2002) Cellular immune response to parasite infection in the drosophila lymph gland is developmentally regulated. Dev Biol 243:65–80

Tanaka K, Tsuzuki S, Matsumoto H, Hayakawa Y (2003) Expression of Cotesia kariyai polydnavirus genes in lepidopteran hemocytes and SF9 cells. J Insect Physiol 49:433–440

Teng ZW, Xu G, Gan SY, Chen X, Fang Q, Ye GY (2016) Effects of the endoparasitoid Cotesia chilonis (hymenoptera: Braconidae) parasitism, venom, and calyx fluid on cellular and humoral immunity of its host Chilo suppressalis (Lepidoptera: Crambidae) larvae. J Insect Physiol 85:46–56. https://doi.org/10.1016/j.jinsphys.2015.11.014

Uçkan F, Ergin E, Rivers DB, Gençer N (2006) Age and diet influence the composition of venom from the endoparasitic wasp *Pimpla turionellae* L. (hymenoptera: Ichneumonidae). Arch Insect Biochem Physiol 63:177–187

Ujita M, Katsuno Y, Kawachi I, Ueno Y, Banno Y, Fujiim H, Hara A (2005) Glucan binding activity of silkworm 30-kDa apolipoprotein and its involvement in defense against fungal infection. Biosci Biotechnol Biochem 69:1178–1185

Vinson SB (1977) Insect host responses against parasitoids and the parasitoid's resistance: with emphasis on the Lepidoptera-Hymenoptera association. In: Bulla LA, Cheng TC (eds) Comparative pathobiology, vol 3. Springer, Boston. https://doi.org/10.1007/978-1-4615-7299-2_6

Wago H (1995) Host defense reactions of insects. Jpn J Appl Entomol Zool 39:1–13

Wang S-S, Wang L-L, Pu Y-x, Liu JY, Wang M-x, Zhu J, Shen Z-y, Shen X-j, Tang Sm (2023) *Exorista sorbillans* (Diptera: Tachinidae) parasitism shortens host larvae growth duration by regulating ecdysone and juvenile hormone titers in *Bombyx mori* (Lepidoptera: Bombycidae). J Insect Sci 23:6. https://doi.org/10.1093/jisesa/iead034

Wertheim B (2022) Adaptations and counter-adaptations in Drosophila host–parasitoid interactions: advances in the molecular mechanisms. Curr Opin Insect Sci 51:100896

Wertheim B, Kraaijeveld AR, Schuster E, Blanc E, Hopkins M, Pletcher SD, Strand MR, Partridge L, Godfray HC (2005) Genome-wide gene expression in response to parasitoid attack in *Drosophila*. Genome Biol 6:R94

Williams MJ (2007) Drosophila hemopoiesis and cellular immunity. J Immunol 178:4711–4716

Wu M-L, Ye G-y, Zhu J-y, Chen X-X, Hu C (2008) Isolation and characterization of an immunosuppressive protein from venom of the pupa-specific endoparasitoid *Pteromalus puparum*. J Invertebr Pathol 99:186–191

Xia Q, Cheng D, Duan J, Wang G, Cheng T, Zha X, Liu C, Zhao P, Dai F, Zhang Z, He N, Zhang L, Xiang Z (2007) Microarray-based gene expression profiles in multiple tissues of the domesticated silkworm, *Bombyx mori*. Genome Biol 8:R162

Xia Q, Zhou Z, Lu C, Chengm D, Dai F, Li B, Zhao P et al (2004) Biology Analysis Group. A draft sequence for the genome of the domesticated silkworm (*Bombyx mori*). Science 306:1937–1940. https://doi.org/10.1126/science.1102210

Xu W, Huang FS, Hao HX, Duan JH, Qiu ZW (2006) Two serine proteases from *Anopheles dirus* haemocytes exhibit changes in transcript abundance after infection of an incompatible rodent malaria parasite, *Plasmodium yoelii*. Vet Parasitol 139:93–101

Yang J, Xu Q, Shen W, Jiang Z, Gu X, Li F, Li B, Wei J (2024) The toll/IMD pathways mediate host protection against dipteran parasitoids. J Insect Physiol 153:104654. https://doi.org/10.1016/j.jinsphys.2024.104614

Zhang YZ, Chen J, Niem ZM, Lü ZB, Wang D, Jiang CY, He PA, Liu LL, Lou YL, Song L, Wu XF (2007) Expression of open reading frames in silkworm pupal cDNA library. Appl Biochem Biotechnol 136:327–343

Zhang G, Schmidt O, Asgari S (2004) A novel venom peptide from an endoparasitoid wasp is required for expression of polydnavirus genes in host hemocytes. J Biol Chem 279:41580–41585

Zhang G, Schmidt O, Asgari S (2006) A calreticulin-like protein from endoparasitoid venom fluid is involved in host hemocyte inactivation. Dev Compar Immunol 30:756–764

Zhu JY, Ye GY, Hu C (2008) Research progress on venom proteins of parasitic hymenoptera. Acta Phytophy Sin 35:270–278

Biotechnological Approaches for the Diagnosis of Silkworm Diseases

Mihir Rabha, Khasru Alam, K. Rahul, and A. R. Pradeep

Abstract

Incidences of silkworm diseases are one of the major stresses, which make considerable economic crop losses and reduce profitability every year in India. Silkworms are highly prone to several innumerable virulent pathogens including bacteria, viruses, fungi, and parasitic microbes like microsporidia. There are four major diseases of silkworms, viz., grasserie, flacherie, muscardine, and pebrine. All these diseases are infectious and spread rapidly to the healthy worms. The infection in the early stage (first to third stage) always leads to the outbreak of the diseases at the later stages. Several reports are available on crop loss due to silkworm diseases, which varies from 15 to 40% in India. For better management of silkworm diseases, early diagnosis of pathogens plays a vital role in reducing the spread of pathogens. In general, silkworm diseases are always diagnosed through visible symptoms and microscopic examinations, which is a labor-intensive, less efficient, less sensitive, and time-consuming process. Several advanced molecular techniques and approaches were already employed for rapid and accurate diagnosis of numerous silkworm pathogens. These techniques are based on nucleic acid hybridization, polymerase chain reaction, serological techniques (antibody based techniques), and biosensors. However, these advanced diagnostic techniques are not feasible at all time as they have their own limitations in many aspects like involvement of high initial cost, requirement of infrastructure, sensitivity, accuracy, and reliability. There is always a demand for the simple, rapid, accurate, and cost-effective techniques for early diagnosis of silkworm diseases. In this chapter, information regarding different advanced techniques employed for silkworm disease diagnosis along with their advantages and limitations will be provided.

M. Rabha (✉) · K. Alam · K. Rahul · A. R. Pradeep
Central Sericultural Research and Training institute, Berhampore, India

R. V Suresh et al. (eds.), *Biotechnology for Silkworm Crop Enhancement*,
https://doi.org/10.1007/978-981-97-5061-0_16

Keywords

PCR · Serological test · Flacerie · Grasserie · Muscardine · Pebrine · Nucleopolyhedrosis · Cytoplasmic polyhedrosis

Introduction

Silkworms are highly susceptible to numerous pathogens including virus, bacteria, fungus, and parasitic microbes like microsporidia. The major diseases of silkworms are grasserie, flacherie, muscardine, and pebrine. Flacherie is known to be caused by multiple pathogens infections (including virus and bacteria), grasserie by nucleopolyhedrosis (NPV) cytoplasmic polyhedrosis (CPV) virus, and muscardine by different species of fungus (*Beauveria* spp., *Metarhizium* spp.), whereas pebrine is another major serious disease caused by microsporidia (*Nosema* spp.) in silkworms that transmits vertically. Generally, silkworm diseases were diagnosed by visible symptoms followed by microscopic examination, which is a laborious, less sensitive, and time-consuming process. For better and efficient management of silkworm diseases, early and advance diagnosis of pathogens plays a vital role in reducing the spread of diseases and to minimize the crop loss. There are numerous molecular techniques and approaches which were already employed for rapid diagnosis and fast detection of pathogens in silkworms, including the techniques based on PCR, DNA hybridization, biochemical assays, antibody (serological test), biosensors, etc. However, these techniques and approaches have their own limitations and advantages like involvement of cost, sensitivity, specificity, accuracy, and time taken to perform the whole process. Different molecular techniques employed for the efficient diagnosis of silkworm diseases along with their limitations and advantages will be provided in this chapter.

Major Diseases of Silkworms

Silkworm (*Bombyx mori* L.) is a monophagous, highly domesticated lepidopteran insect of great economic importance for its silk and is vulnerable to a diverse array of microorganisms (bacteria, fungi, microsporidia, and viruses), resulting in crop losses to the farming fraternity practicing sericulture. The early symptoms of most of the silkworm diseases are very difficult to identify and remain unnoticed till the worms show late symptoms, which is generally too late to manage the disease. For proper effective rearing, silkworms need to be better protected from the pathogens. In this regard, early periodical diagnosis before appearing the disease symptoms needs to be practiced for early detection of the disease pathogens for better management techniques. Once the pathogen entered into the body of silkworms, it becomes almost impossible to save the larvae. The incidences of the silkworms were also highly influenced by different environmental factors like temperature, humidity, and wind (Rahmathulla 2012). Most of the viral and bacterial diseases were favored by the high temperature and moisture, whereas fungal disease like muscardine is favored by the low temperature and high humidity.

Grasserie Grasserie is caused by an occluded virus called *Bombyx mori.* Nucleopolyhedrosis virus (BmNPV) belongs to the family *Baculoviridae*, which is a double stranded DNA virus (dsDNA). Mostly the infection occurs through feeding of contaminated mulberry leaves (Chopade et al. 2021). The early symptoms are not clear and specific. Usually the worms become restless and move on the reams of tray, skin appears shiny, milky white and become fragile, swelling of intersegment, worms loose clasping power of forelegs and finally die with hanging heads. The disease generally appears in hot and humid climate.

Flacherie Flacherie is caused by both viral and bacterial infection. Viruses involved are non-occluded, viz., infectious flacherie virus (IFV) and densonucleosis virus (DNV). IFV is a single-stranded RNA virus (ssRNA), and DNV is a single-stranded DNA virus (ssDNA). The major bacteria involved in flacherie are *Streptococcus*, *Staphylococcus*, and *Bacillus* spp. (*Bacillus thuringiensis*, *Bacillus sotto*, *Serratia marcescens*) (Hukuhara 2014; Rahmathulla 2012). The infected larvae become weak, stop feeding, vomit, become lethargic and inactive, and finally die.

Muscardine In silkworm, white muscardine and green muscardine were commonly observed. White muscardine is caused by entomopathogenic fungus *Beauveria bassiana* (white) and *Metarhizium anisopliae* (green). Both frequently appear during the winter season when there is some unexpected rain. The climatic conditions of low temperature along with high humidity favor the disease outbreaks that can cause considerable crop losses (Hossain et al. 2017; Samson et al. 1990). The infected larvae stop feeding and become inactive and lethargic. Within 10 h after death, the soft corpse gradually hardens and mummifies within 24 h.

Pebrine Pebrine is also called "pepper disease" caused by microsporidia; an obligate intracellular parasite called *Nosema bombycis* (Naegeli) was discovered by Louis Pasteur in 1865. The parasite transmits vertically (transovarial transmission) from mother moth to egg and thereby to next generation larvae. The perbrine infected larvae become lethargic, slow growth, unequal size, irregular moulting, loss of appetite and appear black spots on larval integument. In most of the cases, either larva failed to spin or produce flimsy cocoon. The available method to control the disease is quarantine and complete destruction of infected larvae and mother moths. This devastating disease inflicted extensive damage to the sericultural industry in Europe and Asia during the mid-nineteenth century and still remains one of the impending challenges for the progress of sericulture across the world.

Diagnostic Methods

For successful silkworm cocoon production, proper diagnosis of silkworm diseases is necessary. The conventional diagnostic methods of silkworm diseases are based on microscopic observations employing different types of dyes. However, these methods are laborious, less sensitive, and time-consuming, and several alternative immunological assays and molecular approaches have been designed which include PCR, multi-primer PCR, LAMP, real-time quantitative PCR, DNA hybridization, ELISA, immunoblot, and antibody based fluorescent techniques. Immunoassays are based on the antigen antibody reaction that involves development of antibodies against particular silkworm pathogens. Several immunoassays were developed and employed for diagnosis of silkworm diseases. The advantage of the immunoassays is that the results are visible with the naked eyes. Most of the immunological diagnostic methods employed for the diagnosis of silkworm diseases are immunoblotting, immunodiffusion, latex agglutination, immunofluorescence, dipstick assays, and enzyme-linked immunosorbent assay. These techniques were extensively used for detection of both occluded (BmCPV and BmNPV) and non-occluded virus (BmIFV and BmDNV) including microsporidia. However, these techniques have certain limitations such as requirement of sophisticated laboratory equipment and infrastructure, intricate protocols, technical skilled personnel, and discriminatory issues.

Microscopic Techniques

Conventional microscopic techniques are based on the use of simple microscopy which are laborious and time-consuming in nature and have less sensitivity and specificity. Apart from this, conventional microscopic observation (simple microscopy) can only detect the spores of the pathogens and viral polyhedral occluded bodies, where non-occluded viruses (BmIFV, BmBDV) remain undetectable under simple microscopy. Advanced microscopic techniques including florescent microscopy, phase contrast microscopy, scanning electron microscopy, and transmission electron microscopy were also developed for diagnosis of silkworm diseases (Cali et al. 1991; Iwano and Ishihara 1991; Orenstein 2003). However, these advance microscopy techniques met limited success in its usage for diagnosis of silkworm diseases as they are expensive and sophisticated and require skilled personnel and high cost for maintenance.

Immunological Techniques

Immunodiagnostic techniques are based on antibody-antigen (pathogens) reaction employed to quantify or target the presence of specific pathogens of interest. Immunodiagnostic techniques are generally rapid, accurate, and simple to use. Immunological techniques that are used in silkworm disease diagnosis are enzyme-linked immunosorbent assay (ELISA), immunodiffusion, fluorescence assay, and latex agglutination test.

Enzyme-Linked Immunoassay (ELISA)

ELISA is one of the most popular and widely used immunodiagnostic assays for the diagnosis of silkworm diseases. This technique is based on the use of enzyme-conjugated antibodies which upon reaction with antigens produce a typical measurable color signal that specifies the presence of the target antigens/pathogens. Based on the type of ELISA immunoassay, requirements of primary and/or secondary detection antibody, analyte/antigen, buffer, and substrate/chromogen may vary. The primary antibodies are specific antibody that only binds to the specific target molecules, where secondary antibodies are enzyme-conjugated antibody that binds to the primary antibody and produce color in the presence of the enzyme substrate. The change of the color occurs either due to hydrolysis of phosphate group of alkaline phosphatase (AP) or due to oxidation of substrates by horseradish peroxidase (HRP).

Generally there are four main steps to complete an ELISA immunoassay:

1. Coating of polystyrene ELISA plates (96 wells) with either antigen or antibody followed by washing with phosphate buffered saline (PBS) and a nonionic detergent to remove the unbound molecules. Phosphate buffered saline helps to maintain a constant pH.
2. Blocking of ELISA to reduce nonspecific bindings and block any unbound sites on the ELISA plate improves the signal-to-noise ratio of the assay. Generally blocking was performed either with bovine serum albumin (BSA) or casein and then washing with phosphate buffered saline (PBS).
3. Detection is carried out by adding a substrate that can generate a color. The most commonly used substrate is horseradish peroxidase (HRP) and alkaline phosphatase (AP).
4. Final read based on the change in color absorbance based on the target molecules is measured and analyzed by the ELISA reader.

There are four major types of ELISA: direct ELISA, indirect ELISA, sandwich ELISA, and competitive ELISA (Table 1). There are numerous factors which can interfere ELISA assay at any steps of the testing process including, from the beginning of the sample collection process, shape and quality of the ELISA assay plate, pH of the buffer used for washing, specificity and affinity of primary and secondary antibody, incubation time and temperature, influence and concentration of the blocking buffer, types of target antigen/molecules, cross-reactivity and types of enzyme conjugate, frequency and duration of washing, quality of substrate, and instrumental error for signal detection. The sensitivity and specificity of ELISA technique is higher than that of latex agglutination test (Sivaprasad et al. 2021), but it is not suitable for field application as it requires well-equipped laboratories. In the previous study, indirect ELISA was used for diagnosis of pebrine disease of silkworm caused by *Nosema bombycis* (Wongsorn et al. 2017).

Table 1 Advantages & disadvantages of different types of ELISA techniques

Types of ELISA	Principles	Advantage	Disadvantage
Direct ELISA	The enzyme conjugated primary antibody binds directly to the target antigens/molecules/protein followed by washing and blocking. Final steps involve addition of chromophore/substrate (e.g. alkaline phosphates, or horseradish peroxidase) for colour change	Simplest form of ELISA does not involve use of secondary antibody and hence eliminates secondary antibody cross-reactivity, rapid and simple compared to indirect ELISA	Lower sensitivity compared to the other types of ELISA and its high cost of reaction
Indirect ELISA	It requires two types of antibodies, a primary antibody that binds specifically to the target antigens/molecules/protein of interest and an enzyme conjugated secondary antibody complementary to the primary antibody	The higher sensitivity and less expensive of than that of direct ELISA. It has more flexibility as many possible primary antibodies can be employed	It has high risk of cross-reactivity between the secondary antibodies
Sandwich ELISA	This technique termed as 'Sandwich' as the antigen is captured in between the antibodies. At first the antibody is fixed on the ELISA plates and then the target antigen is added. Then the enzyme conjugated secondary antibody is added. Each of the steps is followed by the washing (to remove unbound excess molecules) and blocking (to block the unbound site). In final step substrate/chromophore is added to produce colour	Sandwich ELISA has the highest sensitivity among all the ELISA techniques	It's a time consuming, expensive and involves complex sample processing steps Requires "matched pair" secondary antibodies (divalent/multivalent antigen)
Competitive ELISA	Competitive ELISA is employed for the detection of antibody specific to an antigen. This technique requires two specific antibodies, first an enzyme-conjugated antibody and the other is antibody present in the test sample/serum (if the serum is positive). Both the antibodies were simultaneously allowed to compete for binding the antigens. The change in the color will depicts the test is negative as the enzyme-conjugated antibody will bound to the antigens, whereas the absence of color will indicates the presence of antibodies in the test serum (positive test)	Require less sample purification, it has low variability Can measure a large range of antigens in a given sample, it can be used for small antigens	Lower specificity and sensitivity which cannot be used in dilute samples

Latex Agglutination Test

One of the most commonly employed immunoassay for disease diagnosis which is based on antigen-antibody reactions results in the formation of visible clumping of particulates (agglutinates) or pathogens. Agglutination test is a semiquantitative, rapid, highly sensitive, and simple user-friendly method. The factors that may influence agglutination are type of antigens and its concentration, avidity of both antigens and antibodies, pH of the buffers, etc. Latex agglutination assay based on monoclonal antibody is used in detection of BmNPV as well as *Nosema bombycis* (Hayasaka and Ayuzawa 1987; Shamim et al. 1997) in mulberry silkworm *Bombyx mori*.

Immunodiffusion Assay

Immunodiffusion is based on the movement of antigen and antibody precipitin through a diffusion supported medium like agarose gel. This is generally a quantitative approach employed for the quantification of antigen in a solution based on the size of the radius formed by the antigen-antibody precipitin complex. Radial immunodiffusion method was successfully employed for the detection of the infectious flacherie virus (IFV) in silkworm based on the formation of precipitin ring of IFV antigen and antibody reaction (Seki and Sekijima 1976).

Immunofluorescence Assay

Immunofluorescence assay is based on the use of fluorescent-labeled antibodies that bind to the target antigens which helps to detect and localize the site of infections by the pathogens. It requires experts and sophisticated fluorescence microscope to observe. Krywienczyk and Sohi (1967) employed this technique to study the crystallization of polyhedra in the nuclei of the infected cells. Later on, the technique was also used to study the site of infection of different silkworm pathogens including *Nosema bombycis*.

Molecular Techniques

During the past two decades, molecular techniques have revolutionized the microbial detection due to the copious advantages they possess, viz., accuracy, aiding in early detection, high sensitivity, greater specificity, rapidity, and ease in analysis and interpretation.

PCR Based Techniques

The tradition polymerase chain reactions technique relies on use of a set of target specific primers (forward and reverse primer) along with the buffer containing dNTPs and taq-polymerase enzyme to amply the target sequence. PCR is a highly robust, specific, and sensitive molecular technique which principle is applied to develop numerous advanced PCR based techniques like RT-PCR, qPCR, multiplex PCR, etc. The multiplex PCR is one of the efficient tools for the early diagnosis of silkworm disease. However, each of these techniques has been designed based on requirements and objective purpose. The conventional PCR was successfully developed and employed for detection of almost all types of major pathogens of silkworms. The principles, advantages, and limitations of all the PCR based techniques were mentioned in Table 2.

Nucleic Acid Lateral Flow Assay (NALFA)

Nucleic acid lateral flow assay is based on the combination of lateral flow technology and immunoassay principles where the antibodies bind with the target analytes (nucleic acid) and produce the color for detection. This technique is very simple, easy to use, rapid, specific, sensitive, and cost-effective compared to other nucleic acid based detection techniques (Table 2). This technique was successfully developed and employed for detection of *Nosema bombycis*, where NALFA is combined with conventional PCR targeting *LSU rDNA* and lateral flow assay with anti-FAM rabbit monoclonal antibody conjugated with gold nanoparticle strip for detection of *N. bombycis* in silkworm eggs (He et al. 2019). However, the sensitivity of the NALFA was found higher than that of agarose gel.

TaqMan Assay

The principle of the TaqMan assay is based on the 5′ exonuclease activity of the Taq polymerase enzyme. TaqMan assay includes the use of forward primer, reverse primer, and one or more TaqMan probes that specifically target a gene sequence. The TaqMan probes are labeled with a fluorophore at the 5′ end and a quencher molecule at the 3′ end. When these fluorophore and quencher molecules remain in the close proximity, the quencher molecule inhibits the fluorescence signals. The probes bind specifically to the complementary region of the template DNA. When the Taq polymerase starts synthesizing new strands and reaches the site of TaqMan probe employing unlabeled forward and reverse primers, it cleaves the probe and separate the fluorophore from quencher molecule leading to emission of the fluorescence signal. This technique was reported by Kary Mullis in 1991, and later on, this technology was developed by the Hoffman-La Roche for diagnostic assays. As the TaqMan assay is based on the PCR, it is relatively simple to implement the technique efficiently (Table 2). Recently, this technique was successfully employed to

Table 2 Advantages & disadvantages of available molecular techniques employed for diagnosis of silkworm diseases

Techniques	Principles	Advantages	Disadvantages
Conventional PCR	Polymerase chain reaction that amplifies the DNA sequences based on enzymatic reaction of Nucleic acid replication	• Early detection is possible before symptom appears • Simple to understand • Differentiation at species level • High sensitivity (10 spores or 0.1 spore DNA in a PCR; Fu et al. 2016)	• Quantification of pathogen i.e. level of infection cannot be determined • Sensitivity of a PCR could be influenced by the technique used for DNA extraction • Require thermal cycle and gel documentation system • Post processing step (multiple step process) • Costly and time consuming and require skilled person. • Require huge funds • Specificity depends on primer sequence designed
qPCR/ quantitative PCR/real time PCR	In this technique, total RNA/mRNA is transcribed into complementary DNA (cDNA) and then cDNA is used as the template for qPCR, where the amount of amplified product is measured in each PCR cycle using fluorescence	• High sensitivity • Quantitative, useful for monitoring of microbial load • More reliable results • No post-PCR processing step • Higher sensitivity compared to conventional PCR • Real time monitoring of amplification	• Require expert for result interpretation • Require sophisticated high cost equipments • Costly chemicals • Need skilled person for preparation and handling of chemicals • Need fluorescent probes • Time consuming process • Prone to temperature and inhibitors • Lab based technique • Specificity depends on primers and probe designed
Multiplex PCR	Multiplex PCR is used to amplify several DNA target sequences simultaneously in a single reaction. It uses multiple primers that amplifies different DNA sequences	• Relatively higher sensitivity than that of conventional PCr • Multiple target sequences can be amplified in a single reaction • Numerous pathogens may be detected in a single reaction • High efficiency	• Critical primer design for same annealing temperature with different product size • Can create a cross-reaction between primer pairs (false positive) • Unspecific amplification • Optimization is difficult • High cost • Mixing different primers can cause some interference in the amplification

(continued)

Table 2 (continued)

Techniques	Principles	Advantages	Disadvantages
LAMP	Loop mediated isothermal reaction is based on *Bst* polymerase enzyme strand displacement activity in addition with replication activity under isothermal condition	• Highly sensitive • Process is simple, rapid and easy to use • Low cost set up compared to conventional PCR • Thermal cycler is not required • Naked eye visual monitoring	• Laboratory based technique • Complicated prime design • Requires multiple primers of high purity • Require strand-displacing DNA polymerase • Require an instrument to maintain a constant temperature (water bath) • Background interference
TaqMan assay	Based on the 5' exonuclease activity of the taq-polymerse enzyme that cleave the taqman probes and release the fluorophore molecule to emit the signal	• High specificity & sensitivity (10^2 copies) • Less cross reactivity • Post PCR processing is not required	• Require skilled person & lab based technique • Complex pre-processing of sample and designing of probe • Costlier chemical (reporter dye) • Costlier equipments • Time consuming process
Aptamer	Chemically synthesized short single stranded nucleic molecules (ssDNA or ssRNA) which can bind to the target molecules with high specificity and affinity by folding into tertiary structures	• Wide ranges of targets • High sensitivity and specificity • High stability • Ease of use and low cost of production • Simple modification and labeling compatibility with different diagnostic approaches	• Difficulties for generating for some molecules such as small molecules • Problems toward commercialization • Sensitivity to nuclease existed in serum and real samples • Changes in affinity and specificity in real situations • Low chemical diversity of natural oligonucleotide

diagnose pebrine disease of silkworm caused by the pathogen *N. bombycis*, whereas the probes and primers targeting *β-tubulin* gene were employed with a limit of detection (LOD) of 6.9×10^2 copies of target gene per reaction (Jagadish et al. 2021).

Loop-Mediated Isothermal Amplification (LAMP)

Loop-mediated isothermal amplification (LAMP) is based on the use of two or three sets of primers along with *Bst* polymerase enzyme from *Geobacillus stearothermophilus* involved in the strand displacement activity in addition to replication activity under isothermal condition. Generally, three types of primers are designed for the target sequence—internal primers (forward and reverse), external primers (forward and reverse), and loop primers (forward and reverse)—to facilitate subsequent amplification. The procedure of the technique is very simple and user-friendly as the whole technique is performed in a single tube for amplification of target DNA sequence in 60–65 °C. It is also one of the low-cost techniques for diagnosis of certain diseases which was invented at the University of Tokyo, in 2000. As the technique uses different sets of primers to amplify the distinct regions of the target sequence, it increases the specificity. Another advantage of LAMP is naked eye visual detection of the resulting amplicon based on hybridization and aggregation of complementary gold nanoparticle-bound (AuNP) single-stranded DNA (ssDNA) that leads to change in blue/purple color instead of normal red color. LAMP is less versatile than the conventional PCR, as LAMP can't be used for cloning and other PCR mediated molecular techniques (Table 2). Primer-primer interaction may occur as it uses multiple primers and may contribute to false result. LAMP requires thermostat for maintaining elevated heat of 60–65 °C. Loop-mediated isothermal reaction in combination with lateral flow assay was developed for diagnosis of BmNPV for visual sensitive detection. However, a set of four primers targeting six unique regions of BmNPV gp41 gene along with labelled probe were used. The technique was successfully developed and employed with detection limit of 0.2 pg of DNA (Zhou et al. 2015). This technique is also employed (Esvaran et al. 2018; Liu et al. 2015; Yan et al. 2014; Yang et al. 2017) for detection and identification of *N. bombycis* based on primers designed to amplify target specific genes (*LSU-rRNA*, *EB-1*, *SWP*, *PTP-1*).

Aptamer-Based Techniques

Aptamers are also known as "chemical antibodies"; they are generally short, single-stranded nucleic acid (ssDNA or ssRNA) sequences which can specifically bind to the target molecules, including peptides, toxins, carbohydrates, small molecules, and even whole live cells. Aptamers are assumed as alternative of antibodies, and the synthesis of aptamers is much cheaper and easier than that of antibodies (Table 2). Ellington and Szostak (1990) introduced the term aptamers to the world in the late twentieth century which is derived from the Latin word *aptus* (fitting) and the Greek word *meros* (part). Aptamers appropriately termed as chemical antibodies are small, single-stranded oligonucleotide (DNA or RNA) molecules that bind to the target molecules with high specificity and affinity by folding into tertiary structures. Aptamers can recognize diverse target molecules which include ions, amino acids, peptides, cofactors, enzymes, immunoglobulins, nucleotides, cell surface receptors, drugs, and organic dyes (Yu et al. 2018). The molecular interactions that exist between an aptamer and its target may be either van der Waals forces, hydrogen bonds, or electrostatic interactions. Recently, an aptamer-based electrochemical sandwich assay in combination with LAMP technique was developed to detect the *Nosema bombycis* (Xie et al. 2018). The PTP1 DNA sequence was targeted by the LAMP, and the pyrophosphate (PPi) produced as by-product during LAMP amplification was further converted into ATP (through addition of adenosine 5′-phosphosulfate (APS) and ATP sulfurylase). Further, the converted ATP was detected by the electrochemical aptasensor.

Critical Issues in the Diagnosis of Silkworm Diseases

Large scale samples In sericulture, silkworm seed production activities involve production of disease-free laying eggs which needs testing of a large number of samples under microscope. Till date, microscopy is the only technique employed for testing of pupa or mother moth in silk worm seed production units. However, this technique is a tedious, labor-intensive, and time-consuming process. Apart from this, lower concentrations of spore remain undetected with large number of samples that reduce the sensitivity of testing for pathogens. Therefore, minimal processing with simple and rapid detection techniques is a prerequisite for the detection of pathogens in sericulture.

Specificity and clarity in diagnosis Lack of clarity regarding sensitivity and accuracy is the major issue with the available techniques. The clarity in the diagnosis also depends on the selection criteria for the samples to be investigated and the intensity of the spores inside the mother moth. Apart from this limitations, techniques based on molecular and biochemical approaches are always found to encounter cross-reactivity that leads to false results and influences the specificity. Whereas microscopic observations are less sensitive and almost impossible to diagnose the pathogen with minimal infections or detection in early stage is a challenge.

Lack of user-friendly techniques Most of the techniques were based on either PCR or antibody based approaches. PCR based techniques were highly sensitive and specific and have the ability to differentiate the strains. But, the PCR based techniques techniques have certain limitations such as requirement of sophisticated equipment, costly chemicals, skilled person to operate and analyze the results and totally lab a based techniques. Contrarily, antibody based techniques are also highly sensitive, easy to operate, portable, and rapid, and the result can be interpreted through the naked eye but the specificity is very low. However, antibody based techniques cannot differentiate the strains.

Perspectives

Regardless of the significant progress in the molecular diagnostics during the last two decades toward development in diagnosis of silkworm disease and pathogens, their practical applicability is extremely limited due to a multitude of factors. All these techniques are complicated and involve multiple steps which require experienced and skilled person, infrastructure, sophisticated equipment, and huge funds. These are the major drawbacks of existing molecular techniques; due to which, these techniques are not transferred to the farmer's level. Researchers across the globe should focus on the development of simple, user-friendly, rapid, sensitive, and most importantly cost-effective techniques for the diagnosis of silkworm diseases.

References

Cali A, Orenstein JM, Kotler DP, Owen R (1991) A comparison of two microsporidian parasites in enterocytes of AIDS patients with chronic diarrhea. J Protozool 38(6):96S–98S

Chopade P, Raghavendra GC, Mohana KS, Bhaskar RN (2021) Assessment of diseases in *Bombyx mori* silkworm - a survey. Global Trans Proc 2(1):133–136

Ellington AD, Szostak JW (1990) *In vitro* selection of RNA molecules that bind specific ligands. Nature 346:818–822

Esvaran VG, Gupta T, Mohanasundaram A, Ponnuve KM (2018) Development of isothermal amplification assay for detection of *Nosema bombycis* infection in silkworm *Bombyx mori* targeting polar tube protein 1 gene. Invertebr Surviv J 15:352–361

Fu Z, He X, Cai S, Liu H, He X, Li M, Lu X (2016) Quantitative PCR for detection of*N osema bombycis* in single silkworm eggs and newly hatched larvae. J Microbiol Methods 120:72–78

Hayasaka S, Ayuzawa C (1987) Diagnosis of microsporidians *Nosema bombycis* and closely related species by antibody sensitized latex. J Sericult Sci Jpn 56:167–170

He Z, Ni Q, Song Y, Wang R, Tang Y, Wu Y, Liu L, Bao J, Chen J, Long M, Wei J, Li C, Li T, Zhou Z, Pan G (2019) Development of a nucleic acid lateral flow strip for rapid, visual detection of *Nosema bombycis* in silkworm eggs. J Invertebr Pathol 164:59–65

Hossain Z, Chakraborty S, Gupta SK, Saha AK, Bindroo BB (2017) Silkworm disease incidence trends during the years 1992-2011 in the Murshidabad district of West Bengal, India. Int J Trop Insect Sci 37(4):259–270

Hukuhara T (2014) The etiology of flacherie, one of the great scourges of sericulture. J Insect Biotechnol Sericol 83:25–31

Iwano H, Ishihara R (1991) Dimorphism of spores of *Nosema* spp. in cultured-cell. J Invertebr Pathol 57:211–219

Jagadish A, Khajje D, Tony M, Nilsson A, Miranda JR, Terenius O, Dubey H, Mishra RK, Ponnuvel KM (2021) Development and optimization of a TaqMan assay for *Nosemabombycis*, causative agent of pébrine disease in *Bombyx mori* silkworm, based on the *β-tubulin* gene. J Microbiol Methods 186:106238

Krywienczyk J, Sohi SS (1967) Immunofluorescence studies of *Bombyx mori* ovarian tissue cultures infected with nuclear-polyhedrosis virus. J Invertebr Pathol 9(4):568–570

Liu JP, Cheng W, Yan YW, Wei JY, Yang JL (2015) Detection of pebrine disease in *Bombyx mori* eggs with the loop-mediated isothermal amplification (LAMP) method based on *EB1* gene. Acta Ecol Sinica 58(08):846–855

Orenstein JM (2003) Diagnostic pathology of microsporidiosis. Ultrastruct Pathol 27(3):141–149

Rahmathulla VK (2012) Management of climatic factors for successful silkworm (*Bombyx mori* L.) crop and higher silk production: a review. Psyche 2012:1–12

Samson MV, Baig M, Sharma SD, Balavenkatasubbaiah M, Sasidharan TO, Jolly MS (1990) Survey on the relative incidence of silkworm diseases in Karnataka, India. Indian J Seric 29:248–254

Seki H, Sekijima Y (1976) Detection of the specific antigen of the infectious flacherie virus in the silkworm, *Bombyx mori* L., by the single radial immunodiffusion method. J Sericult Sci Jpn 45:13–18

Shamim M, Ghosh D, Baig M, Nataraju B, Datta RK, Gupta SK (1997) Production of monoclonal antibodies against *Nosema bombycis* and their utility for detection of pebrine infection in *Bombyx mori* L. J Immunoass 18(4):357–370

Sivaprasad V, Rahul K, Makwana P (2021) Methods in silkworm microbiology: chapter 2 - Immunodiagnosis of silkworm diseases. Methods Microbiol 49:24–46

Wongsorn D, Sirimungkararat S, Hongprayoon R (2017) The efficiency of ELISA technique in *Nosema bombycis* N. of mulberry silkworm (*Bombyx mori* L.) detection (Thai). Asia Pac J Sci Technol 15(7):622–635

Xie S, Tang Y, Tang D (2018) Converting pyrophosphate generated during loop mediated isothermal amplification to ATP: application to electrochemical detection of *Nosema bombycis* genomic DNA PTP1. Biosens Bioelectron 102:518–524

Yan W, Shen Z, Tang X, Xu L, Li Q, Yue Y, Xiao S, Fu X (2014) Detection of *Nosema bombycis* by FTA cards and loop-mediated isothermal amplification (LAMP). Curr Microbiol 69(4):532–540

Yang D, Pan L, Peng P, Dang X, Li C, Li T, Long M, Chen J, Wu Y, Du H, Luo B, Song Y, Tian R, Luo J, Zhou Z, Pan G (2017) Interaction between SWP9 and polar tube proteins of the microsporidian *Nosema bombycis* and function of SWP9 as scaffolding protein contribute to polar tube tethering to the spore wall. Infect Immun 85:3e00872-16

Yu X, Chen F, Wang R, Li Y (2018) Whole-bacterium SELEX of DNA aptamers for rapid detection of *E. coli* O157:H7 using a QCM sensor. J Biotechnol 266:39–49

Zhou Y, Wu J, Lin F, Chen N, Yuan S, Ding L, Gao L, Hang B (2015) Rapid detection of *Bombyx mori*nucleopolyhedrovirus (BmNPV) by loop-mediated isothermal amplification assay combined with a lateral flow dipstick method. Mol Cell Probes 29(6):389–395

Implications of Bioassay in Biotechnology with Relevance to Silkworm Breeding

M. S. Ranjini, N. Chandrakanth, G. R. Manjunath, K. Suresh, K. B. Chandrashekar, and S. Gandhi Doss

Abstract

Bioassay is considered as one of the important assays that contributes valuable information with respect to biotechnological approach, framing successful outcome in silkworm breeding program in the development of abiotic and biotic resistant/tolerant breeds. Bioas say is essential in evaluating consistency and stability of results obtained through biotechnological approaches. Data obtained through bioassay is crucial in all the developmental stages of silkworm life cycle from initial research approach to final inference. Other than the physicochemical and biochemical procedures, bioassay conforms precise validation of biotechnological results. It offers and extends the possibility of efficient trait detection as well as selection of individual with desirable features. Bioassay can be performed through qualitative and quantitative method for assessing the physical effects of a substance that may not be quantified, such as abnormal development or deformity, and estimation of concentration/potency of a substance by measurement of the biological response, respectively. Bioassay confers the detection of physiological responses, viz., mortality, larval growth, development, fecundity, and behavioral responses, viz., feeding deterrence, oviposition deterrence, etc. Determination of LC_{50}, LD_{50}, and LT_{50} is considered as important parameter in bioassay study which can be statistically analyzed through probit regression, and it will be explained in detail with example. Likewise, the survival analysis, viz.,

M. S. Ranjini (✉) · N. Chandrakanth · K. B. Chandrashekar · S. Gandhi Doss
Central Sericultural Research & Training Institute, Central Silk Board, Mysuru, India

G. R. Manjunath
Central Silk Board, Research Coordination Section, Central Office, Bengaluru, India

K. Suresh
Central Sericultural Research & Training Institute, Central Silk Board, Berhampore, West Bengal, India

R. V Suresh et al. (eds.), *Biotechnology for Silkworm Crop Enhancement*, https://doi.org/10.1007/978-981-97-5061-0_17

log-rank test, will be explained in detail which is a well-known tool that compares the entire survival experience among the groups and determines whether a significant effect on survival is present or is nonsignificant. The chapter will provide detailed information on the magnitude of bioassay mediated with biotechnological approaches to undertake silkworm breeding program efficiently.

Keywords

Bioassay · Breeding · Lethal · Probit · Survival

Introduction

Bioassay is a scientific experiment, through which the effect of any type of natural or artificially induced stimulus (abiotic/biotic/induced chemicals) on an organism can be observed, as whether there is impact of the stimulus on the organism body? If so whether it is qualitative or quantitative? In this chapter, attempt has been made on how to calculate LC_{50}/LD_{50} or LT_{50} through probit regression in context to silkworm.

In silkworm breeding, whether it is conventional or molecular approach, demand for breeds conferring resistance to diseases is the need of the hour in order to succeed in the crop. In a broad spectrum, resistance to diseases can be studied and screened through bioassays. The inoculum doses of different viral/fungal/bacterial can be applied to the silkworm/tissues through oral inoculum or injection. Any inoculum experiment requires the calculation of LC_{50}/LD_{50} which is the medial lethal concentration and statistically calculated through probit analysis. A bioassay can be either quantal or quantitative, direct or indirect. If the measured response is binary, the assay is quantal, and if not, it is quantitative. A typical bioassay involves a stimulus (e.g., inoculum containing fungal, viral, or bacterial doses) applicable to the experimental silkworms. It implies similarly to the experiment on abiotic stress responses and deriving the breeding lines based on the tolerance or susceptible levels.

Survival Analysis

Kaplan-Meier Analysis

Kaplan-Meier analysis (Kalpan & Meier 1958) is used to compare the survival of two or more groups, and log-rank test is used to compare the survival distribution and the survival curves, showtime, or age on X-axis and the portion of all individuals surviving on Y-axis. These analyses could be done by using MedCalc software and other statistical software.

Log-Rank Test

Survivorship (lx) is a measure of the proportion of individuals which survive to the beginning of age category x, and it is estimated as lx = nx / n0, where nx is the number of individuals in the study population which survive to the beginning of age category x, and n0 = N (the total population size) (http://mathworld.wolfram.com/LifeExpectancy.html).

The survival analysis is one of the foremost important criteria when one has to take up study on generation of breeding lines with respect to any type of resistance/tolerance to abiotic or biotic stresses.

Probit Analysis, LC_{50}, LD_{50}, and LT_{50}

Probit analysis is a type of regression which is used to analyze variables with binomial response. In 1934, Chester Ittner Bliss explained about probit analysis—he was interested to know how the insects respond to different concentrations of pesticide in way of working on finding an effective pesticide to control insects causing damage on grape leaves (Greenberg 1980). He explained how the sigmoid dose-response curve transforms to a straight line. Median lethal concentration LC_{50} for liquids and median lethal dose LD_{50} for solids are mostly used which represent concentration and dose to which 50% of the individuals of a population would respond. LT_{50} is the median lethal time which is widely utilized to quantify the amount of a stressor which is required to kill an organism exposed from beginning of the time interval.

Design of the Experiment and Statistical Calculations

For example, objective is to screen for resistance/tolerance of silkworm breeding lines against any stress inducing pathogen/pesticide, then different breeding lines have to be exposed to different concentrations of the pathogen load/pesticide concentration accordingly. LC_{50} may be calculated, and it is shown below systematically using Excel which had been explained comprehensively by Paltiyan.

Breeding Line 1

Step 1: Open a new Microsoft Excel sheet.

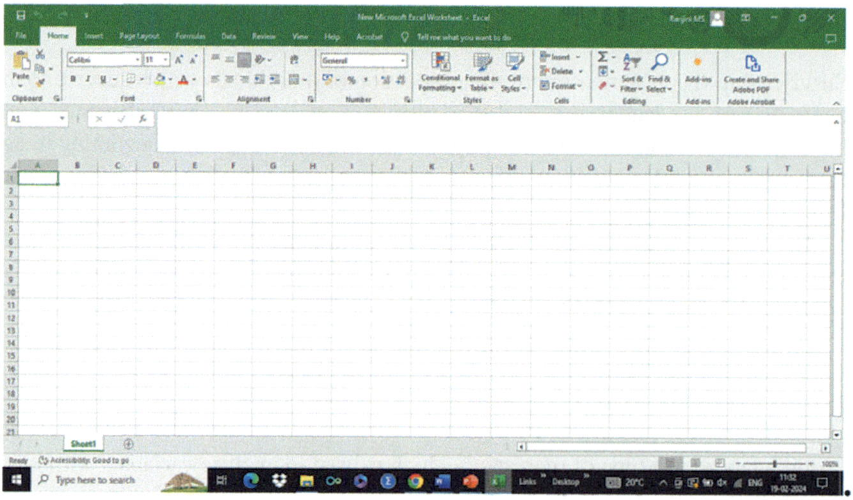

Step 2: Enter different concentrations of the chemical used in the bioassay.

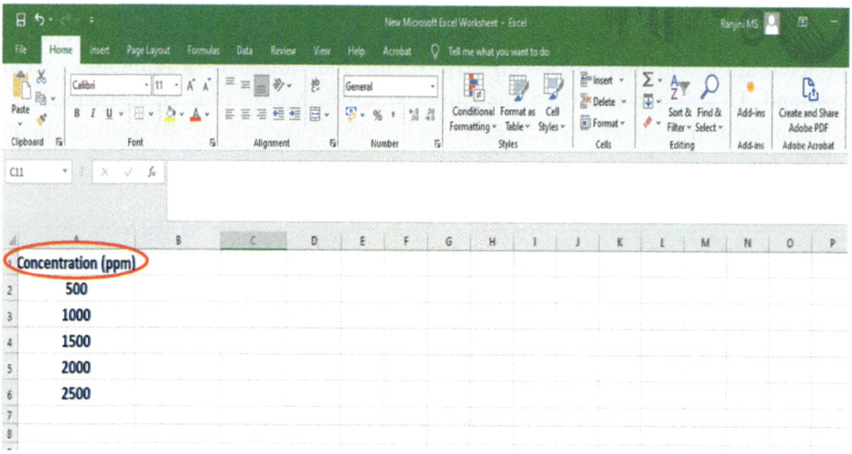

Step 3: Enter the log10 concentration by using the formula = log10(A2) by selecting column A row 2, and drag up to the end of the concentration.

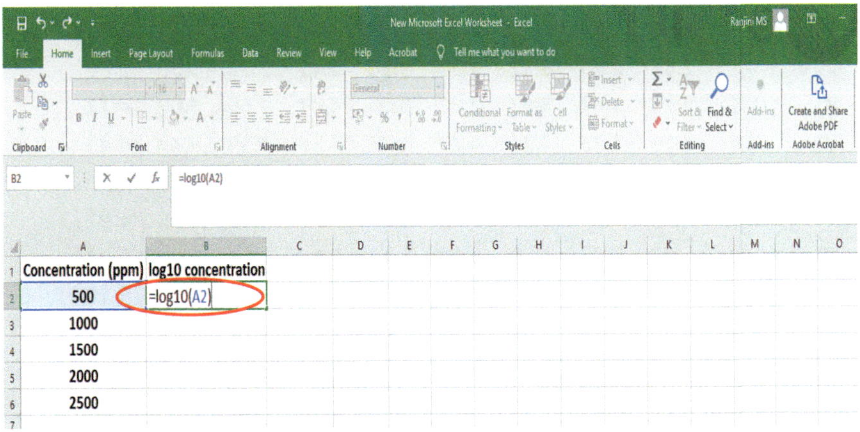

Step 4: Enter the mortality or dead percentage of the individuals exposed to chemical concentrations.

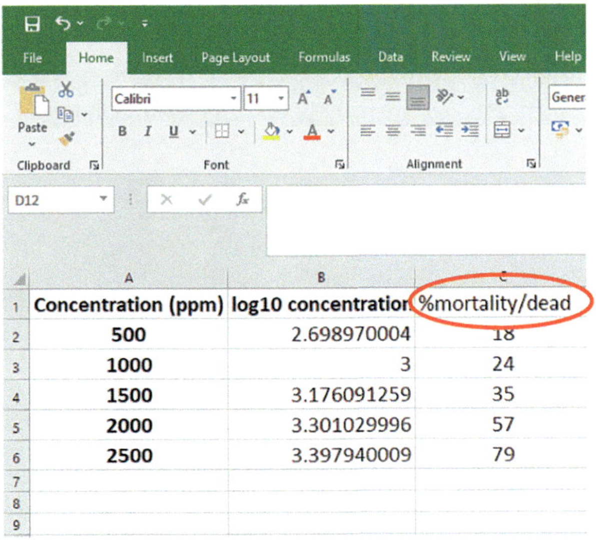

Step 5: Enter the probit in the next column of the mortality by using the Finney's table (Finney 1952)—the probit transformation for 18% mortality (in the mortality column) is 4.08 according to the Finney's table—likewise, enter for all the % mortality.

%	0	1	2	3	4	5	6	7	8	9
0	-	2.67	2.95	3.12	3.25	3.36	3.45	3.52	3.59	3.66
10	3.72	3.77	3.82	3.87	3.92	3.96	4.01	4.05	4.08	4.12
20	4.16	4.19	4.23	4.26	4.29	4.33	4.36	4.39	4.42	4.45
30	4.48	4.50	4.53	4.56	4.59	4.61	4.64	4.67	4.69	4.72
40	4.75	4.77	4.80	4.82	4.85	4.87	4.90	4.92	4.95	4.97
50	5.00	5.03	5.05	5.08	5.10	5.13	5.15	5.18	5.20	5.23
60	5.25	5.28	5.31	5.33	5.36	5.39	5.41	5.44	5.47	5.50
70	5.52	5.55	5.58	5.61	5.64	5.67	5.71	5.74	5.77	5.81
80	5.84	5.88	5.92	5.95	5.99	6.04	6.08	6.13	6.18	6.23
90	6.28	6.34	6.41	6.48	6.55	6.64	6.75	6.88	7.05	7.33

Finney's table (Finney 1952).

Step 6: Go to data analysis on the Excel sheet and select regression, and enter the regression value by selecting probit column as input Y range and log10 concentration column as input X range.

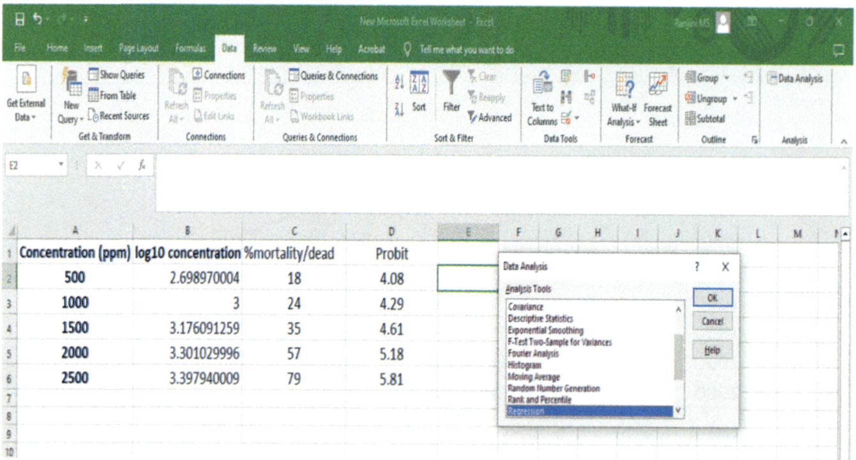

Step 7: In the output summary, the regression coefficient intercept value and X variable value are taken which are −2.392 and 2.307, respectively.

Intercept − 2.392840178.

X variable 1 2.307315317.

The formula Y = ax+b is adopted, where a = x variable value and b = intercept value.

Y is the standard 5 for LC_{50} calculation according to the Finney's table.

LC_{50}: x = (5 + 2.39)/2.30.

LC_{50} = 10^3.21.

LC_{50} = 1633.05 ppm.

Finally, the LC_{50} value of the breeding line 1 is 1633.05 ppm.

Likewise, the LC_{50} value of other breeding lines has to be calculated which gives exact picture of susceptibility and resistance of the breeding lines against a particular pathogen/pesticidal concentration. As mentioned above, the same strategy of calculations would be made to calculate the LT_{50} also wherein the time exposure is taken instead of concentrations.

Conclusion

Biological experiments are very interesting and fascinating; however, the results obtained/observed would be considered reliable only when the data is examined statistically. In particular about bioassay, it plays a crucial role in context with silkworm breeding research, and calculation of median lethal concentration/dose/time is essentially important to validate the inmost character of an individual of the population of different breeding lines. It becomes pavement for considering or characterizing the breeding lines as resistance/tolerance or susceptible to any stress stimulus. Once the probit regression is done, the breeding lines would then be taken further to analyze either conventionally or through molecular markers to develop any resistant/tolerant breeding lines against any biotic/abiotic or chemical/pesticidal stressors.

References

Finney DJ (ed) (1952) Probit analysis. Cambridge, Cambridge University Press

Greenberg BG (1980) Chester I. Bliss, 1899-1979. Int Stat Rev 8(1):135–136

Kalpan EL, Meier P (1958) Nonparametric estimation from incomplete observations. J Am Stat Assoc 53(282):457–481. (http://mathworld.wolfram.com/lifeexpectancy.html)

Application of Sericin in Food Industries and Coating of Fruits and Vegetables

M. A. Ravindra, Azad Gull, Dhaneshwar Padhan, N. Chandrakanth, V. Sobhana, Amit Kumar, Y. Thirupathaiah, and S. Gandhi Doss

Abstract

The cocoon formed by the silkworm *Bombyx mori* contains 25–30% sericin and 70–75% fibroin proteins. Sericin is a natural silk globular protein that is typically eliminated as a biological waste material after separating the fibroin from the silk cocoon to make silk products. The sericin protein is widely employed as a coating material as well as in pharmaceutical, nutraceutical, cosmetic, and therapeutic applications due to its antioxidant, antibacterial, biocompatible, and biodegradable nature. The use of sericin in edible coatings or films not only provides anti-browning, antibacterial, and antioxidant qualities but also improves the nutritional properties of the food product. Thus, using sericin in food applications provides various benefits and expands its scope in the food sector. Furthermore, fruits are an important part of a person's daily diet. Fruit and vegetable postharvest losses range from 15 to 40% due to inappropriate preservation/storage and marketing. Because fruits are perishable in nature, deterioration happens quickly, and different coating materials such as wax, cellulose, and starch are employed to extend the shelf life of the fruit, which are either non-digestive or primary goods. However, sericin, a silk protein that is often eliminated as waste in the sericulture and textile industries, has the potential to be used as a coating material. Approximately 50,000 tons of sericin are typically wasted as reeling waste. Sericin from cocoon reeling waste provides a sustainable and cheaper supply of protein sources and can serve as fat-free edible coating material. It is highly desired to choose coating materials that are easily digestible and have no negative effects on human health. Sericin can be used as a coating material to prevent fruit deterioration by increasing shelf life. As a result, using sericin

M. A. Ravindra (✉) · A. Gull · D. Padhan · N. Chandrakanth · V. Sobhana · A. Kumar · Y. Thirupathaiah · S. Gandhi Doss
Central Sericultural Research & Training Institute, Central Silk Board, Ministry of Textile Government of India, Mysuru, Karnataka, India

R. V Suresh et al. (eds.), *Biotechnology for Silkworm Crop Enhancement*,
https://doi.org/10.1007/978-981-97-5061-0_18

in food packaging and extending the shelf life of fruits are both environmentally beneficial and commercially viable. The features of sericin materials and their possible applications in food-sector enterprises are addressed in-depth in this chapter.

Keywords

Sericin · Coating · Packing · Food industry · Shell life · Eco-friendly

Introduction

Silkworm Cocoon Production

Silk is a natural composite fiber produced by the silkworm *Bombyx mori*. Silkworm cocoons are biological composite structures that protect silkworms from environmental influences and physical attacks of natural predators. Silkworm pupae are killed to obtain raw materials in the textile industry. The process of silk spinning and cocoon construction has undergone a long-term natural selection and extensive evolution. Although cocoons are thin and lightweight, they can protect silkworms from various invasions in nature and provide a good place for silkworm metabolism. Global silk production in 2018–2019 was 192,692 metric tons, with China and India being the main producers. In the textile industry, sericin is removed from the raw silk (degumming process), rendering a much finer silk fiber with a better luster and texture, which is used to make yarns and fabrics.

Fibroin and sericin make up the majority of the two types of proteins that make up silk fibers. Sericin, an adhesive component, covers the outer surface of fibroin, which is located at the core of silk fibers. A high-molecular-weight protein called silk sericin comes from a naturally occurring source, by the insect silkworm *Bombyx mori*. This protein, which is a water-soluble glycoprotein, makes up between 25 and 30% of a silk cocoon. Sericin's biological and mechanical characteristics are affected, depending on the extraction method—it can range in molecular weight from low to high, within the range of 10–400 kDa. Sericin inhibited in vitro lipid peroxidation and antityrosinase activity (Aramwit et al. 2012; Kato et al. 1998). After removal of pigments from sericin, extract also shows antityrosinase activity. This indicates that sericin alone has an antityrosinase activity. Sericin and pigment are hence in charge of the antioxidant qualities (Aramwit et al. 2012). The technique used to extract sericin has an impact on their antioxidant effectiveness. The molecular weight of sericin can be controlled by the extraction conditions. Sericin may also be easily functionalized (mixed with other natural and synthetic polymers) and is FDA-approved, making it more favorable to utilize than other natural biomaterials.

Sericin

A naturally occurring compound called sericin serves as an adhesive to fuse two fibroin filaments together to create silk yarn. With a molecular weight ranging from 20 to 400 kDa and 18 amino acids including essentials, the molecule is extremely hydrophilic. The polar groups (carboxyl, hydroxyl, and amino groups) found in the side chains of amino acids, along with their organic makeup, solubility, and structural organization, allow for cross-linking, copolymerization, and combinations with other polymers. These processes collectively give sericin its special qualities as an antioxidant, moisturizing agent, healing agent, antibacterial agent, antimicrobial agent, and antitumor agent. During the degumming procedure, only the fibroin is utilized to make silk; sericin is typically removed from the fibroin and discarded. The silk business separates sericin from fibroin to increase the smoothness, luster, lightness, and dyeability of the fibers. Sericin is a key ingredient in raw silk. It is projected that, of the 4.0 lakh tons of dried cocoons produced globally, 50,000 tons of waste sericin are often disposed of as sewage waste, posing a risk to the environment (Aramwit et al. 2012; Lamboni et al. 2015; Silva et al. 2022). Based on the protein composition, its applicability for biological applications, textile enhancement, and cosmetic formulations is assessed. There are 18 amino acids in sericin, including aromatic, polar, nonpolar, positively and negatively charged amino acids. Due to its many biological and industrial uses, including potential use in cosmetics, sericin has attracted a lot of attention because of its diverse biological and industrial applications, such as its prospective usage in cosmetics, textiles, biomaterials, and agriculture (Fig. 1).

Sericin in Food Industry

The Food and Drug Administration (FDA) has approved sericin globular protein and its derivatives for inclusion in the generally recognized as safe list (GRAS notice GRN 1026), with no evidence of causing allergy when administered orally and with no cytotoxicity effects as an ingredient of cosmetics. These applications of silk sericin are evident in developed countries like the USA, Japan, Austria, China, and Romania (Ghosh et al. 2019). The weak mechanical properties of sericin can be overcome with the addition of other materials (Kwak et al. 2018).

More than 40% of freshly packaged food and locally farmed items are wasted, yet there are predictions that over 800 million people are starving worldwide. A just 1-week extension of the global food supply's room temperature, shelf life would have a huge positive impact on the agriculture and food production industries, as well as significantly reducing the amount of food waste that now ends up in retail outlets (30% of which is uneaten). With the growing concern about plastic pollution, natural food-packaging materials that are biodegradable have been identified as a potential replacement for synthetic polymers (Adel et al. 2019).

Synthetic polymers are nonbiodegradable and nonrenewable—they cause environmental pollution (Low et al. 2022). By employing biodegradable polymers,

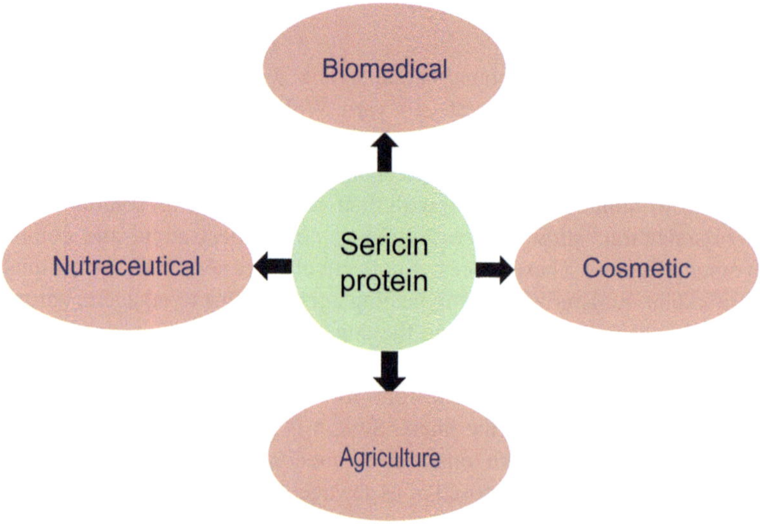

Fig. 1 Application of sericin protein in various fields

Fig. 2 Biological functions of sericin protein

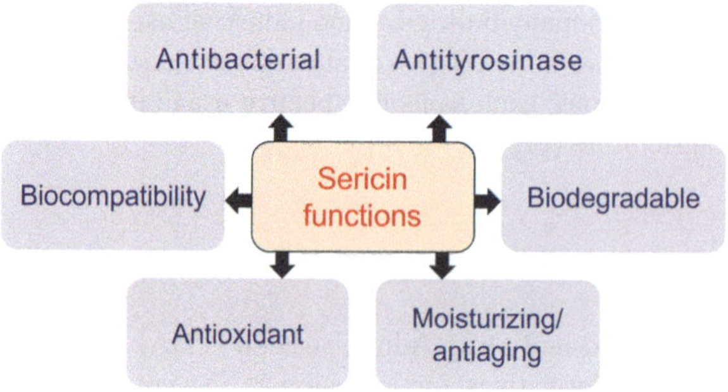

these problems with synthetic polymers can be mitigated (Low et al. 2022). Apart from being biodegradable, biopolymers also have other advantageous qualities, including being readily available, renewable, and nontoxic (Sam et al. 2016). They can also be directly derived from a variety of natural sources (Galgano 2015). The natural basic materials such proteins, lipids, and polysaccharides are used as packing material (McHugh et al. 1993; Mei et al. 2022). Among these, compounds based on proteins can improve food quality and shelf life while also being environmentally friendly (Ma and Song 2005).

It was observed that sericin exhibited strong reactivity and a wide spectrum of biological activities, including antibacterial, antioxidative, biodegradable, and biocompatible properties (Zhang 2002) (Fig. 2).

Because of its accessibility, lack of toxicity, and useful qualities, sericin has been employed mostly as a food additive in the food industry in recent years (Rangi and Jajpura 2015; Takechi and Takamura 2014). The hydrophobicity and hydrophilicity of sericin films, as well as the affinity of sericin, are influenced by its polarity (Meerasri et al. 2022). A sericin film with higher moisture content is formed when the higher polarity type is used (Meerasri et al. 2022). On the other hand, adding sericin hydrolysate to sericin film results in a decrease in both the moisture content and molecular weight—this raises the permeability of water vapor (Meerasri et al. 2022). Furthermore, compared to sericin-only film, which comes from strong acid hydrolysis, the addition of sericin hydrolysate leads to increased antioxidant activity because of the production of polyphenols and alkaloids as well as the total phenolic content (Meerasri et al. 2022). Sericin film's inadequate physical properties can also be addressed by using nano-celluloses, such as cellulose nanofibrils made from bamboo (Kwak et al. 2018). Flexible sericin films are produced by combining glycerol with polysaccharide polymers, such as glucomannan (Sothornvit et al. 2010). Additionally, by reducing the quantity of plasticizer needed to maintain flexibility, sericin can be used in conjunction with other biopolymers to reduce the permeability of films (Sothornvit et al. 2010). The potential use of sericin in food industry is depicted in Fig. 3.

There have been reports that sericin is utilized as a bread component salad dressing (Ghosh et al. 2019; Takechi et al. 2011; Takechi and Takamura 2014). According to Takechi and Takamura (2014), the author found that when sericin is added to flour at a calculated dose of 2–4 g/kg, the bread's height, specific volume, and color tend to decrease, but its internal surface texture and taste remain consistent. The natural food emulsifiers like egg yolk and casein, which can carry a risk of allergic reactions, can be substituted with sericin, which has no immunogenicity potential (Zhang et al. 2006), as an emulsifying agent and acylation with oleic acid can increase sericin's emulsifying activity (Ogino et al. 2006; Takechi et al. 2011).

Fig. 3 Various applications of sericin in food industries

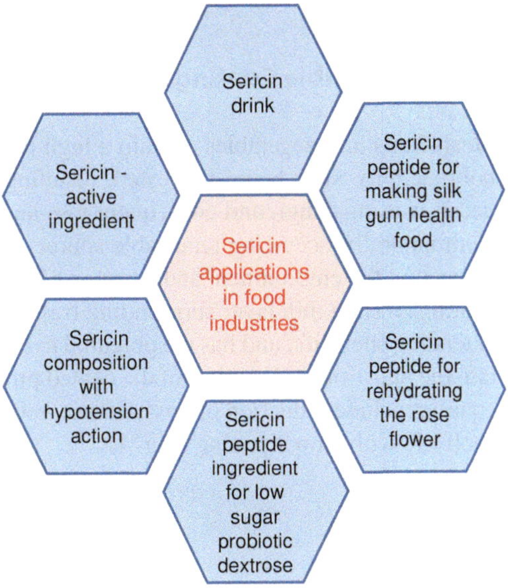

Matran et al. (2023) described the use of sericin in the preparation of jelly deserts. Foods with low energy values, such as those for patients with dysphagia, could be jelled using the sericin glycoprotein. Sericin peptide was added to a high-protein nutrition bar, and by watching this bar harden, it was possible to determine that sericin improved the mobility of water and tiny hydrophilic molecules in the sample, hence reducing the phase separation rate. To stop the protein from self-aggregating, the sericin peptide was added, and this also altered the sample's potential, secondary structure content, and surface hydrophobicity showing an anti-curing agent for the food sector looking to enhance food texture (Zhu et al. 2023a; Zhu et al. 2023b). Improvement in the storage stability of fruits and vegetables was observed with the application of sericin (Puangphet et al. 2018). Furthermore, when it is used as a food component, sericin significantly impacts the texture of food by enhancing its compatibility, mechanical potential, and the shape and size of the food product; for example, it can decrease the hardness and improve the elasticity of steamed potato bread according to Gong et al. (2019). The limited solubility of the powder sericin in solvent and the unpleasant smell of silkworm chrysalises are barriers to its unrestrained use in healthy and functional foods and additives (Wang and Zhang 2023). The glycation method is used to enhance the application of sericin to functional foods (Oh et al. 2011). The solubility in water is increased two to three times when sericin is glycated and also increases its antioxidant capacity (Wang and Zhang 2023). Moreover, the pupal scent of sericin is also specifically removed by glycation and is replaced with a subtly sweet, mellow scent (Wang and Zhang 2023). Therefore, the glycation technique has been proven to have a variety of application opportunities for sericin as a functional food, health food, or food additive, as well as for promoting the sustainable growth of the silk industry (Wang and Zhang 2023). It is reported that sericin, being a natural biomaterial, is safe for biological systems, with low immunogenicity and eliciting almost no allergic responses (Jiao et al. 2017).

Sericin as Edible Coating on Fruits and Vegetables

Fresh fruits and vegetables contain a high percentage of water and continue metabolic activity after being harvested, resulting in ripening, increased sensitivity to decay-causing fungi, and consequent loss and waste. Edible coatings are prepared from naturally occurring renewable sources and can contribute to reducing waste, respecting the environment and consumer health. Sericin, chitosan, and other edible coatings form a thin layer surrounding fresh produce that acts as a protective agent, extending shelf life, and has the potential to control their ripening process and maintain the nutritional properties of the coated product. Functional advantages of edible coating include induction of host defense, reduction in gas exchange, and extending shelf life with slow ripening (Fig. 4).

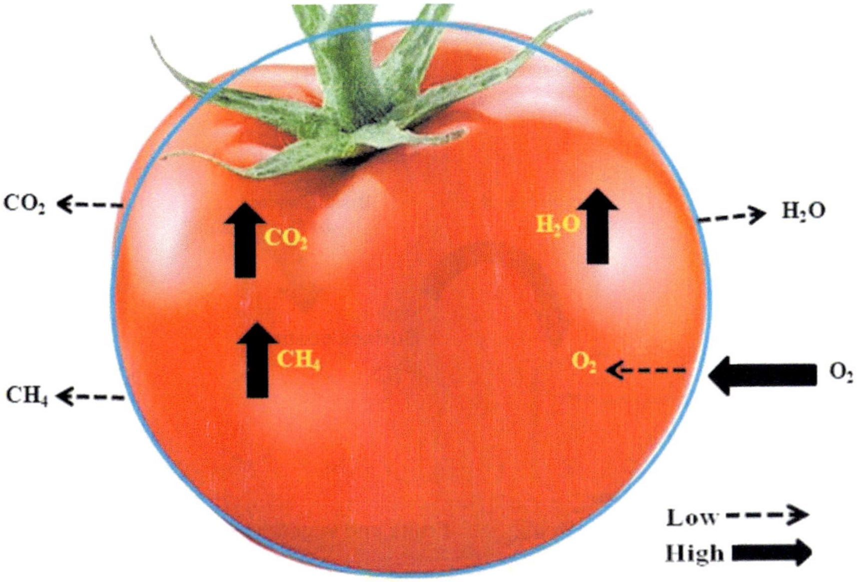

Fig. 4 Level and exchange of gases in the edible coated fruit

Process of Sericin Extraction and Fruit Coating

Cocoon shell was made into small pieces, and distilled water (cocoon 1 g/distilled water 40 ml) was added and placed in a boiling water bath for 1 hr. The filtrate (clear) sample was taken and stored at 4 °C for further use in fruit coating (Fig. 5). Tomatoes at 25 °C and 70% relative humidity may be stored longer if an edible coating substance based on sericin including chitosan, aloe vera, and glycerol is used (Tarangini et al. 2022). It is feasible to preserve the quantity of fruits and avoid aging like postharvest settings when compared to uncoated fruits. The fruit's structure is unaffected by the covering substance (Tarangini et al. 2022). Adding glycine as a plasticizer could result in sericin films with elongation properties. The elasticity of the sericin film is enhanced by the synergistic interactions between glycine and water molecules. A moderate increase in the moisture content and β-sheet structure of the sericin film is also a result of the increased glycine content. According to Yun et al. (2016), glycine plasticization raises the moisture content of films.

Sericin Coating on Fruits

In apple fruits, rice bran extracts effectively inhibited polyphenol oxidase (PPO) activity and browning (Sukhonthara et al. 2016). Fruit is shielded against aging by the sericin layer because of its antioxidant action. These characteristics imply that sericin is a desirable natural dietary element. According to Chimvaree et al. (2019)

Fig. 5 Process of extraction of sericin and coating on fruits

and Thongsook and Tiyaboonchai (2011), a sericin (2%) coating works well to preserve the esthetic appeal of fresh-cut apple fruits while preventing enzymatic browning. Fresh-cut mango fruits can effectively be kept visually appealing by coating them with a 2% sericin solution, which also inhibits the enzymatic browning process (Chimvaree et al. 2019; Thongsook and Tiyaboonchai 2011).

The phenolic content, phenylalanine ammonialyase, polyphenol oxidase, and peroxidase browning enzymatic activity were all reduced by the sericin treatment. However, there was no discernible difference between the treatment and control groups in terms of firmness or ascorbic content. Strawberry and banana coatings with sericin are successful at preserving the fruits (Cuervo Osorio et al. 2021). Pomegranate weight loss was minimized by 10% beeswax and 10% paraffin; gum arabic at 5% preserved fruit respiration and total soluble solid content; fruit peel thickness was significantly preserved by 20% paraffin; chitosan at 1% and 2% significantly affected peroxidase activity. When 2% chitosan was applied, fruit qualities such as firmness, color of the peel, ascorbic acid, and anthocyanin content improved. Browning and decay also decreased. Two weeks of shelf life at 20 °C could be a useful applied treatment to facilitate fruit handling and trade in both domestic and international markets. Pomegranate fruit coated with sericin has not been the subject of any published research. We coated pomegranate fruit with sericin protein, and we discovered that, as compared to untreated fruit, fruit appearance and moisture content did not decrease for up to 10 days. Consequently, sericin can be added to coatings to prolong the shelf life of pomegranate fruits.

Sericin Coating on Vegetables

The primary ingredients of the coating material are 1.5% sericin, 1.0% chitosan, 1.0% aloe vera, and 1.5% glycerol in an aqueous solution, which are applied to tomatoes using a straightforward dip-coating method (Tarangini et al. 2022). On the shelf life and quality of tomatoes during 40 days of storage at 25 °C and 70% relative humidity, sericin-based edible coating materials were found to be effective in reducing weight and firmness losses in the tomatoes under test (Tarangini et al. 2022). The sericin coated tomatoes are stored at room temperature (about 25 °C)—this covering material can keep the tomatoes firm and improve their postharvest quality (Tarangini et al. 2022). The findings show that the new coating material can positively control pH, total soluble sugar, titratable acid, lycopene concentration, and total phenol content (Table 1). It also shows that the coating material did not interfere with the capacity of tomatoes to produce a smooth, impermeable layer without any cracks. It also contains hydrophobic or hydrophilic amino acids, which may be the cause of its antibacterial, anti-inflammatory, anticancer, anti-antityrosinase, and antiaging activities. Sericin has been used in a variety of applications, both alone and in combination with other biomaterials, with potential uses in the food industry (Seo et al. 2023). Comparing the samples to those of the uncoated tomatoes, the titratable acid content increased with the length of storage, whereas the pH, total soluble solid content, total phenol content, total antioxidant content, and lycopene concentration stayed low (Tarangini et al. 2022). Tomatoes' color and ripening process after harvest are influenced by their lycopene concentration, an internal quality attribute. It tends to diminish after reaching the mature stage, when its levels will be higher (Quinet et al. 2019). After being stored for up to 28 days, both coated and untreated tomato fruits had higher lycopene contents. The ripening phases are responsible for this, and on Day 28, the control tomatoes had the greatest lycopene value of 9.7 mg/100 g, compared to the coated tomato. The extended maturity period following harvest is the reason for the drop in lycopene value on Day 35 when compared to the coated tomatoes. When tomatoes are not coated or protected, this behavior is typical (Quinet et al. 2019). Sericin-based coating, on the other hand, slowed down this process and did not show a stationary lycopene saturation threshold.

Edible coating materials based on proteins and polysaccharides typically do not produce good water-vapor barriers and may not have a significant impact on weight loss (Dhall 2013). However, because the sericin-based coating is hydrophobic, it showed a decreased rate of weight loss (Fig. 6). According to multiple publications, fruit postharvest quality can be preserved by protein-based composites (whey,

Table 1 Effect of sericin coating on physical and chemical properties of tomato fruit

	Firmness	pH	Lycopene (mg/100 g)	Sugars (%)	Phenols mg/ml	Titratable acidity (%)
Uncoated	24 N	5.6	10	5.1	251	0.1
Coated	49 N	5.0	7	4.4	196	0.2

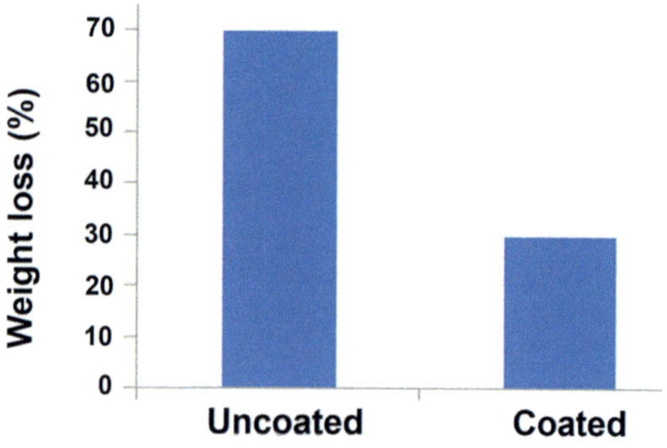

Fig. 6 Effect of sericin coating on weight loss of fruits

Table 2 Commercially available fruits and vegetable coatings (Yadav et al. 2022)

Coating material	Fruits	Vegetables	References
Guar gum and zein	–	Tomato and carrot	Flo chemical (2022)
Esters of sucrose	Apple and pears	–	Kumar et al. (2022)
Natural polymers and nanoparticles of clay	Apples, banana, pears, avocado, grapes, melons	–	Nabaco (2022)
Lipids and glycolipids	Avocado	–	Apeel (2022)
Esters of sucrose	Cherry	–	Pace international (2022)
Carnauba wax	Citrus and tropical fruits	–	Susmita Devi et al. (2023)
Shellac	Apple	–	Alleyne and Hagenmaier (2000)
Carnauba wax	Pomegranate	–	Meighani et al. (2015)
Mineral and vitamin blends	Banana, mango, guava, pear, avocado	Tomato and carrot	Agricoat (2022)
Corn	Mango, cherry	Tomato, pepper, cucumber	Akron (2022)

gelatin, flaxseed protein isolate, etc.) by decreasing stiffness and weight loss (Perez-Gago et al. 2006; Radi et al. 2017; Sharma and Saini 2021). Tarangini et al. (2022) examined how tomato fruit qualities were affected by a coating based on sericin.

According to Khorram et al. (2017), polyphenols generally have an immediate impact on the color, flavor, and scent of fruits in addition to having antibacterial and antioxidant qualities. Sukhonthara et al. (2016) found that rice bran extracts effectively inhibited polyphenol oxidase (PPO) activity and potato browning. When veggies like mushrooms were coated with sericin, there was a roughly 40% reduction in polyphenol oxidase activity, weight loss, and browning (Chimvaree et al. 2019; Thongsook and Tiyaboonchai 2011). Commercially available fruits and vegetable coatings are summarized in Table 2.

There have been reports of sericin having antibacterial activity; however, the antibacterial activity varies according to the different degumming techniques employed to extract sericin. The water degumming method's sericin may be able to inhibit the *Staphylococcus aureus* bacteria, a foodborne disease.

Fruit Spoilage and Preservation

Vegetables and fruits are perishable commodities that might go bad if not handled and stored correctly. Spoilage can be caused by a variety of factors, such as enzymatic activity, microbial development, physical damage, and environmental conditions. The common causes of rotting fruits and vegetables are as follows:

1. Microbial growth: Bacteria, yeast, and mold are naturally present on the surfaces of fruits and vegetables. If not handled carefully or kept in dirty conditions, these bacteria can grow and cause degradation. Microbiological deterioration is indicated by discoloration, sliminess, growth of mold, and odd scents.
2. Enzymatic reactions: Fruits and vegetables include enzymes that can change a food's color, texture, flavor, and nutritional content. In bananas and apples, browning (also known as enzymatic browning) can result from enzyme reactions when fruit flesh is exposed to oxygen. This method may result in a loss of quality and attractiveness.
3. Physical harm: Vegetables and fruits with cuts, bruising, or punctures may let bacteria in and get spoiled. Vegetables should be consumed immediately or thrown aside to prevent further deterioration. Vegetables and fruits can be kept for up to 2 years in the ideal climate and humidity levels.

Most fresh food requires specific humidity and temperature settings for optimal storage. Exposure to high temperatures can accelerate ripening and rotting, while exposure to low temperatures may cause chilling injury or freezing damage. While high humidity may promote the growth of mold, low humidity can cause wilting and drying out. Some fruits and vegetables naturally release ethylene gas, which can accelerate ripening and lead to rotting. Produce that releases ethylene, such as bananas, apples, and tomatoes, should be kept separate from produce that is ethylene-sensitive, like leafy greens.

Conclusion and Future Prospects

Sericin is one of the most important protein components of silkworm cocoon shell which is usually eliminated as reeling waste. Because of its biocompatibility and other biological properties, sericin may be useful as a coating material for enhancing the shelf life of fruits and vegetables without deteriorating their physical and chemical qualities. The extraction process and concentrations of sericin protein play a role in enhancement of shelf life of fruits and vegetables. Furthermore, recently sericin has been employed as a food additive in the food industry due to its accessibility, lack of toxicity, and antioxidant and antimicrobial properties. However, further research needs to be undertaken to find out the exact mechanism and concentration of sericin to be used for enhancing the shelf life of fruits and vegetables and also its use in food industry.

References

Adel AM, Ibrahim AA, El-Shafei AM, Al-Shemy MT (2019) Inclusion complex of clove oil with chitosan/β-cyclodextrin citrate/oxidized nanocellulose biocomposite for active food packaging. Food Packag Shelf Life 20:100307

Agricoat (2022) Nature seal for processors. Online: https://www.agricoat.co.uk/industries/processors/

Akron (2022) AKORN vegetable coatings. Online: https://akorn.tech/our-products/vegetables

Alleyne V, Hagenmaier RD (2000) Candelilla-shellac: an alternative formulation for coating apples. HortScience 35(4):691–693

Apeel (2022) How apeel works. Online: https://www.apeel.com/science

Aramwit P, Siritientong T, Srichana T (2012) Potential applications of silk sericin, a natural protein from textile industry by-products. Waste Manag Res 30(3):217–224

Chimvaree C, Wongs-Aree C, Supapvanich S, Charoenrat T, Tepsorn R, Boonyaritthongchai P (2019) Effect of sericin coating on reducing browning of fresh-cut mango cv.'Nam Dok Mai no. 4'. Agric Nat Resour 53(5):521–526

Cuervo Osorio GA, Murillo Arias YA, Urrea Vélez L (2021) Study of the effect of sericin coatings extracted from Bombyx mori silkworm cocoons on fruit degradation. Rev ION 34(1):15–25

Devi LS, Mukherjee A, Dutta D, Kumar S (2023) Carnauba wax-based sustainable coatings for prolonging postharvest shelf-life of citrus fruits. Sustain Food Technol 1(3):415–425

Dhall RK (2013) Advances in edible coatings for fresh fruits and vegetables: a review. Crit Rev Food Sci Nutr 53(5):435–450

Flo Chemicals (2022) FloZein applications-agriculture. Online: https://www.zeinproducts.com/applications-1

Galgano F (2015) Biodegradable packaging and edible coating for fresh-cut fruits and vegetables. Ital J Food Sci 27(1):1–20

Ghosh S, Rao RS, Nambiar KS, Haragannavar VC, Augustine D, Sowmya SV (2019) Sericin, a dietary additive: mini review. J Med Radiol Pathol Surg 4(2):13–17

Gong S, Yang D, Wu Q, Wang S, Fang Z, Li Y, Wu J (2019) Evaluation of the antifreeze effects and its related mechanism of sericin peptides on the frozen dough of steamed potato bread. J Food Process Preserv 43(8):e14053

Jiao Z, Song Y, Jin Y, Zhang C, Peng D, Chen Z, Wang L (2017) In vivo characterizations of the immune properties of sericin: an ancient material with emerging value in biomedical applications. Macromol Biosci 17(12):1700229

Kato N, Sato S, Yamanaka A, Yamada H, Fuwa N, Nomura M (1998) Silk protein, sericin, inhibits lipid peroxidation and tyrosinase activity. Biosci Biotechnol Biochem 62(1):145–147

Khorram F, Ramezanian A, Hosseini SMH (2017) Shellac, gelatin and Persian gum as alternative coating for orange fruit. Sci Hortic 225:22–28

Kumar L, Ramakanth D, Akhila K, Gaikwad KK (2022) Edible films and coatings for food packaging applications: a review. Environ Chem Lett 20:875–900

Kwak HW, Lee H, Lee ME, Jin HJ (2018) Facile and green fabrication of silk sericin films reinforced with bamboo-derived cellulose nanofibrils. J Clean Prod 200:1034–1042

Lamboni L, Gauthier M, Yang G, Wang Q (2015) Silk sericin: a versatile material for tissue engineering and drug delivery. Biotechnol Adv 33(8):1855–1867

Low JT, Yusoff NISM, Othman N, Wong TW, Wahit MU (2022) Silk fibroin-based films in food packaging applications: a review. Compr Rev Food Sci Food Saf 21(3):2253–2273

Ma YH, Song KB (2005) Physical properties of silk fibroin films treated with various plasticizers. Prev Nutr Food Sci 10(2):187–190
Matran IM, Tarcea M, Rus DC, Voda R, Muntean DL, Cirnatu D (2023) Research and development of a new sustainable functional food under the scope of nutrivigilance. Sustain For 15(9):7634
McHugh TH, Avena-Bustillos R, Krochta JM (1993) Hydrophilic edible films: modified procedure for water vapor permeability and explanation of thickness effects. J Food Sci 58(4):899–903
Meerasri J, Chollakup R, Sothornvit R (2022) Factors affecting sericin hydrolysis and application of sericin hydrolysate in sericin films. RSC Adv 12(44):28441–28450
Mei S, Fu B, Su X, Chen H, Lin H, Zheng Z, Yang DP (2022) Developing silk sericin-based and carbon dots reinforced bio-nano composite films and potential application to litchi fruit. LWT 164:113630
Meighani H, Ghasemnezhad M, Bakhshi D (2015) Effect of different coatings on post-harvest quality and bioactive compounds of pomegranate (*Punica granatum* L.) fruits. J Food Sci Technol 52:4507–4514
Nabaco (2022) NatuWrap helps retain the freshness of a variety of produce. Online: https://www.nabacoinc.com/overview
Ogino M, Tanaka R, Hattori M, Yoshida T, Yokote Y, Takahashi K (2006) Interfacial behavior of fatty-acylated sericin prepared by lipase-catalyzed solid-phase synthesis. Biosci Biotechnol Biochem 70(1):66–75
Oh H, Lee JY, Kim MK, Um IC, Lee KH (2011) Refining hot-water extracted silk sericin by ethanol-induced precipitation. Int J Biol Macromol 48(1):32–37
Pace International (2022) Edible coatings. Available at: https://www.paceint.com/products/coatings/. Accessed 3 Mar 2022
Perez-Gago MB, Serra M, Del Rio MA (2006) Color change of fresh-cut apples coated with whey protein concentrate-based edible coatings. Postharvest Biol Technol 39(1):84–92
Puangphet A, Jiamyangyuen S, Tiyaboonchai W, Thongsook T (2018) Amino acid composition and anti-polyphenol oxidase of peptide fractions from sericin hydrolysate. Int J Food Sci Technol 53(4):976–985
Quinet M, Angosto T, Yuste-Lisbona FJ, Blanchard-Gros R, Bigot S, Martinez JP, Lutts S (2019) Tomato fruit development and metabolism. Front Plant Sci 10:15540
Radi M, Firouzi E, Akhavan H, Amiri S (2017) Effect of gelatin-based edible coatings incorporated with Aloe vera and black and green tea extracts on the shelf life of fresh-cut oranges. J Food Qual 2017:1–10
Rangi A, Jajpura L (2015) The biopolymer sericin: extraction and applications. J Text Sci Eng 5(1):1–5
Sam ST, Nuradibah MA, Chin KM, Hani N (2016) Current application and challenges on packaging industry based on natural polymer blending. In: Natural polymers: industry techniques and applications, pp 163–184
Seo SJ, Das G, Shin HS, Patra JK (2023) Silk sericin protein materials: characteristics and applications in food-sector industries. Int J Mol Sci 24(5):4951
Sharma M, Saini CS (2021) Postharvest shelf-life extension of fresh-cut guavas (*Psidium guajava*) using flaxseed protein-based composite coatings. Food Hydrocoll Health 1:100015
Silva AS, Costa EC, Reis S, Spencer C, Calhelha RC, Miguel SP, Coutinho P (2022) Silk sericin: a promising sustainable biomaterial for biomedical and pharmaceutical applications. Polymers 14(22):4931
Sothornvit R, Chollakup R, Suwanruji P (2010) Extracted sericin from silk waste for film formation. Songklanakarin J Sci Technol 32(1):17–22
Sukhonthara S, Kaewka K, Theerakulkait C (2016) Inhibitory effect of rice bran extracts and its phenolic compounds on polyphenol oxidase activity and browning in potato and apple puree. Food Chem 190:922–927
Takechi T, Maekawa ZI, Sugimura Y (2011) Use of sericin as an ingredient of salad dressing. Food Sci Technol Res 17(6):493–497
Takechi T, Takamura H (2014) Development of bread supplemented with the silk protein sericin. Food Sci Technol Res 20(5):1021–1026

Tarangini K, Kavi P, Jagajjanani Rao K (2022) Application of sericin-based edible coating material for postharvest shelf-life extension and preservation of tomatoes. eFood 3(5):e36

Thongsook T, Tiyaboonchai W (2011) Inhibitory effect of sericin on polyphenol oxidase and its application as edible coating. Int J Food Sci Technol 46(10):2052–2061

Wang HD, Zhang YQ (2023) The glycation of silk sericin to enhance its application to functional foods. LWT 173:114255

Yadav A, Kumar N, Upadhyay A, Sethi S, Singh A (2022) Edible coating as postharvest management strategy for shelf-life extension of fresh tomato (Solanum lycopersicum L.): an overview. J Food Sci 87(6):2256–2290

Yun H, Kim MK, Kwak HW, Lee JY, Kim MH, Lee KH (2016) The role of glycerol and water in flexible silk sericin film. Int J Biol Macromol 82:945–951

Zhang YQ (2002) Applications of natural silk protein sericin in biomaterials. Biotechnol Adv 20(2):91–100

Zhang P, Aso Y, Yamamoto K, Banno Y, Wang Y, Tsuchida K, Fujii H (2006) Proteome analysis of silk gland proteins from the silkworm, *Bombyx mori*. Proteomics 6(8):2586–2599

Zhu HT, Zhang XX, Zhang R, Feng JY, Thakur K, Zhang JG, Wei ZJ (2023a) Silkworm sericin peptides alleviate the hardening of soy protein bars during early storage. J Insects Food Feed 9(6):809–822

Zhu HT, Zhang XX, Zhang R, Feng JY, Thakur K, Zhang JG, Wei ZJ (2023b) Anti-hardening effect and mechanism of silkworm sericin peptide in high protein nutrition bars during early storage. Food Chem 407:135168

Biomedical Applications of Silkworm Sericin

Sayannita Das and Amitava Mandal

Abstract

Biodegradable polymeric membranes are attractive green alternatives to traditionally used artificial polymers, polypropylene suture, etc. and may find varied biomedical applications. Natural composite fiber silk biologically produced by the silkworm named *Bombyx mori* is structurally comprised of chiefly hydrophobic fibroin protein and hydrophilic glycoprotein sericin. Sericin primarily contains four amino acids, threonine, serine, glycine, and aspartic acid. It has two isoforms, sericin A (17% nitrogen) and sericin B (16.8% nitrogen). Sericin shows its attractive bioactivity because of its variable amino acid content. Recent biomedical applications of sericin stem from its altered amino acid composition and presence of different functional groups. Researchers have already reported sericin-based fibers, films, dressings, sponges, hydrogels, foams, and nanoparticles for plausible pharmaceutical and biomedical applications partnering to wound healing, tissue engineering, drug delivery, etc. This chapter provides a concise description about the development of sericin-based hybrid materials for biomedical applications and its future perspectives.

Keywords

Silk Sericin · Antioxidant · Anti-inflammation · Drug delivery · Immunological response

S. Das · A. Mandal (✉)
Molecular Complexity Laboratory, Department of Chemistry, Raiganj University, Raiganj, West Bengal, India

R. V Suresh et al. (eds.), *Biotechnology for Silkworm Crop Enhancement*,
https://doi.org/10.1007/978-981-97-5061-0_19

Introduction

Naturally, two major proteins are necessary for the formation of silkworm cocoon—silk sericin is one of them with the other one being silk fibroin. Sericin protein produced by *Bombyx mori* (a holometabolous insect belonging to the Lepidoptera order and Bombycidae family) provides the ideal conditions for the occurrence of larval metamorphosis to adults (Kundu et al. 2008). Despite its biological functions, sericin has been disregarded in the field of sericulture for a long time as it is considered as a waste product. Sericin possesses some good biological features like biocompatibility, immune-compatibility, biodegradability, anti-inflammatory activity, antibacterial activity, and antioxidant functions (Arango et al. 2021; Liu et al. 2022; Zhang 2002).

This chapter describes some valuable biological properties of silk sericin, which have not found as much priority as fibroin protein. The most notable features of sericin are meticulously discussed, like its tissue engineering, drug delivery, antioxidant, anti-inflammatory, antitumor, anticoagulant, and wound healing effects.

Structure and Properties of Sericin

Structure

Silk fibers (from silkworm cocoons) are mainly consisting of two major proteins: fibroin (65–75%) and sericin (25–35%). The molecule of sericin having a molecular weight ranging from 20 to 400 kDa is highly hydrophilic in nature and contains 18 amino acid residues (Fig. 1) (Kato et al. 1998; Wu et al. 2007).

The presence of different polar groups such as carboxylic acid, hydroxyl, and amino groups of the amino acid side chain and its structural organization with solubility enable cross-linking copolymerizations. Meanwhile, sericin being a globular hydrophilic protein acts as an outer gummy part which connects the fibroin fiber, a hydrophobic stringy protein with a semicrystalline structure for supporting the rigidity and solidity of a fibre. Thus, sericin is regarded as an adhesive joining agent between two filaments of fibroin to build a silk yarn (Fig. 2) (Kumar Sahi et al. 2023; Yao et al. 2022).

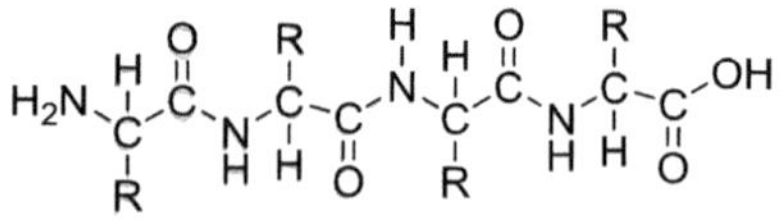

Fig. 1 Structure of silk sericin

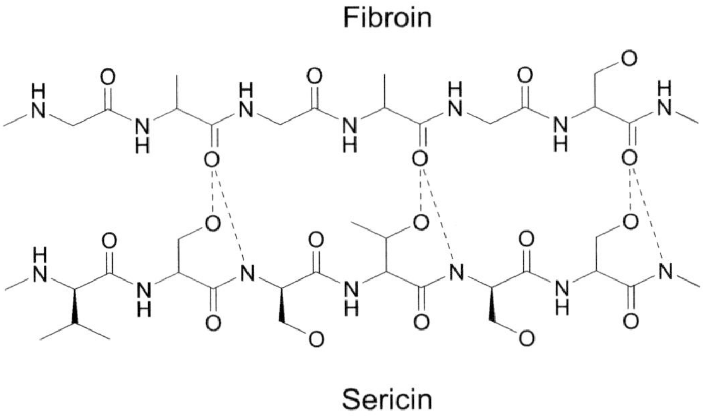

Fig. 2 Complete structure of sericin demonstrating intermolecular hydrogen bonding between sericin and fibroin segments

Properties

Chemical Properties

The glycoprotein sericin contains very high content of hydrophilic amino acids, mostly serine (33%), on its surface, which makes it soluble in water. In addition to that of serine, it contains another 17 amino acids in its structure. These are histidine, glycine, aspartic acid, glutamic acid, threonine, and tyrosine. Sericin consists of random coil and β-sheets that makes it globular in structure. It has carbon (46.5%), oxygen (31%), nitrogen (16.5%), and hydrogen (6%). The structure is sensitive to external stimulus. In the presence of high temperature, mechanical stretching, and/or moisture, it transforms into a β pleated sheet structure, which is the basis of their sol-gel transition. At higher temperature (50–60 °C or even more), sericin adopts its soluble form, whereas, at lower temperatures, its solubility is decreased leading to the conversion of random coil structure into β-sheets that enables the construction of a three-dimensional network resulting in a sericin gel (Kataoka 1977; Zhu et al. 1996, 1998).

Physical Properties

Silk sericin comprises of random coils and β-sheets along with some β-turns, which match to the amorphous and crystalline regions, respectively. Due to high content of random coils, sericin is an amorphous and fragile material in the dry state. Its crystalline property and physical stability can be increased upon treatment with ethanol (EtOH) which could encourage the aggregation of protein, where the transition from random coils to β-sheets may occur due to dehydration and fractionation. Addition of cross-linking agents like glutaraldehyde facilitates the formation of stable β-sheets. Glutaraldehyde makes bonds with the free amines and considerably changes the molecular structure of sericin by increasing its crystalline property. Thus, natural sericin possessing a high molecular weight and β-sheet structures can form gels in solution state. If the solutions are treated at high temperature, sericin could no longer form gels due to denaturation and degradation. Thus, the stability of hydrogen bonds among various hydroxyl groups is reduced by high temperatures, which favors their interaction with water and enables the formation of β-sheet (Dash et al. 2007, 2009).

Extraction Methods

Sericin is separated from fibroin protein through the process called "degumming" of the cocoons in textile industry. During the degumming process of silk, the peptide bonds are broken by the hydrolysis of sericin, and subsequently, it gets detached from fibroin. The extraction process of sericin from silk is possible because of its hydrophilic nature which leads to a higher solubility in water than fibroin, being hydrophobic and insoluble in nature.

Sericin can also be extracted by some chemical methods where the cocoons are boiled in a solution of sodium bicarbonate (Na_2CO_3) and Marseille soap at atmospheric pressure. This is a conventional method of extraction called soap-alkaline degumming process. This method effectively removes sericin from cocoons and isolated the fibroin that the textile industry can use. However, in this process, the molecular weight of recovered sericin is reduced, and some of its functional properties are also lost. Moreover, it is a complex process to separate soap and sericin. Subsequently, there remain some traces of soap which limit its use for biomedical and pharmaceutical purposes (Freddi et al. 2003; Yun et al. 2013).

Various degumming methods for the extraction of sericin have been developed which include heat, chemical, and enzymatic methods as shown in Fig. 3.

In chemical method, several carboxylic acids like citric acid, succinic acid, tartaric acid, etc. or bases like sodium carbonate, sodium silicate, sodium phosphate, etc. are commonly used for sericin extraction from silk. These chemicals hydrolyze the sericin leading to the cleavage of peptide bonds of the amino acid to small molecules and release sericin into the acidic or alkaline solution, where it has high solubility. For instance, sodium carbonate (Na_2CO_3) converts the carboxyl group ($-COOH$) to $-COONa^+$ in sericin molecules, thereby increasing its water solubility through strong hydration of sodium cation (Dou and Zuo 2015; Rangi and Jajpura 2015). Proteolytic enzymes such as degummase, alcalase, savinase, trypsin, etc. are also used for the extraction. These enzymes hydrolyze the peptide bonds between the $-COOH$ groups of lysine or arginine and the $-NH_2$ groups of the adjoining amino acid residues (More et al. 2018).

Another extraction method is the heating method where the cocoons are heated or boiled in water under high pressure by autoclaving. The high temperature and pressure make the hydrogen bonds between the hydroxyl groups more unstable and allow water to combine with the hydroxyl groups of polar amino acids leading to the separation of sericin and fibroin (Silva et al. 2012).

Among all the methods, the most commonly used method for sericin extraction is the heating method. Although this method also causes some degradation of sericin, as high temperatures are applied, sericin retains its noteworthy properties.

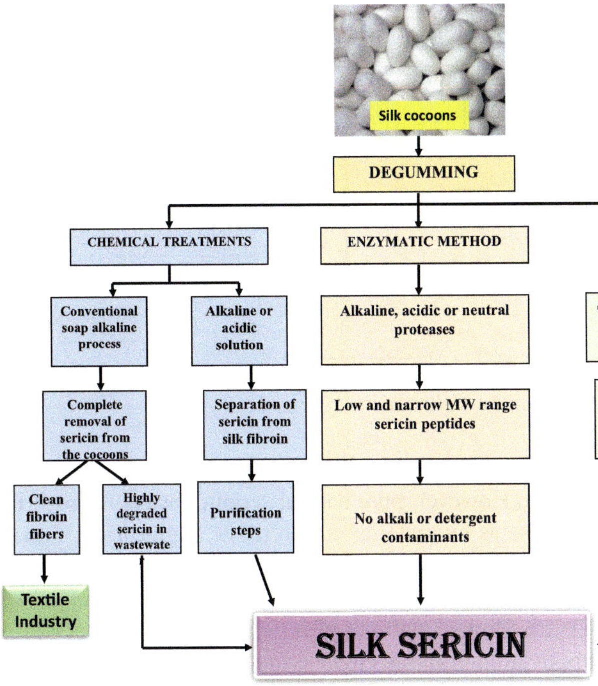

Fig. 3 Different extraction method of silk sericin after degumming

Moreover, the silk is heated in hot distilled water without adding other chemicals so that the obtained sericin does not have any impurities(Chirila et al. 2016).

New technologies have been developed to extract sericin from silk in a greener, more effective and sustainable way in recent years, some of them are—infrared heating, carbon dioxide supercritical fluid extraction, and ultrasounds. However, these techniques demand the use of extra equipment (Wang et al. 2019).

Biomedical Applications of Sericin

Silk sericin has various applications mainly in the biomedical and pharmaceutical fields. Figure 4 describes the main applications of sericin.

In Tissue Engineering

Sericin has a number of remarkable applications in biomedical and tissue engineering. People have blended sericin amid polyvinyl alcohol (PVA) films and silver nanoparticles. The formed matrix has favorable antimicrobial and tissue healing properties. Teramoto et al. showed that if only 10% alcohol is added in sericin

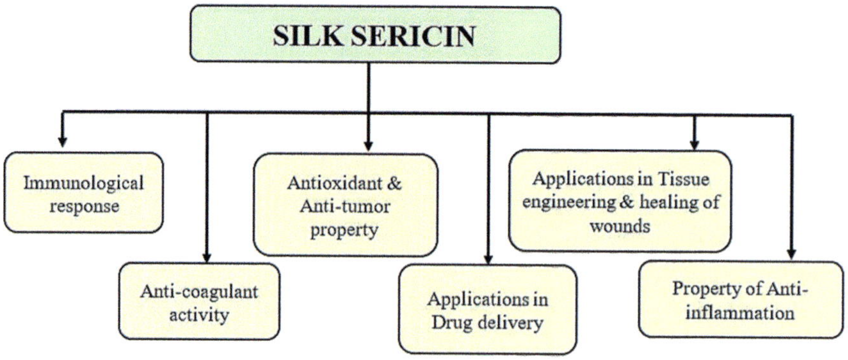

Fig. 4 Different biomedical applications of silk sericin

solution, it forms a hydrogel, which acts as a natural biomaterial without any chemical cross-linking or irradiation. However, pure natural sericin always forms brittle films posing difficulty in its use as biomaterial during tissue engineering. Time to time different approaches has been developed to enhance its physical properties (Teramoto et al. 2005). Sericin-based creams can treat skin tissues in a faster growth.

In addition, 2D and 3D cross-linked matrices of sericin-gelatin have been proficiently employed in the application of tissue engineering. Mandal et al. formulated a sericin—gelatin blended 3D scaffolds and 2D films using *A. mylitta* sericin, and glutaraldehyde was used as a cross-linking agent. The sericin—gelatin combination structure has a uniform pore distribution as well as homogeneous morphology with better mechanical strength and high swell ability. Nayak et al. fabricated some sericin-based 3D porous matrices to construct a possible tissue-engineered skin replacement(Mandal et al. 2009; Nayak et al. 2013).

Besides, sericin blended with the scaffolds of polyvinyl acetate leads to the formation of hydrogel which has been clinically used in skin grafting and effective wound dressing inventions. Later, a highly concentrated sericin solution was proposed to overcome the issue of the poor mechanical strength of sericin. This is known as robust sericin hydrogel, having good cytocompatibility and high resistance. It promotes a versatile platform for tissue engineering and renewing medicinal applications (Zhang et al. 2022).

In Drug Delivery Applications

Sericin has polar side chains and hydrophobic spheres; thus, it possesses an amphiphilic nature. It can easily bind hydrophobic and hydrophilic drugs as well as charged therapeutic molecules which makes it a good carrier of drugs. Additionally, sericin shows high capacity for moisture absorption and desorption and also has long in vivo half-life. These properties favor its drug delivery applications. Besides, sericin-based hydrogels, synthesized by the method of cross-linking, ethanol precipitation, and/or mixing with different polymers, are suitable drug delivery vehicle (Lamboni et al. 2015; Nishida et al. 2011).

Wang et al. reported the development of an injectable highly cross-linked 3D sericin gel which encourages cell adhesion and high survival properties. The excellent physicochemical properties of these 3D sericin gel show good drug release capability that could be utilized in multifunctional ways for tissue repair during cell therapy (Wang et al. 2014).

Recently, Wang and his coworkers developed one sericin/dextran hydrogel for modern drug delivery approaches which could carry small drug molecules and some macromolecular systems. This hydrogel is also injectable and could be used as an optical material to follow the applied drug molecules in malignant melanoma through its photoluminescence action (Liu et al. 2016).

Most recently, in 2022, Xia et al. described an effective approach to engineer a silk sericin nanosphere for recombinant human lactoferrin delivery for ulcerative colitis treatment. They developed some genetically modified silk fibers from transgenic silkworms. These modified silk fibers were utilized for developing recombinant human lactoferrin by ethanol precipitation. These functional materials are spherical in shape with promising therapeutic effect (Xu et al. 2022).

Oral administration of drugs is suitable over any other mood. A sericin-based pH-responsive oral drug delivery system was reported by Oh et al. They prepared sericin beads by using LiCl/DMSO system. They studied the release performance of nonsteroidal anti-inflammatory drug, diclofenac, at different pH. At higher pH, the release efficacy is more. According to their report, these sericin beads can find applications in agricultural fields (Oh et al. 2007).

Wound Healing

Sericin catalyzes the migration, propagation, and formation of collagen. Therefore, multiple studies evidenced the wound curative properties of sericin. Aramwit and coworkers developed one potent antibiotic cream compositionally silver sulfadiazine blended with 8% sericin in a scientific study to treat second-degree burn wounds. The developed antibiotic cream was applied on 29 volunteers who got a second-degree burn. For all cases, sericin cream cured the wound within 5–7 days, and for every case, epithelialization at the burn surface happened readily. Overall, the new method has decreased patients' hospitalization increasing the life expectancy (Aramwit et al. 2013).

In another study, Nagar et al. used small sericin (30 kDa) for corneal lesion treatment of type 2 diabetes mellitus. They showed that diluted sericin in saline accelerated the healing process of corneal damage. Further, addition of sericin to the culture media of human corneal epithelial cell line effectively increased the cell propagation and adhesion(Nagar et al. 2009). Besides, corneal injuries and diabetes mellitus can cause degeneration on the central and peripheral nervous system.

Sericin also exhibited promising activity in reducing diabetes mellitus. Using sericin orally at a dose of 2.4 g/kg for 35 days reduced the blood glucose levels in animal model and upheld the formation of neurofilament protein in nerves. It also increased the nerve growth factor in spinal ganglion and in anterior horn cells. These results show that sericin guards the sciatic nerve and the nerve cells related against injuries caused by diabetes mellitus (Song et al. 2013).

Antioxidant Activity

Sericin also has effective antioxidant properties which result from its scavenging activity of reactive oxygen species, as well as lipid peroxidation inhibition and anti-tyrosinase and anti-elastase properties. Li et al. observed that sericin might increase the properties of some antioxidant enzymes like superoxide catalase, glutathione, and peroxidase dismutase (Li et al. 2008).

The antioxidant properties of sericin are associated with its high content of serine and threonine. The involved hydroxyl groups chelate different trace elements like copper and iron. Besides, the pigment molecules like flavonoids and carotenoids gathered in sericin layers. This may be one of the causes of its antioxidant and anti-tyrosinase properties.

Aramwit et al. showed that sericin found from cocoons during pigment extraction has anti-tyrosinase activity, which is greater than the sericin attained from cocoons having pigments (Aramwit et al. 2010). These studies, thus, propose the practice of sericin as benign adjuvant in cosmetic industries.

Anticoagulants

Sericin exhibits anticoagulant property when sulfated as heparin (a sulfated polysaccharide). Sulfation of sericin with chlorosulfonic acid was carried out in pyridine at 80 °C for 8 h with stirring. FT-IR (Fourier-transform infrared) spectroscopy and ^{1}H-NMR (nuclear magnetic resonance) spectroscopy clearly showed that sericin binds considerable amounts of sulfate chiefly through the hydroxyl groups of serine when treated with chlorosulfonic acids in pyridine. Appearance of IR peak at 935 cm^{-1} confirmed the generation of covalent sulfates through cross-linking. The final molecular weight of the sulfated serine determines its anticoagulant strength (Monti et al. 2007; Tamada et al. 2004).

Anti-Inflammatory Property

Inflammation in the body represents the main healing process. Through the process of inflammation, phagocytosis of all contaminants and necrotic tissues at the site of wound occurs. Besides, at the stage of inflammation, inflammatory cells emit cytokines that take on the cells. These are responsible for new tissue formation. However, this strategy required to be restrained as the uncontrolled countenance of inflammatory cytokines encourages metalloproteinases, which are responsible for the dilapidation of extracellular matrix. Thus, any biomaterials developed to treat wounds should have the property to minimize inflammation (Anderson et al. 2008; Koh and DiPietro 2011).

Overall, analysis to determine anti-inflammatory property is frequently based on the evaluation of the release of inflammatory cytokines like interleukin 1 (IL—1) and tumor necrosis factor-alpha (TNF-α). In literature, it is found that IL-1 and TNF-α are the most important mediators of inflammation. They also encourage the release of adhesion molecules that are crucial for the proliferative phase. Different in vitro and in vivo studies demonstrated that sericin dictates the synthesis of inflammatory cytokines IL-1 and TNF-α (Anderson et al. 2008).

Antitumor Activity

In cancer treatment, chemotherapy is used to treat cancer affecting the normal as well as neoplastic cells, thus reflecting its limit toward clinical applications. Zhaorigetu et al. studied the effect of supplementation of 30% sericin of an animal model against colon tumorigenesis. Sericin had no effect on body weight and pattern of food consumption. However, it decreased the prevalence of colon adenoma resulting in lowered level of rate of cell propagation, oncogene formation, and oxidative stress. The mechanism of action of the antitumor activity of sericin is unclear. In mechanistic study, Zhaorigetu et al. reported that the supplemented sericin always decreased the number aberrant crypt foci on the intestine. Inside the body, ingested sericin remains as undigested that induces low stomach tumorigenesis, oxidative stress, etc. in the colon (Zhaorigetu et al. 2001, 2007). The same group also determined the anti-skin tumor activity of sericin. At first, the rats were induced with 12-O-tetradecanoylphorbol-13-acetate (TPA) and 7,12-dimethylbenz (a) anthracene (DMBA) to generate skin tumor. Application of sericin topically in dose dependent manner (2.5 mg/dose) delayed the progression of tumor significantly. Fifteen weeks of sericin application resulted in a small tumor proving the significance of sericin to inhibit tumorigenesis and skin tumor numbers in mice (Zhaoringetu et al. 2023).

Other Applications

Besides the abovementioned biomedical applications of silk sericin, it has a number of potential applications in the field of cosmetology. Due to its unique chemical and physical properties, it can also be used as an ingredient for skin and hair care cosmetic preparations. According to the reports of Expert Panel for Cosmetic Ingredient Safety, only sericin is safe among all other silk proteins. This may be the reason of using sericin in the cosmetic products (Johnson et al. 2020). There are so many benefits of using sericin in different cosmetic products like sunscreens and shampoos. It retains skin hydration and elasticity for longer time. The products became lower irritant during cleaning and also have anti-wrinkle and antiaging properties.

The skin whitening potential of sericin-based creams is due in part to its antioxidant properties favorable for use in hair care treatment (Camargo et al. 2022; Wang et al. 2021), to protect hair from damage due to hair bleaching or coloring.

Conclusion

Silk protein sericin (a water-soluble glycoprotein) is a natural polymer generated and secreted by the insect *B. mori*. It is composed of 18 amino acids, with strong polar side groups (amino groups, carboxyl and hydroxyl). It can be extracted by various methods after degumming. Its potential use is because of its excellent biochemical and biophysical properties. Sericin is a material of enormous interest in industry and has great promise to generate new functional materials for advanced biomedical applications.

References

Anderson JM, Rodriguez A, Chang DT (2008) Foreign body reaction to biomaterials. Semin Immunol 29:86–100

Aramwit P, Damrongsakkul S, Kanokpanont S, Srichana T (2010) Properties and anti-tyrosinase activity of sericin from variousextraction methods. Biotechnol Appl Biochem 55:91–98

Aramwit P, Palapinyo S, Srichana T, Chottanapund S, Muangman P (2013) Silk sericina meliorates wound healing and its clinical efficacy in burn wounds. Arch Dermetol Res 305:585–594

Arango MC, Montoya Y, Peresin MS, Bustamante J, Álvarez-López C (2021) Silk sericin as a biomaterial for tissue engineering: a review. Int J Polym Mater 70:1115–1129

Camargo FB, Minami MM, Rossan MR, Magalhães WV, Ferreira VTP, Campos PMBGM (2022) Prevention of chemically induced hair damage by means of treatment based on proteins and polysaccharides. J Cosmet Dermatol 21:827–835

Chirila TV, Suzuki S, McKirdy NC (2016) Further development of silk sericin as a biomaterial: comparative investigation of the procedures for its isolation from Bombyx mori silk cocoons. Prog Biomater 5:135–145

Dash R, Ghosh SK, Kaplan DL, Kundu SC (2007) Purification and biochemical characterization of a 70 kDa sericin from tropical tasar silkworm, Antheraea mylitta. Comp Biochem Physiol B: Biochem Mol Biol 147:129–134

Dash BC, Mandal BB, Kundu SC (2009) Silk gland sericin protein membranes: fabrication and characterization for potential biotechnological applications. J Biotechnol 144:321–324

Dou H, Zuo B (2015) Effect of sodium carbonate concentrations on the degumming and regeneration process of silk fibroin. J Text Inst 106:311–319

Freddi G, Mossotti R, Innocenti R (2003) Degumming of silk fabric with several proteases. J Biotechnol 106:101–112

Johnson W, Bergfeld WF, Belsito DV, Hill RA, Klaassen CD, Liebler DC, Marks JG, Shank RC, Slaga TJ, Snyder PW (2020) Safety assessment of silk protein ingredients as used in cosmetics. Int J Toxicol 39:127–144

Kataoka K (1977) The solubility of sericin in water. Nippon Sanshigaku Zasshi 46:227–230

Kato N, Sato S, Yamanaka A, Yamada H, Fuwa N, Nomura M (1998) Silk protein, sericin, inhibits lipid peroxidation and tyrosinase activity. Biosci Biotechnol Biochem 62:145–147

Koh TJ, DiPietro LA (2011) Inflammation and wound healing: the role of the macrophage. Expert Rev Mol Diagn 13:23–26

Kumar Sahi A, Gundu S, Kumari P, Klepka T, Sionkowska A (2023) Silk-based biomaterials for designing bioinspired microarchitecture for various biomedical applications. Biomimetics 8:55–58
Kundu SC, Dash BC, Dash R, Kaplan DL (2008) Naturalprotective glue protein, sericin bioengineered by silk worms: potential for biomedical and biotechnological applications. Prog Polym Sci 33:998–1012
Lamboni L, Gauthier M, Yang G, Wang Q (2015) Silk sericin: A versatile material for tissue engineering and drug delivery. Biotechnol Adv 33:1855–1867
Li YG, Ji DF, Lin TB, Zhong S, Hu GY, Chen S (2008) Protective effect of sericin peptide against alcohol-induced gastric injury in mice. Chin Med J 121:2083–2087
Liu J, Qi C, Tao K, Zhang J, Zhang J, Xu L, Jiang X, Zhang Y, Huang L, Li Q (2016) Sericin/dextran injectable hydrogel as an optically trackable drug delivery system for malignant melanoma treatment. ACS Appl Mater Interfaces 8:6411–6422
Liu J, Shi L, Deng Y, Zou M, Cai B, Song Y, Wang Z, Wang L (2022) Silk sericin-based materials for biomedical applications. Biomaterials 287:121638
Mandal BB, Priya AS, Kundu SC (2009) Novel silk sericin/gelatin 3D scaffolds and 2D films: fabrication and characterization for potential tissue engineering application. Acta Biomater 5:3007–3020
Monti P, Freddi G, Arosio C, Tsukada M, Arai T, Taddei P (2007) Vibrational spectroscopic study of sulphated silk proteins. J Mol Struct 202:834–836
More SV, Chavan S, Prabhune AA (2018) Silk degumming and utilization of silk sericin by hydrolysis using alkaline protease from beauveria Sp. (MTCC 5184): A green approach. J Nat Fibers 15:373–383
Nagar N, Murao T, Ito Y, Okamoto N, Sasaki M (2009) Enhancing effects of sericin on corneal wound healing in Otsuka Long-Evans Tokushima Fatty rats as a model of human type 2 diabetes. Biol Phram Bull 32:1594–1599
Nayak S, Dey S, Kundu SC (2013) Skin equivalent tissue-engineered construct: co-cultured fibroblasts/keratinocytes on 3D matrices of sericin hope cocoons. PLoS One 8:74779
Nishida A, Yamada M, Kanazawa T, Takashima Y, Ouchi K, Okada H (2011) Sustained-release of protein from biodegradable sericin film, gel and sponge. Int J Pharm 407:44–52
Oh H, Lee JY, Kim A, Ki CS, Kim JW, Park YH, Lee KH (2007) Preparation of silk sericin beads using LiCl/DMSO solvent and their potential as a drug carrier for oral administration. Fibers Polym 8:470–476
Rangi A, Jajpura L (2015) The biopolymer sericin: extraction and applications. Int J Text Sci 5:1–5
Silva VR, Ribani M, Gimenes ML, Scheer AP (2012) High molecular weight sericin obtained by high temperature and ultrafiltration process. Procedia Eng 42:833–841
Song CJ, Yang ZI, Zhong MR, Chen ZH (2013) Sericin protects against diabetes-induced injuries in sciatic nerve and related nerve cells. Neural Regen Res 8:506–513
Tamada Y, Sano M, Niwa K, Imai T, Yoshino G (2004) Sulfation of silk sericin and anticoagulant activity of sulfated sericin. J Biomater Sci Polym Ed 15:971–980
Teramoto H, Nakajima KI, Takabyashi C (2005) Preparation of elastic silk sericin hydrogel. Biosci Biotechnol Biochem 69:845–847
Wang WH, Lin WS, Shih CH, Chen CY, Kuo SH, Li WL, Lin YS (2021) Functionality of silk cocoon (Bombyx mori l.) sericin extracts obtained through high-temperature hydrothermal method. Materials 14:5314–5317
Wang W, Pan Y, Gong K, Zhou Q, Zhang TQ, Li A (2019) Comparative study of ultrasonic degumming of silk sericin using citric acid, sodium carbonate and papain. Color Technol 135:195–201
Wang Z, Zhang Y, Zhang J (2014) Exploring natural silk protein sericin for regenerative medicine: an injectable, photo-luminescent, cell-adhesive 3D hydrogel. Sci Rep 4:7064
Wu JH, Wang Z, Xu SY (2007) Preparation and characterization of sericin powder extracted from silk industry waste water. Food Chem 103:1255–1262
Xu S, Yang Q, Wang R, Tian C, Ji Y, Tan H, Zhao P, Kaplan DL, Wang F, Xia Q (2022) Genetically engineered pH-responsive silk sericin nanospheres with efficient therapeutic effect on ulcerative colitis. Acta Biomater 144:81–95

Yao X, Zou S, Fan S, Niu Q, Zhang Y (2022) Bioinspired silk fibroin materials: from silk building blocks extraction and reconstruction to advanced biomedical applications. Mater Today Bio 16:100381–100384

Yun H, Oh H, Kim MK, Kwak HW, Lee JY, Um IC, Vootla SK, Lee KH (2013) Extraction conditions of Antheraea mylitta sericin with high yields and minimum molecular weight degradation. Int J Biol Macromol 52:59–65

Zhang YQ (2002) Applications of natural silk protein sericin in biomaterials. Biotechnol Adv 20:91–100

Zhang Y, Tangfeng W, Shen C, Xu G, Chen H, Yan H, Xiong M, Zhang G (2022) A robust sericin hydrogel formed by a native sericin from silkworm bodies. Fibers Polym 23:1826–1833

Zhaorigetu S, Sasaki M, Kato N (2007) Consumption of sericin suppresses colon oxidative stress and aberrant crypt foci in 1,2-dimeth hydrazine-treated rats by colonundigestedsericin. J Nutr Sci Vitaminol 53:297–300

Zhaorigetu S, Sasaki M, Watanabe H, Kato N (2001) Supplemental silk protein, sericin, suppresses colon tumorigenesisin 1,2-dimethylhydrazine-treated mice by reducing oxidative stress and cell proliferation. Biosci Biotech Bioch 65:2181–2186

Zhaoringetu S, Yanaka N, Sasaki M, Watanabe H, Kato N (2023) Silk protein, sericin, suppresses DMBA-TPA-induced mouse skin tumorigrnrsis by reducing oxidative stress, inflammatory responses and endogeneous tumor promote TNF-α. Oncol Departs 10:537–543

Zhu LJ, Arai M, Hirabayashi K (1996) Sol-gel transition of sericin. Nippon Sanshigaku Zasshi 65:270–274

Zhu LJ, Yao J, Youlu L (1998) Structural transformation ofsericina dissolved from cocoon layer in hot water. Zhejiang Nongye Daxue Xuebao 24:268–272

9789819750603